Antimicrobial D

The Wound Care Applications

Developments in Applied Microbiology and Biotechnology

Antimicrobial Dressings
The Wound Care Applications

Edited by

Raju Khan

*CSIR-Advanced Materials and Processes Research Institute (AMPRI),
Bhopal, Madhya Pradesh, India*

Sorna Gowri

*CSIR-Advanced Materials and Processes Research Institute (AMPRI),
Bhopal, Madhya Pradesh, India*

ELSEVIER

ACADEMIC PRESS
An imprint of Elsevier

Academic Press is an imprint of Elsevier
125 London Wall, London EC2Y 5AS, United Kingdom
525 B Street, Suite 1650, San Diego, CA 92101, United States
50 Hampshire Street, 5th Floor, Cambridge, MA 02139, United States
The Boulevard, Langford Lane, Kidlington, Oxford OX5 1GB, United Kingdom

Notices
Knowledge and best practice in this field are constantly changing. As new research and experience broaden our
understanding, changes in research methods, professional practices, or medical treatment may become necessary.

Practitioners and researchers must always rely on their own experience and knowledge in evaluating and using any
information, methods, compounds, or experiments described herein. In using such information or methods they
should be mindful of their own safety and the safety of others, including parties for whom they have a professional
responsibility.

To the fullest extent of the law, neither the Publisher nor the authors, contributors, or editors, assume any liability
for any injury and/or damage to persons or property as a matter of products liability, negligence or otherwise, or
from any use or operation of any methods, products, instructions, or ideas contained in the material herein.

ISBN: 978-0-323-95074-9

For Information on all Academic Press publications
visit our website at https://www.elsevier.com/books-and-journals

Publisher: Stacy Masucci
Acquisitions Editor: Kattie Washington
Editorial Project Manager: Billie Jean Fernandez
Production Project Manager: Sajana Devasi P. K.
Cover Designer: Greg Harris

Typeset by MPS Limited, Chennai, India

Contents

List of contributors

Cátia Alves
Centre for Textile Science and Technology (2C2T), University of Minho, Guimarães, Portugal

Komal Attri
School of Chemistry and Biochemistry, Thapar Institute of Engineering and Technology, Patiala, Punjab, India; BioX, Centre of Excellence for Emerging Materials, Thapar Institute of Engineering and Technology, Patiala, Punjab, India

Geetanjali Baruah
Department of Biotechnology, School of Health Sciences, The Assam Kaziranga University, Jorhat, Assam, India

Jejiron M. Baruah
Department of Chemistry, North Lakhimpur College (Autonomous), Lakhimpur, Assam, India

Subhashini Bharathala
Department of Emergency Medicine, All India Institute of Medical Sciences, New Delhi, New Delhi, India

Sanjeev Bhoi
Department of Emergency Medicine, All India Institute of Medical Sciences, New Delhi, New Delhi, India

Manish Biyani
Department of Bioscience and Biotechnology, Japan Advanced Institute of Science and Technology, Nomi, Ishikawa, Japan

Diptiman Choudhury
School of Chemistry and Biochemistry, Thapar Institute of Engineering and Technology, Patiala, Punjab, India; BioX, Centre of Excellence for Emerging Materials, Thapar Institute of Engineering and Technology, Patiala, Punjab, India

Bodhisatwa Das
Department of Biomedical Engineering, Indian Institute of Technology Ropar, Ropar, Punjab, India

Mehrez E. El-Naggar
Pretreatment and Finishing of Cellulosic-Based Fibers Department, Textile Research and Technology Institute, National Research Centre, Dokki, Cairo, Egypt

Marta Fernandes
Centre for Textile Science and Technology (2C2T), University of Minho, Guimarães, Portugal

Sonu Gandhi
DBT-National Institute of Animal Biotechnology (NIAB), Hyderabad, Telangana, India

Satyabrat Gogoi
Department of Chemistry, School of Basic Sciences, The Assam Kaziranga University, Jorhat, Assam, India

Ayush Gupta
Department of Microbiology, AIIMS Bhopal, Madhya Pradesh, India

Pratik Kolhe
DBT-National Institute of Animal Biotechnology (NIAB), Hyderabad, Telangana, India

Lakshmi Kanth Kotarkonda
Department of Emergency Medicine, All India Institute of Medical Sciences, New Delhi, New Delhi, India

Akshay Kumar
Department of Emergency Medicine, All India Institute of Medical Sciences, New Delhi, New Delhi, India

Nalinee Kumari
Department of Zoology, DPG Degree College, Gurugram, Haryana, India

Tarun Kumar Kumawat
Department of Botany, University of Rajasthan, Jaipur, Rajasthan, India; Department of Biotechnology, Biyani Girls College, Jaipur, Rajasthan, India

Varsha Kumawat
Biyani Institute of Pharmaceutical Sciences, Jaipur, Rajasthan, India

Liliana Melro
Centre for Textile Science and Technology (2C2T), University of Minho, Guimarães, Portugal

Mohit
Department of Biomedical Engineering, Indian Institute of Technology Ropar, Ropar, Punjab, India

Sagnik Nag
Department of Biotechnology, Vellore Institute of Technology (VIT), Vellore, Tamil Nadu, India

Bhingaradiya Nutan
Department of Biosciences and Bioengineering, Indian Institute of Technology Bombay, Mumbai, Maharashtra, India

Jorge Padrão
Centre for Textile Science and Technology (2C2T), University of Minho, Guimarães, Portugal

Anjali Pandit
Department of Biotechnology, Biyani Girls College, Jaipur, Rajasthan, India

Huda R.M. Rashdan
Chemistry of Natural and Microbial Products Department, Pharmaceutical and Drug Industries Research Institute, National Research Centre, Dokki, Cairo, Egypt

Ana Isabel Ribeiro
Centre for Textile Science and Technology (2C2T), University of Minho, Guimarães, Portugal

Jayanta K Sarmah
Department of Chemistry, School of Basic Sciences, The Assam Kaziranga University, Jorhat, Assam, India

Maitri Shah
DBT-National Institute of Animal Biotechnology (NIAB), Hyderabad, Telangana, India

Deepinder Sharda
School of Chemistry and Biochemistry, Thapar Institute of Engineering and Technology, Patiala, Punjab, India

Bhoomika Sharma
Department of Biotechnology, Biyani Girls College, Jaipur, Rajasthan, India

Vishnu Sharma
Department of Biotechnology, Biyani Girls College, Jaipur, Rajasthan, India

Vijay Pal Singh
CSIR-Institute of Genomics and Integrative Biology, New Delhi, New Delhi, India

Arvind K. Singh Chandel
Center for Disease Biology and Integrative Medicine, Graduate School of Medicine, The University of Tokyo, Bunkyo-Ku, Tokyo, Japan

Tej Prakash Sinha
Department of Emergency Medicine, All India Institute of Medical Sciences, New Delhi, New Delhi, India

Marcela Slovakova
Department of Biological and Biochemical Sciences, Faculty of Chemical Technology, University of Pardubice, Pardubice, Czech Republic

Xinyu Song
Key Laboratory of Marine Chemistry Theory and Technology, Ministry of Education, Ocean University of China, Qingdao, Shandong, P.R. China

Amit Tyagi
Institute of Nuclear Medicine and Allied Sciences, Defence Research & Development Organisation, Timarpur, New Delhi, New Delhi, India

Liangmin Yu
Key Laboratory of Marine Chemistry Theory and Technology, Ministry of Education, Ocean University of China, Qingdao, Shandong, P.R. China

Andrea Zille
Centre for Textile Science and Technology (2C2T), University of Minho, Guimarães, Portugal

Overview and summary of antimicrobial wound dressings and its biomedical applications

1

Tarun Kumar Kumawat[1,2], Varsha Kumawat[3], Vishnu Sharma[2], Anjali Pandit[2], Bhoomika Sharma[2], Sagnik Nag[4], Nalinee Kumari[5] and Manish Biyani[6]

[1]*Department of Botany, University of Rajasthan, Jaipur, Rajasthan, India* [2]*Department of Biotechnology, Biyani Girls College, Jaipur, Rajasthan, India* [3]*Biyani Institute of Pharmaceutical Sciences, Jaipur, Rajasthan, India* [4]*Department of Biotechnology, Vellore Institute of Technology (VIT), Vellore, Tamil Nadu, India* [5]*Department of Zoology, DPG Degree College, Gurugram, Haryana, India* [6]*Department of Bioscience and Biotechnology, Japan Advanced Institute of Science and Technology, Nomi, Ishikawa, Japan*

List of abbreviations

AgNPs	silver nanoparticles
AuNPs	gold nanoparticles
CS	chitosan
ECDC	European Centre for Disease Prevention and Control
ECM	extracellular matrix molecules
EO	essential oils
EPS	extracellular polymeric substances
IV	intravenous
MRSA	methicillin-resistant *Staphylococcus aureus*
PHMB	polyhexamethylene biguanide
PVPI	povidone iodine
ROS	reactive oxygen species
TTC	tetracycline

1.1 Introduction

The skin accounts for approximately 15% of total body mass and is composed of three layers: epidermis, dermis, and hypodermis [1]. The skin senses temperature and moisture to govern the body's interior environment [2]. The most important functions are preventing desiccation and protecting internal organs and other structural components from environmental damage [3]. Burns, stitches, and accidental cuts cause the most damage to the skin [4]. Every day, a wide range of surgical operations are performed all over the world. The majority of these treatments leave some kind of wound that must be treated in order to prevent further health problems [5]. A wound is defined as any change in the structural or functional characteristics of the skin [6]. Many different factors can be used to classify a wound, including its origin, location, symptoms, severity, and amount of tissue

Antimicrobial Dressings. DOI: https://doi.org/10.1016/B978-0-323-95074-9.00004-X

lost. It can also be classified as clean, polluted, diseased, or colonized [7]. Wounds can be acute or chronic in nature. A chronic wound has impaired physiology, whereas an acute wound has normal physiology and healing stages [8].

When microbes enter a skin breach, they cause wound contamination. Infection slows wound healing and may result in death. Life-threatening wounds can be fatal both immediately and later. The former is caused by severe exsanguinations or underlying organ damage, whereas the latter by infection and sepsis. As a result, the primary issue with wound care is wound infection [9]. The most invasive infections in burn wounds worldwide are caused by Methicillin-resistant *Staphylococcus aureus* (MRSA) and *S. aureus* (MRSA). *Pseudomonas aeruginosa* and multidrug-resistant *Acinetobacter baumanii* infections are more common in tropical climates [10]. As a result, wound care that aims to reduce bacterial infection has received a lot of attention around the world. Dressings have been used to cover wounds for a long time because they aid in wound healing by acting as a physical barrier against infection, keeping the surrounding area moist, and absorbing any exudates. The following are the requirements that an optimal wound dressing should meet:

1. absorb and eradicate the toxins and exudate,
2. permit gaseous exchange,
3. maintain a high moistness,
4. prevent secondary infection,
5. provide thermal insulation,
6. not be cytotoxic [11].

Wound healing continues to be the most difficult problem in wound therapy [12]. It is widely accepted that providing a moist, warm environment for a wound by covering it with an optimal dressing membrane is an effective way to promote healing. This environment also encourages the growth and colonization of microorganisms that require sophisticated medical treatments. Antimicrobial wound dressings reduce bacterial biofilm formation and the spread of microbial infestations [13,14]. The healthcare industry's focus on wound care has shifted globally in order to provide a more supportive environment for healing [15]. Wound infections were often difficult to treat, but patients now have access to wound dressings containing silver, iodine, polyhexamethylene biguanide, and octinidine [16]. The alarming global development of multidrug-resistant bacteria, driven by antibiotic abuse and misuse, is also increasing the need in the wound care industry for sophisticated antimicrobial chemicals and wound dressings with a broad spectrum of action [17]. These dressings deliver antimicrobials in a controlled manner, achieving antibacterial action while maintaining therapeutic concentrations in healing tissues [18].

1.2 Pathophysiology of wound and healing process

Wounds form when the cellular cohesion of tissue is disrupted by structural, physiological, or metabolic disorders [19]. Skin wounds are characterized as either acute or chronic based on their healing duration [20,21]. Acute wounds are defined by a disruption of tissue integrity that occurs suddenly, recovers quickly, and has no long-term consequences [22]. Platelets, keratinocytes, immunological surveillance cells, and fibroblasts are among the cell types involved in tissue repair [23].

The dermal and epidermal integrity is compromised, resulting in chronic wounds that do not heal in a reasonable amount of time and are linked to predisposing factors [24]. Bacterial colonization is common in chronic wounds and is thought to be a primary source of inflammation [23]. The majority of chronic wound patients, such as the elderly and diabetic patients with lower body (leg and foot) ulcers, have poor blood circulation in their lower limbs, rendering conventional oral and intravenous (IV) antibiotics ineffective [21].

In a series of perfectly synchronized processes, the wound healing process involves both local and mobile cells, as well as extracellular matrix elements and hydrophilic mediators [6]. Hemostasis, inflammation, migration, proliferation, and finally remodeling are the stages of skin wound healing [21,25−28] (Fig. 1.1). The production of fluids, proteins, and dead cells by metabolically active damaged tissue in the early stages of wound healing reduces the risk of bacterial infection [29]. It is critical to create a controlled environment in the wound area, such as aeration, temperature, and the availability of trace elements, vitamins, and minerals, in order to promote complex cellular activity during the recovery process [30,31].

1.3 Wound dressings

Every year, thousands of people are burned by hot water, fires, accidents, and boiling oil. Such mishaps frequently result in disabilities during therapy, excessive medical costs, and even death.

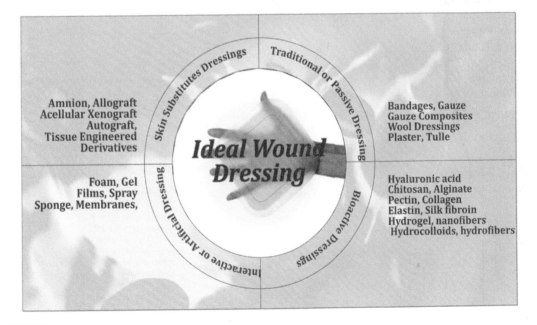

FIGURE 1.1

Classification of dressings for wounds.

Keeping wounds clean and moist while preventing bacterial and fungal infections are difficult aspects of wound care management [9]. Skin wound management focuses on fast and effective healing. This goal can be achieved by dressing the wound. The wound dressing is a critical component and clinical protocol of care for the healing process [32] which have an essential function to play in the control of exudate, as well as infections [33]. Wound dressing materials have a variety of applications. As a result, wound dressings are classified into four types based on their interaction with biological tissue: dressings that are traditional or passive, skin substitutes, artificial or interactive dressings, and bioactive dressing [34−37] (Fig. 1.2).

From 2200 BC, one of the earliest medical journals describes wound healing as "three therapeutic gestures": washing, plastering, and bandaging. Although wound healing technology appears to be unchanged, it has advanced [38]. Wound dressing and healing research was largely ignored until the mid-1960s. It was previously assumed that keeping the wound dry and exposed aids in faster and more efficient healing [39,40]. The ideal conditions for wound healing were initially proven by Winter [41] using the first generation of polymeric materials used in wound dressings.

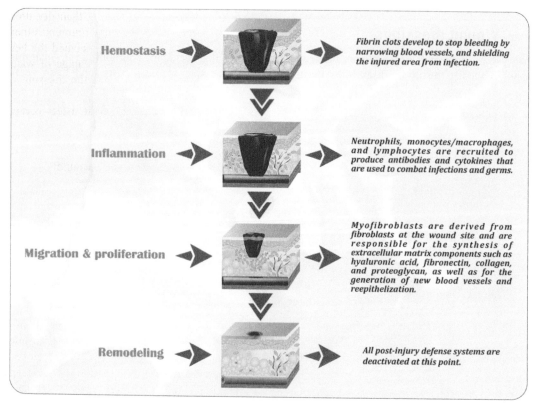

FIGURE 1.2

The healing procedure for a wound.

This discovery has paved the way for wound dressing to go from inert to functionally active materials, radically altering how wounds are treated.

A wound dressing should facilitate self-healing and shield damaged skin from external pressures, dust, and pathogens [42]. Dressings should be cytocompatible, soft, flexible, mechanically robust, gas permeable, and able to handle wound exudates [43]. Chronic and nonhealing wounds have a significant societal and economic impact. When wounds become infected with bacteria, they heal more slowly or not at all. As a result, as the prevalence of chronic wounds rises, so does the demand for more sophisticated therapeutic wound care. Current wound dressings fall short of ideal criteria, necessitating research into new, more effective dressings [44]. The wound care industry has recently been aware of a rising need for cutting-edge antimicrobial wound dressings [18].

1.4 Ideal wound dressing requirements

The best wound dressing maintains wound moisture, allows oxygen to pass through, absorbs wound exudate, accelerates re-epithelialization for wound closure, and reduces healing time, pain, and infection [45]. Moist dressings have been shown to promote faster wound healing than dry dressings. This is because skin regeneration and repair can only occur in a moist environment; otherwise, eschars or inflammation will form. As a result, wet or moist dressings were deemed the best choice for skin repair and wound dressing. Furthermore, they contain a high percentage of water and are naturally permeable. Furthermore, oxygen must be allowed to permeate the dressing in order to nourish repairing cells [46].

In general, the best way to treat slow-healing wounds is with a dressing that meets certain requirements. These requirements are shown in Fig. 1.3:

1. a physical barrier to stop further harm and exposure;
2. a carrier for a bactericidal chemical to get rid of bacterial infections around the wound;
3. an elimination mechanism for metabolites;
4. a wound-healing environment conditioner [29].

Several different approaches may be used to provide the most effective skin wound dressings available.

1.5 Antimicrobial wound dressing

Antimicrobial dressings contain an antibacterial agent, a biocide that inhibits germ growth in a wound or on clear skin. As technology has advanced, several antiseptic treatments that are far more effective at killing bacteria without harming surrounding healthy tissue have been developed. Antiseptics include silver, titanium oxide, zinc oxide, and iodine. Antimicrobial antiseptic dressings have proven effective in preventing microbiological infections. Antibiotic resistance in bacteria is the leading cause of the need for antimicrobial dressings [47,48]. The primary types of wound dressings that retain moisture, include films, sponges, foams, hydrocolloids, alginates, and hydrogels (Fig. 1.4; Table 1.1).

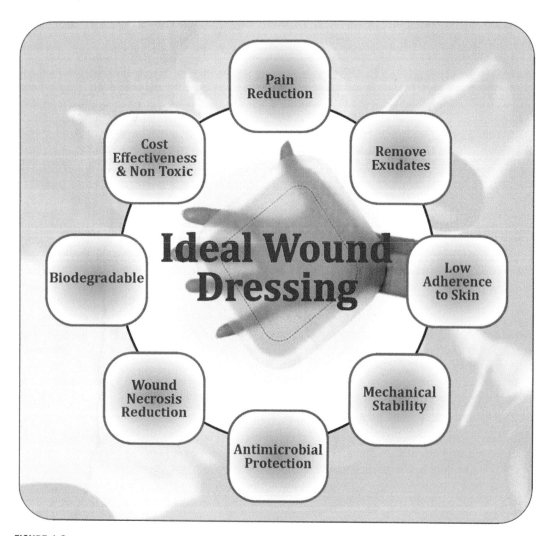

FIGURE 1.3

Ideal wound dressing requirements.

The ideal antimicrobial dressing must have a broad range of antimicrobial action against any pathogens responsible, be non-allergenic and non-toxic to the host, remove waste and maintain a humid environment in the wound area, release medications quickly and consistently, eliminate odor, and be reasonably priced [60,61]. The two most common types of antimicrobial wound dressing are antiseptic dressing and antibiotic dressing. Antiseptic dressings have numerous applications, including infection prevention and treatment caused by bacteria, fungi, protozoa, viruses, and prions [62]. Certain antiseptic dressings, on the other hand, cause cytotoxicity in host cells based on the number of doses, such as keratinocytes, fibroblast cells, and white corpuscles [63,64].

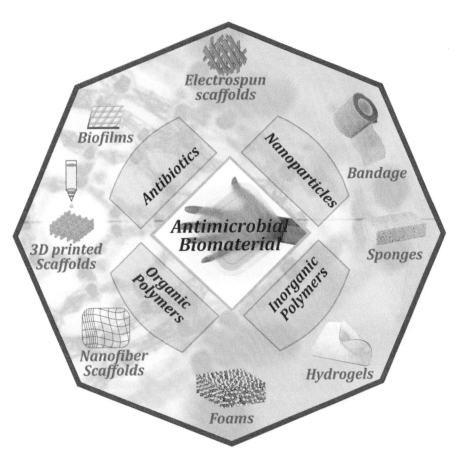

FIGURE 1.4

Wound dressing type.

Fibroblasts and keratinocytes are highly intoxicated by povidone-iodine concentrations larger than 0.05% and 0.004%, respectively [65]. Antibiotic dressings are harmless and can successfully treat target locations while causing no harm to the host [64].

1.6 Antimicrobial agents for wound dressing

Wound dressings are meant to prevent and/or control infection in the wound and surrounding areas. To meet these expectations, researchers have created wound dressings that use antibacterial agents in a variety of ways. Antimicrobials should be effective against a wide range of infections while causing minimal toxicity or sensitivities [66]. A high bioburden will impair normal wound healing and may result in the formation of nonhealing wounds. Consequently, antimicrobial-containing

Table 1.1 Wound dressings with antimicrobial agents.

S. No.	Wound dressing type	Polymeric material	Antibiotic drug	References
1	Hydrogels	poly(2-hydroxyethyl methacrylate) S-Nitrosothiol Polyvinylalcohol	Nitric Oxide	Masters et al. [49–51]
2	Microsphere	Gelatin Chitosan	Doxycycline Levofloxacin	Adhirajan et al. [52,53]
3	Films	PVA/sodium alginate PVA/dextran	Clindamycin and nitrofurazone Gentamicin	Kim et al. [54–56]
4	Nanofiber mats	Polyurethane/dextran PVA/poly(vinyl acetate)	Ciprofloxacin	Unnithan et al. [57]
5	Scaffolds	Chitosan/polyethylene glycol Collagen	Ciprofloxacin Doxycycline	Adhirajan et al. [58,59]

dressings are frequently employed [67]. Natural products, antibiotics, and nanoparticles are used nowadays as antimicrobial agents.

1.6.1 Natural products for wound dressing

Because they are safer, naturally derived antimicrobials are becoming more popular in antimicrobial dressings. Biochemically produced bacteriocins, plant extracts, and enzymes are used in the antimicrobial dressing [68]. Dressings are now made from a wide range of polymers, both synthetic and natural. Poly (methacrylates), polyvinyl pyrrolidine, and other synthetic dressings are examples. These dressings are difficult to handle due to their low mechanical strength. As a result, considerable effort has been expended in developing dressings based on natural polymers [21]. Essential oil (EO), honey, and turmeric are used as natural products for wound dressing.

1.6.1.1 Essential oil dressing

Dressing with EO is high in bioactive components that have anti-allergic, antioxidant, antiviral, antibacterial, and rejuvenating properties [69]. Bioactive wound dressings with slow-release EOs keep wounds moist. This creates an ideal environment for wound healing [70]. Environmental factors such as the latitude at which plants are grown, the stage of growth at which they were selected, the drying techniques used, and the conditions under which they were stored all have an impact on the EO content. As a result, it appears that using pure chemicals found in EO composition is reasonable [71]. Antimicrobial properties of herbal EOs have been known for millennia. When it comes to antimicrobial composites, they have a lot more potential, particularly in terms of resistance to bacterial strains [72]. Many EOs, including thyme, peppermint, cinnamon, rosemary, tea tree, eucalyptus, lavender, and lemongrass, have antibacterial properties. Wound antibacterial properties of eugenol and limonene deposited in nanofluid-based magnetit [73]. Fermentation, expression, and solvent extraction are all common processes. Because of their extraordinary benefits, they have piqued the interest of many industries, including food, perfume, aromatherapy, and pharmaceuticals [74]. The health industry is becoming more accepting of plant EOs and their components due to their antioxidant and antibacterial properties [75]. It is a recent development to incorporate EOs into electrospun fibers for wound dressings [76].

1.6.1.2 Honey dressing

Honey is a naturally existing sweetener and is effective as a topical wound care treatment due to its broad-spectrum antibacterial action as well as its healing effects [77–79]. It can be stored for a long time and is easily assimilated even after a significant amount of time, making it an excellent choice for antimicrobial therapy due to the lack of toxicity or adverse effects, as well as the low cost of maintenance and easy availability [80]. According to studies, honey's bioactivities boost the immune system's response, reduce inflammation, and promote rapid autolytic debridement [79]. Despite being acidic, honey may provide the best environment for fibroblasts to function, making bacterial survival more difficult. For partial thickness burns and pressure ulcers, it outperforms amniotic membrane, silver sulfadiazine, and ethoxy-diamino-acridine with nitrofurazone [81]. Honey's high osmolarity causes water to be evacuated from the wound when applied topically, diluting the honey and stimulating the glucose-oxidase enzyme. It forms a protective layer over the wound, keeping it moist and promoting wound healing [82]. It was reused for wound care and as a wound-prevention dressing [78]. Honey promotes collagen production and angiogenesis, which aids in wound granulation and epithelialization after debridement. The sugar, amino acids, vitamins, and minerals in honey may promote cell growth. The acidity of honey promotes the release of oxygen from hemoglobin into tissues [83]. Honey also contains enzymes such as catalase, which can aid in healing. Honey is used to treat burns, infected wounds, and decubitus ulcers. It is used to prevent bacterial colonization and to speed wound healing after vulvectomy. Gram-positive and Gram-negative bacteria and fungi are inhibited by honey [84].

1.6.2 Antibiotics and other antiseptics for wound dressing

Various studies have also shown that certain bacteriostatic or bactericidal medications can aid in the healing of wounds. Despite the fact that a wide range of antibiotics have been shown to be extremely effective against infection-causing bacteria, antibacterial wound dressings have only used tetracyclines (TTC), quinolones, cephalosporins, and aminoglycosides. Although several antibiotics are effective in treating wound infections, overuse and/or misuse of these medications may result in infection resistance [18]. Thousands of antibiotics exist, but only about 1% are currently in clinical use due to toxicity concerns or host cell absorption limitations. Antibiotics can interfere with bacterial function or structure, as well as metabolic processes [20]. The current crop of topical antibacterial agents is rounded out by antiseptics such as polyhexamine and silver compounds such as silver sulfadiazine and ionic silver-impregnated dressings [60].

1.6.2.1 Iodine dressing

Iodine has been used by doctors to keep wounds from becoming infected for nearly 150 years. Povidone-iodine has been used as a topical therapy since the 1950s [85]. Povidone iodine (PVPI) is a wound antiseptic. PVPI is a water-soluble combination of iodine and polyvinylpyrrolidone that kills bacteria, viruses, fungi, protozoa, and yeasts [86]. PVPI preparations have been accessible without a prescription in many countries for decades and are widely regarded as effective antiseptics [87]. Iodine's primary function in wound treatment is as an antibacterial agent [88]. Iodine's microbicidal effects include oxidizing nucleotides, fatty/amino acids, and respiratory chain enzymes, rendering them inert, and inhibiting bacterial cellular processes and structures [89]. In addition to promoting healing, antibacterial agents like cadexomer iodine have been proven to be effective against biofilm-forming *P. aeruginosa* and *S. aureus* [90,91].

1.6.2.2 Tetracycline dressing

Tetracycline (TTC) antibiotics are a class of antibiotic that are widely used in veterinary medicine, human medicine, and agriculture. These residues encourage the growth of antibiotic-resistant bacteria, which can be harmful to human health and increase disease risk [92]. These antibiotics can be applied topically to effectively destroy bacteria [93]. It is commonly used to treat skin infections such as acne, periodontal infections, and urinary tract infections. TTC was tested for antimicrobial activity against a variety of pathogens and was discovered to have a low minimum inhibitory concentration [94]. They are effective against Gram-positive and Gram-negative bacteria, have clinical safety, intravenous (IV) and oral forms, and are tolerated by the majority of group members. Their antibacterial properties, as well as the specificity of antimicrobial drugs, are influenced by their intrinsic antibiotic resistance pathways [95]. Despite their solubility in water, TTCs are highly miscible with alcohols, dimethyl sulfoxide, and dimethyl formamide. In polar solvents such as chloroform, dichloromethane, and ethyl acetate, all TTCs are very slightly soluble [96]. Currently, TTC-containing ointments are also utilized for treatment, which can cause skin sensitivity in rare cases. TTCs are broad-spectrum antibiotics having bacteriostatic properties. This antibiotic is sensitive to *Streptococci, Listeria, pneumococci, Vibrio cholerae, Campylobacter jejuni*, and *Treponema pallidum*, among other microorganisms. Interfering with protein synthesis in bacterial cells by inhibiting transfer-RNA binding to the m-RNA-ribosome complex is part of their antibacterial strategy. This prevents protein synthesis from taking place. TTCs are used locally in amounts ranging from 2% to 3%. Furthermore, it has the potential to cause photosensitization of the skin (pigmentation) [97].

1.6.2.3 Polyhexamethylene biguanide dressing

Polyhexamethylene biguanide (PHMB), also known as polyhexanide, has been around for a while and is commonly used as a preservative in pool sanitizers, contact lens solutions, and cosmetics. As a wound antibacterial, PHMB is available in solutions, gels, non-adherent bacterial barriers, and biocellulose dressings [98]. PHMB is an efficient, non-toxic antimicrobial agent with an antibacterial action against Gram-negative and Gram-positive bacteria [99]. It has demonstrated efficacy against a vast array of pathogens, including *Escherichia coli, Staphylococcus epidermidis*, and even *Acanthamoeba castellanii* [100]. The agent's method of action is caused by the interaction between the positive PHMB group and the negative phospholipids in bacterial cell walls [101]. When applied to the skin, eyes, epithelium of the nose, or wounds, there have been no reports of PHMB causing any adverse effects. This substance is considered to be reasonably safe [102].

1.6.3 Nanoparticles for wound dressing

The discovery of new materials for wound dressings is a tough but necessary task [103]. Nanomedicine technologies, namely nanoparticles, are a viable strategy for the creation of potential antimicrobial medicines [104]. Recent consideration has been given to bioactive wound dressings that are activated by nanoparticles such as polymeric, silver, gold, glass, and zinc oxide to increase wound dressing potentials above traditional ones [105]. The incorporation of nanoparticles into scaffolds reflects the novel notion of "nanoparticle dressing," which has recently received a lot of interest for wound healing [106].

1.6.3.1 Silver nanoparticles

Antibiotic-resistant bacteria have reignited interest in silver and other non-antibiotic therapies that had previously been abandoned following the discovery of penicillin and other antibiotics [107]. With the advent of nanotechnology, a new therapeutic modality for the use of AgNPs in wounds has emerged [108]. AgNPs have demonstrated significant promise in a wide range of applications, including detection and diagnosis, medication transport, biomaterials and device coating, novel antimicrobial agents, and regeneration materials [48]. Silver-based lotions, ointments, and AgNPs-based wound dressings are commercially accessible [109,110].

Silver ions kill bacteria through a variety of mechanisms, including binding to the outer membrane and the bacterial cell wall, which changes the permeability and, by extension, the function of the bacterial cell membrane. Silver metal and its compounds were thus able to prevent wound infection [111]. Burns, wounds, and bacterial illnesses are treated with metallic silver, silver nitrate, and silver sulfadiazine. Antibiotic resistance in bacteria and the discovery of antibiotic-resistant species have sparked interest in silver solutions for chronic wound treatment. Ionic silver is a powerful antibacterial agent whose use in medicine has increased as a result of antibiotic resistance [108]. AgNPs are antibacterial compounds that can be used to inhibit the growth of a wide range of microorganisms. AgNPs are thus useful in a wide range of medical devices and antimicrobial control systems [112]. In response to AgNPs, reactive oxygen species and free radicals were produced, which could lead to apoptosis-like reactions, cell membrane disintegration, and DNA damage. As is well known, the antibacterial activity of AgNPs varies greatly depending on their size, shape, and surface characteristics [113].

1.6.3.2 Gold nanoparticles

AuNPs were created in collaboration with other chemicals to have progressive antibacterial and injured-tissue healing activity. A variety of AuNPs are frequently used to improve cutaneous wound dressings. AuNPs can cause cytotoxicity on either side during production and distribution in vivo. Because uncoated AuNPs are sensitive to pH, temperature, electrolytes, and solvents, they accumulate [114]. AuNPs have been used to augment the antibacterial efficacy of a conventional antibiotics [115]. Gram-negative bacteria can be killed by antibacterial intermediate-treated AuNPs. Furthermore, AuNP-containing electrospun polymeric fibers can be used as wound dressings [116]. AuNPs cause membrane damage and reduce energy metabolism in bacteria. They have a multivalent interaction with the bacterial cell membrane. In the future, AuNPs could be used to treat wound infections [117].

The AuNPs have outstanding antimicrobial properties against both laboratory and clinical MDR isolates, and they are non-toxic to human cells. AuNPs are intriguing antibacterial agents due to their nontoxicity, polyvalent effects, and combined accessibility of functionalization [118]. Small-molecule capped AuNPs containing mercaptan, amine, hydrosulfonyl, or phosphonic groups have antibacterial effects [119]. AuNPs were used as a therapeutic therapy for chronic wound healing because they decrease the manufacture of pro-inflammatory cytokines such as TNF - α, IL -12, and

IL -6, as well as wound repair proteins [120]. AuNPs serve two purposes: they preserve the wound site and act as a replacement space for tissue formation. AuNPs are a novel type of nanoparticle that could be used to treat skin abnormalities. AuNPs were investigated for use as wound dressing material. AuNPs' antibacterial properties, hydrophilicity, and mechanical strength were improved, resulting in a more favorable environment for cell proliferation and adhesion [121].

1.6.4 Polymer-based antimicrobial dressing

European Centre for Disease Prevention and Control (ECDC) reported approximately 33,000 antibiotic-resistant bacteria-related deaths in [122]. Antimicrobial resistance is thus a global issue. Thus, natural and synthetic molecules are a valuable resource for developing novel antimicrobials. Polymers that inhibit microorganism growth are the next big thing in antimicrobials [123]. Natural polymers, such as polysaccharides, are of particular interest to researchers because they are abundant in nature, do not harm human tissues, and have ideal physical and chemical properties. Polysaccharides and their derivatives can do more than simply heal wounds [1]. Alginate, collagen, chitosan (CS), gelatin, and hyaluronic acid were utilized to produce most wound dressings because of their biocompatibility and biodegradability [14] (Fig. 1.5).

CS is a biopolymer that is widely used in medicine, particularly in wound dressings. This is because it aids wound healing by preventing bleeding, inhibiting bacterial growth, being biocompatible and biodegradable, preventing scarring, and rapidly releasing drugs from the matrix [15]. CS is a form of chitin that has been partially or completely deacetylated. Chitin is a natural cationic polysaccharide made of $(1 \rightarrow 4)$-2-amino-2-deoxy-β-d-glucan [124]. CS is an easy-to-make,

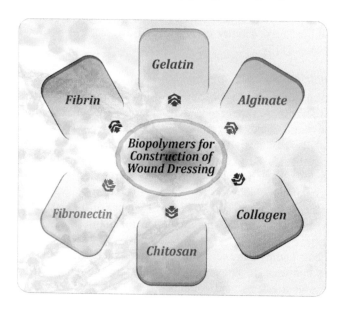

FIGURE 1.5

Biopolymers for construction of wound dressings.

affordable, long-lasting, biodegradable, non-toxic, and pathogen-killing natural cationic polysaccharide [125]. Biocellulose is widely used biomedically which is a trendy novel wound dressing lacking antibacterial characteristics [126]. The antibiotics and antiseptics povidone-iodine (PI) and polihexanide (PHMB) were included in biocellulose dressings with antibacterial activity incorporated into them due to the biopolymer's nanostructured design and intended for use in the fight against infection [45,127,128]. Bacteria live a biofilm lifestyle, in which they cling to surfaces and are protected from host defenses and biocides by a matrix of self-produced Extracellular Polymeric Substances (EPS). Biofilm containing antibiotic-resistant microorganisms poses a threat to public health. By destroying biofilms, antibiotics and other antimicrobials can be made more effective [129].

Natural sodium alginate is used to treat wounds. This polymer is being researched for biomedical applications as an extracellular matrix material and scaffold for tissue engineering. Although sodium alginate has low heat stability, high hydrophilicity, and poor mechanical properties, it also has disadvantages depending on its application [130]. Cui et al. [131] developed pH-controllable CuO_2-loaded gelatin sponge dressings for wound healing. pH-responsive dressings release Cu_{2+} and H_2O_2 in bacterial-infected wounds. After that, a Fenton-like reaction between the breakdown products should produce poisonous •OH, disrupting the bacterial barrier and causing bacterial death without harming surrounding tissues.

1.7   Conclusions and future perspectives

People may die as a result of wound dehydration, severe infection, and insufficient mechanical stress protection for damaged tissue. Almost everyone has had an open skin wound as a result of an accident, a burn, a condition (such as diabetes), or a surgical procedure. Some environmental infections, mucosal bacteria, and skin microflora from nearby skin can easily infect these wounds. There are over 3000 different types of bandages available, allowing doctors to treat any wound. Chronic wounds are notoriously difficult to treat, such as venous leg ulcers, diabetic wounds, and pressure ulcers. Bacterial infections, on the other hand, can slow wound healing and even endanger the patient's life. Incorporating bioactive compounds and antimicrobial agents into a dressing can improve its biocompatibility and biological properties, allowing for faster healing. This may have an effect on the dressing's exudate absorption and mechanical properties. Antiseptics may help prevent resistance, but they are toxic to human cells. Medication delivery methods are critical in these situations. As a result, more research is needed to overcome or reduce negative effects associated with biomaterial alterations utilizing bioactive chemicals, such as mechanical property degradation or a decrease in cell proliferation. Wound dressings of the future should be multifunctional devices (sensors and medicinal substances) that promote healing, prevent infection, and monitor wound condition. To summarize, significant effort has gone into developing effective treatment options for infected wounds.

Competing interests declaration

The authors state that they have no conflicts of interest.

Acknowledgments

We thank the Director, R&D of Biyani Girls College, Jaipur; Head, Department of Botany, University of Rajasthan, Jaipur, and Chairman and Vice Chairman of DPG Degree College, Haryana, India for their support and encouragement.

References

[1] M. Rahimi, E.B. Noruzi, E. Sheykhsaran, B. Ebadi, Z. Kariminezhad, M. Molaparast, et al., Carbohydrate polymer-based silver nanocomposites: recent progress in the antimicrobial wound dressings, Carbohydr. Polym. 231 (2020) 115696.

[2] Z. Muwaffak, A. Goyanes, V. Clark, A.W. Basit, S.T. Hilton, S. Gaisford, Patient-specific 3D scanned and 3D printed antimicrobial polycaprolactone wound dressings, Int. J. Pharm. 527 (1−2) (2017) 161−170.

[3] A. Moeini, P. Pedram, P. Makvandi, M. Malinconico, G. Gomezd'Ayala, Wound healing and antimicrobial effect of active secondary metabolites in chitosan-based wound dressings: a review, Carbohydr. Polym. 233 (2020) 115839.

[4] S. MacNeil, Progress and opportunities for tissue-engineered skin, Nature. 445 (7130) (2007) 874−880.

[5] A. Lumbreras-Aguayo, H.I. Melendez-Ortiz, B. Puente-Urbina, C. Alvarado-Canche, A. Ledezma, J. Romero-Garcia, et al., Poly (methacrylic acid)-modified medical cotton gauzes with antimicrobial and drug delivery properties for their use as wound dressings, Carbohydr. Polym. 205 (2019) 203−210.

[6] T. Velnar, T. Bailey, V. Smrkolj, The wound healing process: an overview of the cellular and molecular mechanisms, J. Int. Med. Res. 37 (5) (2009) 1528−1542.

[7] J.E. Mellerio, Infection and colonization in epidermolysis bullosa, Dermatol. Clin. 28 (2) (2010) 267−269.

[8] Armstrong, D.G., Meyr, A.J., Basic principles of wound management - UpToDate. UpToDate 1−50, 2016.

[9] E. Shams, H. Yeganeh, H. Naderi-Manesh, R. Gharibi, Z. Mohammad Hassan, Polyurethane/siloxane membranes containing graphene oxide nanoplatelets as antimicrobial wound dressings: *in vitro* and *in vivo* evaluations, J. Mater. Sci. Mater. Med. 28 (5) (2017) 75.

[10] C. Dhand, M. Venkatesh, V.A. Barathi, S. Harini, S. Bairagi, E. Goh Tze Leng, et al., Bio-inspired crosslinking and matrix-drug interactions for advanced wound dressings with long-term antimicrobial activity, Biomaterials. 138 (2017) 153−168.

[11] Y. Zhong, H. Xiao, F. Seidi, Y. Jin, Natural polymer-based antimicrobial hydrogels without synthetic antibiotics as wound dressings, Biomacromolecules. 21 (8) (2020) 2983−3006.

[12] R. Gharibi, S. Kazemi, H. Yeganeh, V. Tafakori, Utilizing dextran to improve hemocompatibility of antimicrobial wound dressings with embedded quaternary ammonium salts, Int. J. Biol. Macromol. 131 (2019) 1044−1056.

[13] J.S. Boateng, K.H. Matthews, H.N.E. Stevens, G.M. Eccleston, Wound healing dressings and drug delivery systems: a review, J. Pharm. Sci. 97 (8) (2008) 2892−2923.

[14] S. Hu, X. Cai, X. Qu, B. Yu, C. Yan, J. Yang, et al., Preparation of biocompatible wound dressings with long-term antimicrobial activity through covalent bonding of antibiotic agents to natural polymers, Int. J. Biol. Macromol. 123 (2019) 1320−1330.

[15] S. Anjum, A. Arora, M.S. Alam, B. Gupta, Development of antimicrobial and scar preventive chitosan hydrogel wound dressings, Int. J. Pharm. 508 (1−2) (2016) 92−101.

[16] D.J. Leaper, G. Schultz, K. Carville, J. Fletcher, T. Swanson, R. Drake, Extending the TIME concept: what have we learned in the past 10 years? Int. Wound J. 9 (2012) 1−19.

[17] J.A. Lemire, J.J. Harrison, R.J. Turner, Antimicrobial activity of metals: mechanisms, molecular targets and applications, Nat. Rev. Microbiol. 11 (6) (2013) 371−384.

[18] W.A. Sarhan, H.M. Azzazy, I.M. El-Sherbiny, Honey/chitosan nanofiber wound dressing enriched with *Allium sativum* and *Cleome droserifolia*: enhanced antimicrobial and wound healing activity, ACS Appl. Mater. Interfaces. 8 (10) (2016) 6379−6390.

[19] A. Chaudhari, K. Vig, D. Baganizi, R. Sahu, S. Dixit, V. Dennis, et al., Future prospects for scaffolding methods and biomaterials in skin tissue engineering: a review, Int. J. Mol. Sci. 17 (12) (2016) 1974.

[20] D. Simoes, S.P. Miguel, M.P. Ribeiro, P. Coutinho, A.G. Mendonça, I.J. Correia, Recent advances on antimicrobial wound dressing: a review, Eur. J. Pharm. Biopharm. 127 (2018) 130−141.

[21] O. Sarheed, A. Ahmed, D. Shouqair, J. Boateng, Antimicrobial dressings for improving wound healing, Wound Healing-New Insights into Ancient Challenges, InTech, 2016, pp. 373−398.

[22] F. Gottrup, A. Melling, D.A. Hollander, An overview of surgical site infections: aetiology, incidence and risk factors, World Wide Wounds. 5 (2) (2005) 11−15.

[23] S.A. Eming, P. Martin, M. Tomic-Canic, Wound repair and regeneration: mechanisms, signaling, and translation, Sci. Transl. Med. 6 (265) (2014) 265sr6. 265sr6.

[24] K. Jarbrink, G. Ni, H. Sönnergren, A. Schmidtchen, C. Pang, R. Bajpai, et al., Prevalence and incidence of chronic wounds and related complications: a protocol for a systematic review, Syst. Rev. 5 (1) (2016) 152.

[25] A.S. Weyrich, G.A. Zimmerman, Platelets: signaling cells in the immune continuum, Trends Immunol. 25 (9) (2004) 489−495.

[26] P. Agrawal, S. Soni, G. Mittal, A. Bhatnagar, Role of polymeric biomaterials as wound healing agents, Int. J. Low. Extrem. Wounds 13 (3) (2014) 180−190.

[27] P.I. Morgado, A. Aguiar-Ricardo, I.J. Correia, Asymmetric membranes as ideal wound dressings: an overview on production methods, structure, properties and performance relationship, J. Memb. Sci. 490 (2015) 139−151.

[28] J.M. Reinke, H. Sorg, Wound repair and regeneration, Eur. Surg. Res. 49 (1) (2012) 35−43.

[29] Y. Zhang, M. Jiang, Y. Zhang, Q. Cao, X. Wang, Y. Han, et al., Novel lignin−chitosan−PVA composite hydrogel for wound dressing, Mater. Sci. Eng. C. 104 (2019) 110002.

[30] S. Singh, A. Young, C.-E. McNaught, The physiology of wound healing, Surg 35 (9) (2017) 473−477.

[31] E. Handayani, H.T. Damailia, W. Pujiastuti, Factor affecting perineal wound healing, J. Midwifery Sci. Basic. Appl. Res. 2 (2) (2020) 38−42.

[32] B. Atiyeh, J. Ioannovich, C. Al-Amm, K. El-Musa, Management of acute and chronic open wounds: the importance of moist environment in optimal wound healing, Curr. Pharm. Biotechnol. 3 (3) (2002) 179−195.

[33] D.G. Metcalf, D. Parsons, P.G. Bowler, Clinical safety and effectiveness evaluation of a new antimicrobial wound dressing designed to manage exudate, infection and biofilm, Int. Wound J. 14 (1) (2017) 203−213.

[34] A.E. Stoica, C. Chircov, A.M. Grumezescu, Nanomaterials for wound dressings: an up-to-date overview, Molecules. 25 (11) (2020) 2699.

[35] V. Vivcharenko, A. Przekora, Modifications of wound dressings with bioactive agents to achieve improved pro-healing properties, Appl. Sci. 11 (9) (2021) 4114.

[36] S. Alven, X. Nqoro, B.A. Aderibigbe, Polymer-based materials loaded with curcumin for wound healing applications, Polymers. 12 (10) (2020) 2286.

[37] S.P. Ndlovu, K. Ngece, S. Alven, B.A. Aderibigbe, Gelatin-based hybrid scaffolds: promising wound dressings, Polymers. 13 (17) (2021) 2959.

[38] E. Rezvani Ghomi, S. Khalili, S. Nouri Khorasani, R. Esmaeely Neisiany, S. Ramakrishna, Wound dressings: current advances and future directions, J. Appl. Polym. Sci. 136 (27) (2019) 47738.

[39] E.A. Kamoun, X. Chen, M.S. Mohy Eldin, E.R.S. Kenawy, Crosslinked poly (vinyl alcohol) hydrogels for wound dressing applications: a review of remarkably blended polymers, Arab. J. Chem. 8 (1) (2015) 1−14.

[40] M. Kokabi, M. Sirousazar, Z.M. Hassan, PVA−clay nanocomposite hydrogels for wound dressing, Eur. Polym. J. 43 (3) (2007) 773−781.

[41] G.D. Winter, Formation of the scab and the rate of epithelization of superficial wounds in the skin of the young domestic pig, Nature. 193 (4812) (1962) 293−294.

[42] R. Gharibi, H. Yeganeh, Z. Abdali, Preparation of antimicrobial wound dressings via thiol−ene photopolymerization reaction, J. Mater. Sci. 53 (3) (2018) 1581−1595.

[43] I.R. Sweeney, M. Miraftab, G. Collyer, A critical review of modern and emerging absorbent dressings used to treat exuding wounds, Int. Wound J. 9 (6) (2012) 601−612.

[44] T.A. Jeckson, Y.P. Neo, S.P. Sisinthy, B. Gorain, Delivery of therapeutics from layer-by-layer electrospun nanofiber matrix for wound healing: an update, J. Pharm. Sci. 110 (2) (2021) 635−653.

[45] S. Napavichayanun, P. Amornsudthiwat, P. Pienpinijtham, P. Aramwit, Interaction and effectiveness of antimicrobials along with healing-promoting agents in a novel biocellulose wound dressing, Mater. Sci. Eng. C55 (2015) 95−104.

[46] E.A. Kamoun, E.R.S. Kenawy, X. Chen, A review on polymeric hydrogel membranes for wound dressing applications: PVA-based hydrogel dressings, J. Adv. Res. 8 (3) (2017) 217−233.

[47] A. Mohandas, S. Deepthi, R. Biswas, R. Jayakumar, Chitosan based metallic nanocomposite scaffolds as antimicrobial wound dressings, Bioact. Mater. 3 (3) (2018) 267−277.

[48] F. Paladini, M. Pollini, Antimicrobial silver nanoparticles for wound healing application: progress and future trends, Mater. (Basel) 12 (16) (2019) 2540.

[49] K.S.B. Masters, S.J. Leibovich, P. Belem, J.L. West, L.A. Poole-Warren, Effects of nitric oxide releasing poly (vinyl alcohol) hydrogel dressings on dermal wound healing in diabetic mice, Wound Repair. Regen. 10 (5) (2002) 286−294.

[50] G.M. Halpenny, R.C. Steinhardt, K.A. Okialda, P.K. Mascharak, Characterization of pHEMA-based hydrogels that exhibit light-induced bactericidal effect via release of NO, J. Mater. Sci. Mater. Med. 20 (11) (2009) 2353−2360.

[51] Y. Li, P.I. Lee, Controlled nitric oxide delivery platform based on S -nitrosothiol conjugated interpolymer complexes for diabetic wound healing, Mol. Pharm. 7 (1) (2010) 254−266.

[52] N. Adhirajan, N. Shanmugasundaram, S. Shanmuganathan, M. Babu, Functionally modified gelatin microspheres impregnated collagen scaffold as novel wound dressing to attenuate the proteases and bacterial growth, Eur. J. Pharm. Sci. 36 (2−3) (2009) 235−245.

[53] J. Guan, L.Z. Dong, S.J. Huang, M.L. Jing, Characterization of wound dressing with microspheres containing levofloxacin, Proc. 2010 Int. Conf. Inf. Technol. Sci. Manag. 1−2 (2010) 344−348.

[54] M.-R. Hwang, J.O. Kim, J.H. Lee, Y.I. Kim, J.H. Kim, et al., Gentamicin-loaded wound dressing with polyvinyl alcohol/dextran hydrogel: gel characterization and *in vivo* healing evaluation, AAPS PharmSciTech. 11 (3) (2010) 1092−1103.

[55] J.O. Kim, J.Y. Choi, J.K. Park, J.H. Kim, S.G. Jin, S.W. Chang, et al., Development of Clindamycin-loaded wound dressing with polyvinyl alcohol and sodium alginate, Biol. Pharm. Bull. 31 (12) (2008) 2277−2282.

[56] J.O. Kim, J.K. Park, J.H. Kim, S.G. Jin, C.S. Yong, D.X. Li, et al., Development of polyvinyl alcohol–sodium alginate gel-matrix-based wound dressing system containing nitrofurazone, Int. J. Pharm. 359 (1–2) (2008) 79–86.

[57] A.R. Unnithan, N.A.M. Barakat, P.B. Tirupathi Pichiah, G. Gnanasekaran, R. Nirmala, Y.S. Cha, et al., Wound-dressing materials with antibacterial activity from electrospun polyurethane-dextran nanofiber mats containing ciprofloxacin HCl, Carbohydr. Polym. 90 (4) (2012) 1786–1793.

[58] M. Sinha, R.M. Banik, C. Haldar, P. Maiti, Development of ciprofloxacin hydrochloride loaded poly (ethylene glycol)/chitosan scaffold as wound dressing, J. Porous Mater. 20 (4) (2013) 799–807.

[59] N. Adhirajan, N. Shanmugasundaram, S. Shanmuganathan, M. Babu, Collagen-based wound dressing for doxycycline delivery: *in-vivo* evaluation in an infected excisional wound model in rats, J. Pharm. Pharmacol. 61 (12) (2009) 1617–1623.

[60] K. Vowden, P. Vowden, Wound dressings: principles and practice, Surg 35 (9) (2017) 489–494.

[61] K.F. Cutting, Wound dressings: 21st century performance requirements, J. Wound Care. 19 (Sup 1) (2010) 4–9.

[62] G. Gethin, Role of topical antimicrobials in wound management, J. Wound Care (2009) 4–8.

[63] A. Drosou, A. Falabella, R.S. Kirsner, Antiseptics on wounds: an area of controversy, Wounds. 15 (2003) 149–166.

[64] B.A. Lipsky, C. Hoey, Topical antimicrobial therapy for treating chronic wounds, Clin. Infect. Dis. 49 (10) (2009) 1541–1549.

[65] R.I. Burks, Povidone-Iodine solution in wound treatment, Phys. Ther. 78 (2) (1998) 212–218.

[66] W. Zhong, Efficacy and toxicity of antibacterial agents used in wound dressings, Cutan. Ocul. Toxicol. 34 (1) (2015) 61–67.

[67] C. Wiegand, M. Abel, P. Ruth, P. Elsner, U.C. Hipler, *In vitro* assessment of the antimicrobial activity of wound dressings: influence of the test method selected and impact of the pH, J. Mater. Sci. Mater. Med. 26 (1) (2015) 18.

[68] R. Irkin, O.K. Esmer, Novel food packaging systems with natural antimicrobial agents, J. Food Sci. Technol. 52 (10) (2015) 6095–6111.

[69] I. Negut, V. Grumezescu, A. Grumezescu, Treatment strategies for infected wounds, Molecules. 23 (9) (2018) 2392.

[70] K. Lee, S. Lee, Electrospun nanofibrous membranes with essential oils for wound dressing applications, Fibers Polym. 21 (5) (2020) 999–1012.

[71] M. Michalska-Sionkowska, M. Walczak, A. Sionkowska, Antimicrobial activity of collagen material with thymol addition for potential application as wound dressing, Polym. Test. 63 (2017) 360–366.

[72] F. Altaf, M.B.K. Niazi, Z. Jahan, T. Ahmad, M.A. Akram, A. Safdar, et al., Synthesis and characterization of PVA/Starch hydrogel membranes incorporating essential oils aimed to be used in wound dressing applications, J. Polym. Environ. 29 (1) (2021) 156–174.

[73] I. Liakos, L. Rizzello, D.J. Scurr, P.P. Pompa, I.S. Bayer, A. Athanassiou, All-natural composite wound dressing films of essential oils encapsulated in sodium alginate with antimicrobial properties, Int. J. Pharm. 463 (2) (2014) 137–145.

[74] E.P. dos Santos, P.H.M. Nicácio, F.C. Barbosa, H.N. da Silva, A.L.S. Andrade, M.V.L. Fook, et al., Chitosan/essential oils formulations for potential use as wound dressing: physical and antimicrobial properties, Materials (Basel) 12 (14) (2019) 2223.

[75] G. Kavoosi, A. Rahmatollahi, S. Mohammad Mahdi Dadfar, A. Mohammadi Purfard, Effects of essential oil on the water binding capacity, physico-mechanical properties, antioxidant and antibacterial activity of gelatin films, LWT - Food Sci. Technol. 57 (2) (2014) 556–561.

[76] I. Unalan, S.J. Endlein, B. Slavik, A. Buettner, W.H. Goldmann, R. Detsch, et al., Evaluation of electro-spun Poly (ε-Caprolactone)/Gelatin nanofiber mats containing clove essential oil for antibacterial wound dressing, Pharmaceutics. 11 (11) (2019) 570.

[77] A. Bergman, J. Yanai, J. Weiss, D. Bell, M.P. David, Acceleration of wound healing by topical application of honey, Am. J. Surg. 145 (3) (1983) 374−376.

[78] F.D. Halstead, M.A. Webber, M. Rauf, R. Burt, M. Dryden, B.A. Oppenheim, *In vitro* activity of an engineered honey, medical-grade honeys, and antimicrobial wound dressings against biofilm-producing clinical bacterial isolates, J. Wound Care 25 (2) (2016) 93−102.

[79] P. Molan, T. Rhodes, Honey: a biologic wound dressing, Wounds. 27 (6) (2015) 141−151.

[80] T. Md Abu, K.A. Zahan, M.A. Rajaie, C.R. Leong, S. Ab Rashid, N.S. Mohd Nor Hamin, et al., Nanocellulose as drug delivery system for honey as antimicrobial wound dressing, Mater. Today Proc. 31 (2020) 14−17.

[81] S.S. Abou Zekry, A. Abdellatif, H.M.E. Azzazy, Fabrication of pomegranate/honey nanofibers for use as antibacterial wound dressings, Wound Med. 28 (2020) 100181.

[82] A.M. Scagnelli, Therapeutic review: Manuka honey, J. Exot. Pet. Med. 25 (2) (2016) 168−171.

[83] K. Lay-flurrie, Honey in wound care: effects, clinical application and patient benefit, Br. J. Nurs. 17 (Sup 5) (2008) S30−S36.

[84] Z. Vardi, N. Barzilay, H.A. Linder, A. Coh, Local application of honey for treatment of neonatal postoperative wound infection, Acta Paediatr. 87 (4) (1998) 429−432.

[85] M.J. Hoekstra, S.J. Westgate, S. Mueller, Povidone-iodine ointment demonstrates *in vitro* efficacy against biofilm formation, Int. Wound J. 14 (1) (2017) 172−179.

[86] M. Summa, D. Russo, I. Penna, N. Margaroli, I.S. Bayer, T. Bandiera, et al., A biocompatible sodium alginate/povidone iodine film enhances wound healing, Eur. J. Pharm. Biopharm. 122 (2018) 17−24.

[87] B. Globel, H. Globel, C. Andres, [Iodine resorption from PVP-iodine preparations after their use in humans], Dtsch. Med. Wochenschr. 109 (37) (1984) 1401−1404.

[88] P.L. Bigliardi, S.A.L. Alsagoff, H.Y. El-Kafrawi, J.-K. Pyon, C.T.C. Wa, M.A. Villa, Povidone iodine in wound healing: a review of current concepts and practices, Int. J. Surg. 44 (2017) 260−268.

[89] J. Kanagalingam, R. Feliciano, J.H. Hah, H. Labib, T.A. Le, J.-C. Lin, Practical use of povidone-iodine antiseptic in the maintenance of oral health and in the prevention and treatment of common oropharyngeal infections, Int. J. Clin. Pract. 69 (11) (2015) 1247−1256.

[90] L. Danielsen, G.W. Cherry, K. Harding, O. Rollman, Cadexomer iodine in ulcers colonised by *Pseudomonas aeruginosa*, J. Wound Care. 6 (4) (1997) 169−172.

[91] P.M. Mertz, M.F. Oliveira-Gandia, S.C. Davis, The evaluation of a Cadexomer Iodine wound dressing on Methicillin Resistant *Staphylococcus aureus* (MRSA) in acute wounds, Dermatol. Surg. 25 (2) (1999) 89−93.

[92] R. Daghrir, P. Drogui, Tetracycline antibiotics in the environment: a review, Environ. Chem. Lett. 11 (3) (2013) 209−227.

[93] G. Kiaee, M. Etaat, B. Kiaee, S. Kiaei, H.A. Javar, Multilayered controlled released topical patch containing tetracycline for wound dressing, J. Silico Vitr. Pharmacol. 02 (02) (2016).

[94] A.I. Rezk, J.Y. Lee, B.C. Son, C.H. Park, C.S. Kim, Bi-layered nanofibers membrane loaded with titanium oxide and tetracycline as controlled drug delivery system for wound dressing applications, Polymers (Basel) 11 (10) (2019) 1602.

[95] T.H. Grossman, Tetracycline antibiotics and resistance, Cold Spring Harb. Perspect. Med. 6 (4) (2016) a025387.

[96] S. O'Connor, D.S. Aga, Analysis of tetracycline antibiotics in soil: advances in extraction, clean-up, and quantification, TrAC. - Trends Anal. Chem. 26 (6) (2007) 456−465.

[97] I. Lovetinska-Slamborova, P. Holy, P. Exnar, I. Veverkova, Silica nanofibers with immobilized tetracycline for wound dressing, J. Nanomater. 2016 (2016) 1−6.

[98] E. To, R. Dyck, S. Gerber, S. Kadavil, K.Y. Woo, The effectiveness of topical Polyhexamethylene Biguanide (PHMB) agents for the treatment of chronic wounds: a systematic review, Surg. Technol. Int. 29 (2016) 45−51.

[99] S. Napavichayanun, R. Yamdech, P. Aramwit, The safety and efficacy of bacterial nanocellulose wound dressing incorporating sericin and polyhexamethylene biguanide: *in vitro, in vivo* and clinical studies, Arch. Dermatol. Res. 308 (2) (2016) 123−132.

[100] A. Worsley, K. Vassileva, J. Tsui, W. Song, L. Good, Polyhexamethylene Biguanide:Polyurethane blend nanofibrous membranes for wound infection control, Polym. (Basel) 11 (5) (2019) 915.

[101] K. Kaehn, Polihexanide: a safe and highly effective biocide, Skin. Pharmacol. Physiol. 23 (Suppl. 1) (2010) 7−16.

[102] N.F. Kamaruzzaman, R. Firdessa, L. Good, Bactericidal effects of polyhexamethylene biguanide against intracellular *Staphylococcus aureus* EMRSA-15 and USA 300, J. Antimicrob. Chemother. 71 (5) (2016) 1252−1259.

[103] J. Avossa, G. Pota, G. Vitiello, A. Macagnano, A. Zanfardino, M. Di Napoli, et al., Multifunctional mats by antimicrobial nanoparticles decoration for bioinspired smart wound dressing solutions, Mater. Sci. Eng. C. 123 (2021) 111954.

[104] V. Ambrogi, D. Pietrella, A. Donnadio, L. Latterini, A. Di Michele, I. Luffarelli, et al., Biocompatible alginate silica supported silver nanoparticles composite films for wound dressing with antibiofilm activity, Mater. Sci. Eng. C. 112 (2020) 110863.

[105] M. Rahimi, R. Ahmadi, H. Samadi Kafil, V. Shafiei-Irannejad, A novel bioactive quaternized chitosan and its silver-containing nanocomposites as a potent antimicrobial wound dressing: structural and biological properties, Mater. Sci. Eng. C. 101 (2019) 360−369.

[106] K. Kalantari, E. Mostafavi, A.M. Afifi, Z. Izadiyan, H. Jahangirian, R. Rafiee-Moghaddam, et al., Wound dressings functionalized with silver nanoparticles: promises and pitfalls, Nanoscale. 12 (4) (2020) 2268−2291.

[107] N. Beyth, Y. Houri-Haddad, A. Domb, W. Khan, R. Hazan, Alternative antimicrobial approach: nano-antimicrobial materials. Evidence-based complement, Altern. Med. 2015 (2015) 1−16.

[108] R. Singh, D. Singh, Chitin membranes containing silver nanoparticles for wound dressing application, Int. Wound J. 11 (3) (2014) 264−268.

[109] S.S.D. Kumar, N.K. Rajendran, N.N. Houreld, H. Abrahamse, Recent advances on silver nanoparticle and biopolymer-based biomaterials for wound healing applications, Int. J. Biol. Macromol. 115 (2018) 165−175.

[110] A. Paduraru, C. Ghitulica, R. Trusca, V.A. Surdu, I.A. Neacsu, A.M. Holban, et al., Antimicrobial wound dressings as potential materials for skin tissue regeneration, Mater. (Basel) 12 (11) (2019) 1859.

[111] T. Maneerung, S. Tokura, R. Rujiravanit, Impregnation of silver nanoparticles into bacterial cellulose for antimicrobial wound dressing, Carbohydr. Polym. 72 (1) (2008) 43−51.

[112] A. Hebeish, M.H. El-Rafie, M.A. EL-Sheikh, A.A. Seleem, M.E. El-Naggar, Antimicrobial wound dressing and anti-inflammatory efficacy of silver nanoparticles, Int. J. Biol. Macromol. 65 (2014) 509−515.

[113] N. Eghbalifam, S.A. Shojaosadati, S. Hashemi-Najafabadi, A.C. Khorasani, Synthesis and characterization of antimicrobial wound dressing material based on silver nanoparticles loaded gum Arabic nanofibers, Int. J. Biol. Macromol. 155 (2020) 119−130.

[114] S. Al-Musawi, S. Albukhaty, H. Al-Karagoly, G.M. Sulaiman, M.S. Alwahibi, Y.H. Dewir, et al., Antibacterial activity of honey/chitosan nanofibers loaded with capsaicin and gold nanoparticles for wound dressing, Molecules 25 (20) (2020) 4770.

[115] L. Zhou, K. Yu, F. Lu, G. Lan, F. Dai, S. Shang, et al., Minimizing antibiotic dosage through in situ formation of gold nanoparticles across antibacterial wound dressings: a facile approach using silk fabric as the base substrate, J. Clean. Prod. 243 (2020) 118604.

[116] X. Yang, J. Yang, L. Wang, B. Ran, Y. Jia, L. Zhang, et al., Pharmaceutical intermediate-modified gold nanoparticles: against multidrug-resistant bacteria and wound-healing application via an electrospun scaffold, ACS Nano 11 (6) (2017) 5737–5745.

[117] B. Lu, H. Ye, S. Shang, Q. Xiong, K. Yu, Q. Li, et al., Novel wound dressing with chitosan gold nanoparticles capped with a small molecule for effective treatment of multiantibiotic-resistant bacterial infections, Nanotechnology. 29 (42) (2018) 425603.

[118] X. Zhao, Y. Jia, J. Li, R. Dong, J. Zhang, C. Ma, et al., Indole derivative-capped gold nanoparticles as an effective bactericide *in vivo*, ACS Appl. Mater. Interfaces. 10 (35) (2018) 29398–29406.

[119] K. Wang, Z. Qi, S. Pan, S. Zheng, H. Wang, Y. Chang, et al., Preparation, characterization and evaluation of a new film based on chitosan, arginine and gold nanoparticle derivatives for wound-healing efficacy, RSC Adv. 10 (35) (2020) 20886–20899.

[120] KM. Zepon, M.S. Marques, A.W. Hansen, C. Pucci, A.F. Do, F.D.P. Morisso, et al., Polymer-based wafers containing in situ synthesized gold nanoparticles as a potential wound-dressing material, Mater. Sci. Eng. C. 109 (2020) 110630.

[121] L. Wang, J. Yang, X. Yang, Q. Hou, S. Liu, W. Zheng, et al., Mercaptophenylboronic acid-activated gold nanoparticles as nanoantibiotics against multidrug-resistant bacteria, ACS Appl. Mater. Interfaces. 12 (46) (2020) 51148–51159.

[122] ECDC. European Centre for Disease Prevention and Control. <https://www.ecdc.europa.eu/en/news-events/33000-people-die-every-year-due-infections-antibiotic-resistant-bacteria>, 2019 (accessed 16.11.19).

[123] A. Matica, A. Tøndervik, H. Sletta, V. Ostafe, Chitosan as a wound dressing starting material: antimicrobial properties and mode of action, Int. J. Mol. Sci. 20 (23) (2019) 5889.

[124] M. Hosseinnejad, S.M. Jafari, Evaluation of different factors affecting antimicrobial properties of chitosan, Int. J. Biol. Macromol. 85 (2016) 467–475.

[125] E.-R. Kenawy, S.D. Worley, R. Broughton, The chemistry and applications of antimicrobial polymers: a state-of-the-art review, Biomacromolecules. 8 (5) (2007) 1359–1384.

[126] E. Liyaskina, V. Revin, E. Paramonova, M. Nazarkina, N. Pestov, N. Revina, et al., Nanomaterials from bacterial cellulose for antimicrobial wound dressing, J. Phys. Conf. Ser. 784 (2017) 012034.

[127] W. Shao, H. Liu, S. Wang, J. Wu, M. Huang, H. Min, et al., Controlled release and antibacterial activity of tetracycline hydrochloride-loaded bacterial cellulose composite membranes, Carbohydr. Polym. 145 (2016) 114–120.

[128] C. Wiegand, S. Moritz, N. Hessler, D. Kralisch, F. Wesarg, F.A. Muller, et al., Antimicrobial functionalization of bacterial nanocellulose by loading with polihexanide and povidone-iodine, J. Mater. Sci. Mater. Med. 26 (10) (2015) 245.

[129] P.G. Bowler, D. Parsons, Combatting wound biofilm and recalcitrance with a novel anti-biofilm Hydrofiber® wound dressing, Wound Med. 14 (2016) 6–11.

[130] B.S. Vasile, A.C. Birca, M.C. Musat, A.M. Holban, Wound dressings coated with silver nanoparticles and essential oils for the management of wound infections, Materials (Basel) 13 (7) (2020) 1682.

[131] H. Cui, M. Liu, W. Yu, Y. Cao, H. Zhou, J. Yin, et al., Copper peroxide-loaded gelatin sponges for wound dressings with antimicrobial and accelerating healing properties, ACS Appl. Mater. Interfaces. 13 (23) (2021) 26800–26807.

Traditional and modern wound dressings—characteristics of ideal wound dressings

Huda R.M. Rashdan[1] and Mehrez E. El-Naggar[2]

[1]*Chemistry of Natural and Microbial Products Department, Pharmaceutical and Drug Industries Research Institute, National Research Centre, Dokki, Cairo, Egypt* [2]*Pretreatment and Finishing of Cellulosic-Based Fibers Department, Textile Research and Technology Institute, National Research Centre, Dokki, Cairo, Egypt*

2.1 Introduction

A pathogenic case is a skin wound caused by chemical or physical damage or surgical procedures. The healing duration and the source of the damage determine whether the wound is chronic or acute. During certain diseases, such as cancer, diabetes, and vascular diseases, the wound cannot be healed over time and is referred to as a chronic wound that is difficult to heal. Furthermore, the wound healing process varied depending on the depth, size, and level of damage in the dermis and epidermis layers. Other factors that contribute to the prolonged wound healing process include age, immunological or nutritional weakness, and chronic stress [1–6]. There is ample evidence that severe chronic wounds burden healthcare and patients. These skin wounds are mostly exposed to pathogenic microbes, which cause dangerous inflammations that disrupt the healing cycle. Bacterial infections are common affects during wound healing, posing a threat to human health. Would healing is an intricate process in which damaged tissues contract and close to restore the skin's protective function [7–12]. The wound dressing is a substrate that has been designed and manufactured to be in direct contact with the wound. Dressing products containing drugs such as antibiotics, vitamins, and protease inhibitors can be used to speed up the healing process and increase the rate of tissue healing. For treating the infected wounds, biomaterials containing various bioactive and antimicrobial agents are applied to the wound surface. Traditional dry wound dressing or osmotic wound dressing is very common in third-world countries. Traditional methods leave particles at the wound site and lack ideal characteristics. These dressing materials adhere to the wound and dehydrate it, necessitating skilled nursing care to replace it on a regular basis to prevent wound dryness [13–16]. Taking in mind that, these dressing materials should not be applied directly to the wet wound surface. Their application should be restricted to specific cases where the wound is clean and dry, or they may be used as secondary dressing materials [17–20]. Furthermore, water evaporation in the dry dressing cools the wound site, slowing wound healing. Recently, studies have revealed that wound healing is much faster with a wet (modern) dressing than with a dry (traditional) dressing [16,21–23]. The modern dressing promotes wound healing and the formation of granular tissue while also keeping the wound site warm and moist. The primary function of the

Antimicrobial Dressings. DOI: https://doi.org/10.1016/B978-0-323-95074-9.00002-6

ideal dressing is to create a moist healing environment in the wound site by passing appropriate steam and absorbing excess inflammatory secretions in the wound. Furthermore, the ideal dressing should have antioxidant and antimicrobial properties, as well as being biodegradable, biocompatible, and easily detached from the wound surface without causing damage to the tissues surrounding the wound [24–27].

2.2 Wound dressings from passive to smart dressings

According to the manuscripts discovered in the ancient civilization, wound bandaging dates back to 2000 BC. The ancient Egyptians were the first to develop wound dressings, according to archeologists. The ancient wound systems consisted of three main steps, the first of which involved washing the wound, followed by making a plaster and badging. The ancient Egyptians were the first to use honey in wound care and to reverse the adhesion skin wound dressing. Nowadays, the plaster is known as plaster in which a mixture of various materials, such as specific oils, medical herbs, or clay, is used to form the plaster basis. Until now, wound dressings have been used to create an environment that can accelerate and improve wound healing. Because of recent advances in material science and technology, smart wound dressing has emerged as a promising strategy for developing and improving wound healing management. Wound dressings can play an important role in the healing of chronic wounds that cannot be recurred in a timely and normal manner, and they can restore the wound to its normal state. The psychological and financial burdens have compelled researchers to investigate and develop new strategies to improve biomaterials features and fabrication techniques in order to design advanced skin wound dressings that achieve the best healing outcomes.

2.3 Ideal wound dressings requirements

Skin wound healing is one of the most complex and sophisticated pathological healing processes in the human body. Depending on the type of wound, different factors are involved in the healing process. The skin wound healing process went through three major stages: inflammation, proliferation, and remodeling. The ideal wound dressing can provide a vital condition to amplify the healing process by not only protecting the wounds from pathogenic infections but also acting as a supportive structure. The ability to absorb secretions, prevent wound dryness and dehydration, and maintain gas, moisture, and environment exchange permeability in the wound site are the main essential features of proper ideal wound dressings. They should also have adequate mechanical strength. Nontoxic, nonadherent, and biocompatible wound dressings that are also comfortable, easy to remove, and have antimicrobial properties are ideal. In addition to speeding up the healing process, the best wound dressings should help to reduce cicatrix. The severe pain associated with removing the intertwined sections of dressings from the wound surface, which increases inflammations and worsens the wound, was one of the most commonly reported disadvantages of traditional or dry wound dressings. Inflammations as a result of this disrupted the wound healing process. As a result, research has been conducted to create a new generation of dressings aimed at intelligently monitoring the healing process and minimizing the limitations associated with traditional dry wound dressings. The ability to manage moisture is regarded as an essential feature of the best wound dressers. Several research scholars have looked into it. For example, Thomas shed light on the importance of wound environment moisture as an essential feature in the production of wound dressers to prevent wound maceration. In his study, he compared the characteristics of the ideal wound dressing for recovering and protecting the highly vulnerable peri-wound tissue from secondary severe damage caused by dressing removal and replacement caused by prospective infection and poor moisture management. The study demonstrated that advancements in wet modern dressings, which included permeable foam/film and hydrogels with improved fluid handling properties, can significantly reduce the risk of wound maceration formation when compared to dry traditionally used dressings. The moisture level at the wound interface has a significant impact on the wound healing process. Coloumbe and his colleagues compared the rate of re-epithelization and wound closure in wounds with a controlled moisture level and wounds exposed to air. They discovered that wounds with a controlled moisture level improve and accelerate the rate of epithelization. Sweeney et al. demonstrated that excessive moisture, particularly that secreted from chronic wounds, has the ability to damage the macerated peri-wound epidermis and slow the recovery and regeneration process. Another critical significant parameter for an ideal dressing is wound secretion management. The composition of excaudate secreted varies depending on the type of wound. Wound secretions appeared for improving the wound healing and remodeling, whereas in chronic wounds, differentiation and proliferation of wound reconstructing cells slowed due to high levels of various denatured proteins, as well as proinflammatory cytokines and protease. Some wounds are also classified as highly exuding. The ideal wound dressings for chronic wounds are chosen based on the type, size, volume, and viscosity of the secretions. The wound dressings for highly exuding wounds must have demonstrated adequate liquid absorption in relation to their physical dimensions; otherwise, fluid leaks around or throw the dressings. The highly exuding wounds provided an ideal environment for pathogenic microbial growth, slowing down the healing process. Along with the ability to absorb secreted fluids, the ability to hold the absorbed secreted fluids is an important feature of the ideal wound dressings. Other important characteristics of ideal wound dressings include the ability to maintain a moisture balance in the wound's environment. Improving new synthetic composites has the potential to provide significant characteristics similar to natural skin tissue, as well as preventing microbial penetration and balancing the moisture level in the wound environment. A symmetric polymeric membrane, membranes composed of many different layers, each layer with its own unique function, and composite have recently been considered as ideal promising wound dressings. Many different fabrication techniques, such as bioprinting, electrospinning, and dry/wet, have been developed for the development of these wound dressings. Alves et al. created photocrosslinkable symmetric multilayer electrospun nanofibrous networks for skin tissue differentiation and regeneration using polycaprolactone (PCL)/gelatin methacryloyl (GelMA) at the top layer, and GelMA/chitosan methacrylamide (chMA) at the bottom layer. The in vitro studies demonstrated that this prototype has the perfect features for providing a healing protective environment for wounds. Morgado et al. used the supercritical carbon dioxide ($scCO_2$)-assisted phase inversion technique to create interconnected microporous composite poly(vinyl alcohol) (PVA)/chitosan (Ch) asymmetrical membranes with drug-releasing efficacy in another study. The interconnected composite of these membranes can improve and facilitate wound exudate absorption while also delivering encapsulated ibuprofen to the wound site. Despite significant efforts in the development and design of promising ideal wound dressings that must be compatible with the dynamic features of chronic wounds, reaching ideal dressings remains the primary challenge for healthcare systems. The ideal wound dressing materials should be capable of monitoring the wound's progression/deterioration status, which requires advanced types of materials embedded in the wound dressings composite. Ideal smart dressings have been manufactured and used in recent years.

2.4 Traditional wound dressings

Hundreds of wound dressings have been developed over the centuries with the goal of controlling tissue damage and enhancing wound healing. In this regard, the essential main features of wound dressing can be stated as maintaining the wound site's moisture, promoting the elimination of excess wound secretions, supporting and protecting the wound from secondary injury, and preventing microbial infections [28,29]. Recently, the most common types of wound dressings developed for wound healing are hydrogel, gauze/bandage, and foam.

2.5 Types of traditional wound dressings

2.5.1 Bandage and gauze wound dressings

Bandages and gauzes, which have been developed for a long time, are among the most well-known types of dry wound dressings. Cotton gauze is the most commonly used and well-known material used in wound healing due to its hygroscopic property, skin affinity, and biocompatibility [30]. Meanwhile, the cotton gauzes are inert, limiting their ability to promote wound healing. Recently, researchers have developed a variety of cotton gauze modifications to incorporate various function components into them in order to improve their features and endow them with more adorable properties. For example, antimicrobial potency can be introduced into cotton gauze via various modification methods such as nanoparticle loading, polymerization with antimicrobial polymers, cationization, and coating [31,32]. Modifying and improving the existing matrix to develop new wound dressings is more convenient and simpler; however, the current matrix itself can become the most significant barrier that limits wound dressing design, as a result of which development cannot significantly alter the basic structure and mechanical properties of the developed wound dressings.

2.5.2 Foam/sponge wound dressings

Foam/sponge wound dressings are designed and manufactured with a porous structure and can be used for wound healing as well as hemostasis. Foam has the ability to absorb excess wound secretions, which aids in wound healing. Polyurethane (PU) foam has been developed with notable properties such as anti-inflammatory, antimicrobial, and promote and enhance re-epithelialization. Furthermore, the sponge revealed a highly connected porous structure, which enhances its ability to absorb excess wound exudates. However, the absorbed exudates and blood in the dressings can form clots in the sponge, resulting in pain and difficulty removing the dressings from the wound's surface [33,34].

2.5.3 **Hydrogel/adhesive wound dressings**

Hydrogels have sparked considerable interest in their use as wound dressings due to their exceptional ability to maintain moisture levels in the wound site as well as their high exudate absorption efficacy [35]. Hydrogel dressings can enhance and promote fibroblast differentiation and proliferation, as well as keratinocyte migration, which is necessary for wound healing and epithelialization. Many different hydrogel dressings have recently been designed and developed with various significant properties such as adhesion property, injectability, antimicrobial potency, self-healing efficacy, drug release property, and antioxidant potency [34,36,37]. Hydrogels are commonly developed and used in wound healing processes, but they currently face several limitations when used as dressings, including their poor mechanical properties, which prevent them from being used for motional wounds.

2.6 **Limitations of traditional wound dressings**

Despite the fact that many different wound dressings have been developed and tested in clinical trials in recent years, the most commonly used wound dressings continue to face unique challenges, restrictions, and limitations. Most well-known wound dressings can aid in wound healing by protecting the wound site from secondary injury, microbial infection, and other external harm. Meanwhile, wound dressings with the ability to facilitate wound healing and closure on their own are difficult to come by [35]. Wound healing is a dynamic process with multiple stages, each with its own unique process at the physiological and molecular levels. Each stage has its own set of requirements. Currently, developed wound dressings can not meet all of the wound's requirements at each stage. These wound dressings are generally not responsive to changes in the external environment. Furthermore, any abnormal changes that occurred during the wound healing process resulted in stalled wound healing and secondary development in chronic wounds. In this regard, modified wound dressings capable of repairing the abnormal changes that occur during the wound healing process which, in turn, would be advantageous in wound management. These wound dressings, known as smart wound dressings, have the ability to reflect wound status and provide timely wound treatment guidance [38,39].

2.7 **Smart wound dressings**

Smart wound dressings have emerged to enable new wound healing techniques. The term "smart wound dressings" refers to wound dressings that can interact with wounds, react and sense changes in the wound environment by utilizing built-in sensors or smart materials such as self-healing materials and stimuli-responsive materials. Meanwhile, many different strategies are involved in the wound healing process, and each strategy has its own set of characteristics, features, and physiological processes [40]. Smart dressings are known to be concerned with each stage's demands and to reveal an ideal proper response to changes in the wound environment. Numerous smart wound dressing categories have emerged to date, including stimuli-responsive wound dressings, self-healing wound dressings for motional wounds, biomechanical wound dressings, and self-removable wound dressings [40,41].

These smart wound dressings demonstrated an exceptional ability to interact with the wound's environment, reflecting changes that occur during the wound healing process. Furthermore, they prevent wound infection, the development of chronic wounds, and the promotion of autonomous wound recovery. Fig. 2.1 demonstrates the different currently developed smart wound dressings.

2.8 Biomechanical wound dressing

Wound dressings can effectively enhance and promote wound healing in a variety of ways. Traditional wound dressings are used to cover wounds to protect them from external harm. Modern wound dressings, on the other hand, are focused on facilitating the wound healing process by introducing one or more steps in the traditional wound healing process. Wound contraction is commonly associated with wound healing. The appropriate rate of wound contraction has been shown to be beneficial in enhancing and promoting wound healing [42]. During the fetal wound healing process, the wound closure process is primarily dependent on the actin cable that forms in the wound leading edges, pulling the wound edges together. In this regard, Blacklow et al. [43] has been developed a thermos-triggered contraction rigid hydrogel to promote the wound healing and closure. The developed hydrogel had two layers: an

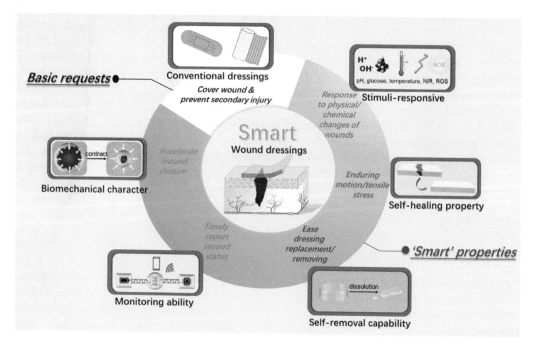

FIGURE 2.1

Significant features of smart wound dressings and smart wound dressings.

Reproduced with permission from Elsevier. Copyright 2021. https://www.sciencedirect.com/science/article/abs/pii/
S1748013221002152.

adhesion surface and a thermosensitive matrix. The contraction property is determined by the thermosensible poly(N-isopropylacrylamide). Silver nanoparticles were encapsulated with alginate, giving this developed hydrogel rigidity and antimicrobial properties. The carbodiimide-mediated reaction between the alginate, chitosan, and proteins in the tissue occurred in the tissue surface layer, providing the hydrogel with adhesiveness to the native tissue of origin. The rigidity or toughness of this developed hybrid hydrogel can reach approximately 500 J/m^2 (fracture energy), the adhesion energy of the developed hybrid hydrogel with the embedded silver particles is approximately around 175 J/m^2. Furthermore, the surface layer adhered to the tissue and the matrix layer generated a contraction stress along with antimicrobial potency, affording a so-called purse-string-like manner of contraction to the wounds as illustrated in Fig. 2.2A. The in vivo results showed that the developed hydrogel can effectively reduce wound closure time and improve granulation tissue formation. Despite the fact that this developed hydrogel demonstrated promising wound closure results, it lacked bioactivity. Furthermore, the reaction to achieve tissue adhesion requires the presence of catalyst reagents, which are usually toxic to the native tissue. To avoid the disadvantages and drawbacks of this biomedical dressing, researchers created a contractible nontoxic wound dressing with drug-release and bioactivity efficacy. According to Li et al., a hydrogel wound dressing has thermos-triggered contraction efficacy and thermos-dependent drug release. The PNIPAM was responsible for the dressing contraction efficacy. Because of the presence of polydopamine, this hydrogel wound dressing demonstrated proper adhesion potency (adhesion strength is approximately 9.68 kpa). Once this hydrogel wound dressing applied to the wound site, this hydrogel dressing was first adhere strongly to the native wound tissue and then the temperature of the body would trigger the hydrogel contraction subsequently palling the wound edges together as demonstrated in Fig. 2.2B [44]. Furthermore, PNIPAM provided support for the hydrogel's thermos-responsive drug release feature. The in vivo study results revealed that this developed hydrogel wound dressing could significantly improve and promote wound closure while also facilitating the formation of granulated tissue, vascularization, and collagen deposition. One disadvantage of those who developed PNIPAM-based contractive hydrogels is that they require high temperatures to enhance their contraction because the lower critical solution temperature (LCST) of PNIPAM's is around 32°C. Regardless, the temperature of the human body varies from one location to the next. More future studies on lowering the contraction-triggering temperature become critical in this regard. Shape memory wound dressings were another option for improving and facilitating wound closure. Li et al., created a wound dressing out of polyurethane-urea elastomer, which is a set of copolymers of polyethylene glycol (PEG), polycarpolactone (PCL), and aniline trimer. This elastomer was formed into a film and demonstrated good elastic properties with a high modulus up to approximately 21 MPa. This dressing had a shape memory feature that allowed it to be changed, deformed and recovered to its native cylindrical shape when exposed to body temperature at 37°C. This wound dressing can contract the wound, allowing it to be healed and recovered more quickly (Fig. 2.2C) [45]. With the same concept, Li [46] developed a wound dressing with a multi-responsive shape memory. This wound dressing is effectively responsive to both moisture and thermos changes. Upon raising the temperature to 90°C, the dressing can be returned to its initial shape immediately and when utilized in a humid environment, this wound dressing can be gradually recovered. They used these wound dressings as a dressing ring on the site of the wound, this ring can seal and squeeze the wound when triggered by hot vapor as demonstrated in Fig. 2.2D. Meanwhile, the hot-vapor-triggered contraction may be problematic because the hot vapor may cause skin tissue damage. Shape memory wound dressings can provide biomechanical benefits for wound healing and closure. Because of their shape memory properties, they saw self-removal wound dressings as a future trend.

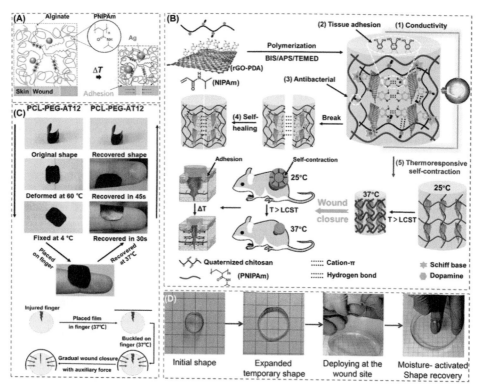

FIGURE 2.2

Biomechanical dressings for wounds that speed up wound healing and closure: (A) by mimicking the embryonic wound closure achieved by actin contraction, a thermo-responsive wound dressing with adhesive capacity on the skin was fabricated, (B) wound dressing that can contract at body temperature, helping wound closure, (C) PCL-PEG-AT shape memory wound dressing that can provide auxiliary force to help wound closure, and (D) shape memory wound dressing ring.

Reproduced with permission from Elsevier. (A) [43] https://s100.copyright.com/CustomerAdmin/PLF.jsp?ref=deab9f03-bf1d-4639-8545-2d7c9139024d; (B) [44] https://s100.copyright.com/CustomerAdmin/PLF.jsp?ref=deab9f03-bf1d-4639-8545-2d7c9139024d; (C) [45] https://s100.copyright.com/CustomerAdmin/PLF.jsp?ref=3bf07032-b63f-4a86-9359-78d0ba87cd4f; (D) [46] https://s100.copyright.com/CustomerAdmin/PLF.jsp?ref=8a3dd1de-987a-44e9-9e3e-a5d3a08a2d70.

2.9 pH responsive wound dressing

The pH of normal healthy skin is around 4—6; this value was altered during the skin infection at the infected chronic wounds. There is widespread agreement that chronic wounds have a pH greater than 7. However, a few studies have suggested that the pH of chronic wounds, particularly ulcers, is around 5.5. In general, the pH of the wound differs from the pH of the intact skin [47] (Fig. 2.3A). Researchers have modified and developed different pH-responsive wound dressings for

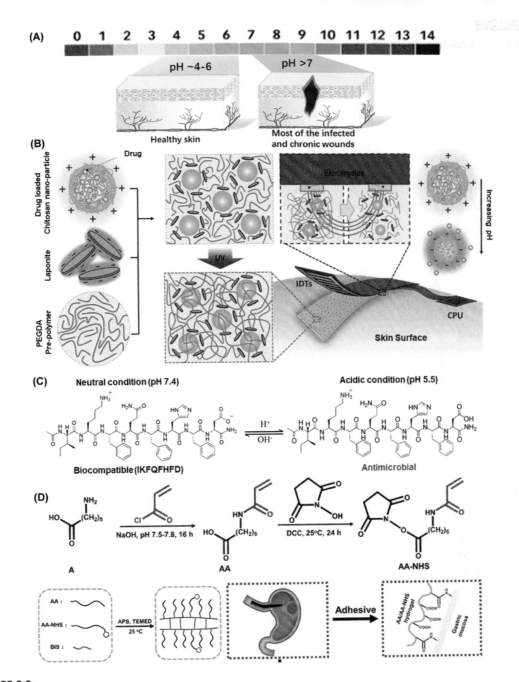

FIGURE 2.3

pH-responsive wound dressings: (A) pH difference between healthy skin and the infected or chronic wounds, (B) the wound dressing is responsive to the controlled pH changing, (C) pH-responsive wound dressing composed of the biocompatible peptide. This peptide showed antibacterial properties at acidic conditions, and (D) pH-responsive wound dressing composed of AA and AA-NHS [48]. *AA*, acryloyl-6-aminocaproic acid; *AA-NHS*, AA-g-*N*-hydroxysuccinimide.

wound management based on this pH difference [48]. The stimuli-responsive materials are divided into two categories: those that respond to local pH changes and those that respond to externally controlled changes. Kiaee et al. [49] controlled the pH of the wound dressings utilizing an electrical voltage. The wound dressing is fabricated using poly (ethylene glycol)-diacrylate/laponite and encapsulated using chitosan nanoparticles as shown in Fig. 2.3B. A current voltage was applied directly to the hydrogel with copper wire as cathode and zinc wire as anode. The pH increased near the anode upon applying the voltage as the + ve charge moving away from the anode (the cathode not involved in the hydrogel) and the redox reaction takes place at the electrode. As the pH increased, the chitosan nanoparticles deprotonated and contract in a basic environment and the electrostatic interaction between chitosan nanoparticles and laponite become weak, that resulting in drug release from the chitosan nanoparticles as shown in Fig. 2.3B. This externally controlled pH change can control the environment pH in a timely and adequate manner to induce drug release from the wound dressing. In any case, the external control is ineffective in practice, and this pH responsive wound dressing requires a very high pH value of around 14 to achieve the promising drug release. Local-pH-change responsive wound dressings that respond to slight alkaline or acidic pH changes are advantageous and desirable. The researchers created an acidic-condition-responsive hydrogel to manage and control drug release in order to protect the wound from external harms and improve wound healing and recovery. Ninan and colleagues, for example, created a carboxylated agarose/tannic acid hydrogel with high sensitivity to low pH (around 5.5). When the temperature is reduced from 70°C to 4°C, a hydrogel can be formed using carboxylated agarose and physical crosslinking. However, zinc ions capable of crosslinking the tannic acid chain and the agarose chain. Under acidic condition, the ionic crosslinking was disrupted and the hydrogel network become loose, allowing the release of the tannic acid from the hydrogel. Furthermore, tannic acid demonstrated potent anti-inflammatory and antimicrobial activities, as well as long-term tannic acid release, which protects chronic wounds from infection. Meanwhile, there is no discernible difference in the rate of tannic acid release between hydrogels containing zinc ions at low pH and hydrogels containing no zinc ions at any other pH. Another study developed an acidic pH-responsive wound dressing after researchers discovered that the pathogenic pH of chronic wounds is around 5.5 [50]. At neutral pH, an acidic pH-switchable peptide was used to create a hydrogel, and due to its instability, the developed hydrogel will release this antimicrobial peptide at acidic pH (Fig. 2.3C). Photothermal agents and other drugs can be loaded into this hydrogel and designed to be released from the developed hydrogel when the acidic pH is applied. In vitro studies revealed that these wound dressings can effectively damage the biofilm while also activating cell differentiation and proliferation. Although the pH at the site of chronic wounds is still debatable, acidic pH-responsive wound dressing may not be acted as effective as expected in this case.

Wounds in the stomach are exposed to the stomach's acidic medium. He et al. [48] created an acidic pH-sensitive hydrogel that can be used as a wound dressing to control gastric hemorrhage. This wound dressing hydrogel effectively promotes gastric wound healing. Hydrogels developed using acryloyl-6-aminocaproic acid (AA) and AA-g-N-hydroxysuccinimide (AA-NHS) were formed through free radical polymerization as illustrated in Fig. 2.3D. Because of its reaction with the amino groups in the native tissue, NHS provided these hydrogels with excellent adhesion properties. Because of the formed hydrogen bonds between the carboxyl groups at low pH, the swelling ratio of those hydrogels decreased sharply when exposed to acidic solution (pH of the gastric juice around 2). The low swelling ratio associated with the acidic medium makes this wound dressing hydrogel stable in the acidic gastric juice. On the swine gastric bleeding model, these developed hydrogels demonstrated excellent hemostatic properties and were also capable of preventing delayed bleeding. The only disadvantage of this development is the use of cytotoxic catalysts in the hydrogel fabrication process.

2.10 Thermo-responsive wound dressing

The temperature at the wound site is also differ from the intact tissue. the temperature at the healthy skin is approximately around 37°C, at the wound site and when the wound infected or inflammations happens, the temperature mostly increased as shown in Fig. 2.4A. Recently, researchers have developed wound dressings with thermos-responsive features that have the ability to deliver drugs and factors that promote and accelerate wound healing and closure. In the study of Montaser et al. [51] a hydrogel has been fabricated using sodium alginate- graft-NIPAM and poly-vinyl alcohol (PVA). Diclofenac sodium (DS) and anti-inflammatory drug, were encapsulated in the hydrogel and revealed sustain release from the hydrogel at various temperatures. The drugs were continuously released from the hydrogel at 25°C. Because of the thermos-responsive feature of the NIPAM, a secondary stage of release was observed when the temperature was raised to 37°C. First, when the temperature is lower than its LCST, NIPAM forms hydrogen bonds primarily with the −OH groups in PVA, resulting in the retention of approximately 30% of the encapsulated drugs in the hydrogel at RT via release (Fig. 2.4B). When the temperature was increased to be 37°C, NIPAM was contract and the hydrogen bonds were broken, resulting in releasing the remained drugs in the hydrogel (Fig. 2.4B). Analogously, owing to the thermos-responsive character of the NIPAM, Lin et al. [52] has been developed a hydrogel capable of step wisely releasing the drugs and other factors. NIPAM was used to create a nanogel gel that was embedded with basic fibroblast growth factor (bFGF), and the LCST of this nanogel is approximately 33°C. Another developed nanogel, made with NIPAM-co-acrylic acid (NIPAM-co-AA), was embedded with DS and its LCST was around 40°C. To be used on the wound surface, all of the developed nanogels were encapsulated in a sodium alginate hydrogel. During the inflammation stage, the temperature at the wound site was around 37°C, which was lower than the temperature of the NIPAM-co-AA nanogel and greater than the LCST of NIPAM nanogel, resulting in the release of the DS, an anti-infection drug that protects wounds from pathogenic microbial infections during the inflammation stage (Fig. 2.4C). The remodeling stage and the formation of new tissue begin with wound healing and closure. At this stage, the wound site temperature is around 25°C, which is lower than the LCTS of the NIPAM nanogel, resulting in bFGF release and increased angiogenesis (Fig. 2.4C). Wound animal models treated with wound dressing composites had higher angiogenesis and lower inflammation than those treated with SA hydrogel embedded directly with bFGF and DS. This thermos-triggered stepwise drug-release technique provided a novel solution for the design of thermos-responsive wound dressings. Furthermore, the thermo-responsive wound dressing with controlled drug release is an excellent method for treating wound infections. Although the temperature of the body's skin varies from site to site, thermos-responsive wound dressings must come into closer contact with the wound area to detect temperature changes, resulting in insensitivity of the thermos-responsive action of the developed wound dressing.

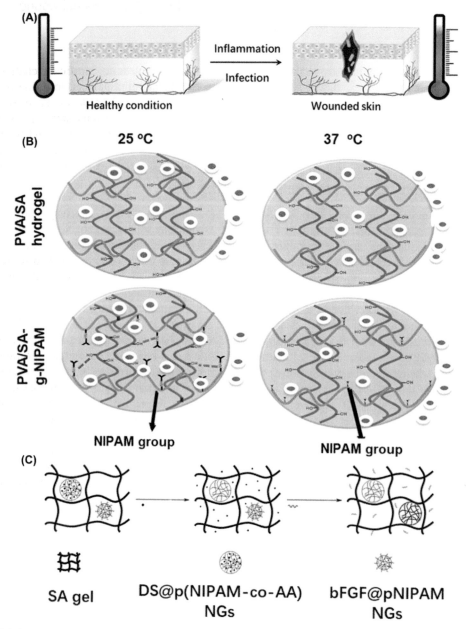

FIGURE 2.4

Drug release from the wound dressing due to the temperature: (A) temperature increased when the wound gets infected or inflammation happens, (B) sodium alginate (SA)-*graft*-NIPAM enabled the thermo-responsive drug release from the PVA/SA-*g*-NIPAM wound dressing, (C) a dual-factor stepwise release system that responds to temperature change [48]. *PVA*, polyvinyl alcohol.

The treatment of chronic wounds, as well as diabetic ulcers, is a significant challenge in wound healing. Diabetes is commonly associated with wound healing defects, resulting in many complications and delayed wound healing. Furthermore, hyperglycemia makes it easier for pathogenic microbes to infect wounds [53,54] (Fig. 2.5A). In this regard, wound dressings with the ability to respond to blood glucose levels will be beneficial to use in diabetic wound treatment. As a result, the development of modified wound dressings capable of responding to changes in glucose levels and managing wound healing via glucose-responsive drug release is a promising idea. Glucose-triggered drug release system utilizing phenylboronate ester and Schiff-base reaction has been designed. Chitosan was created and modified with phenylboronate (CSPBA), which can react with PVA; additionally, benzaldehyde-capped PEG was used as a crosslinker to interact with the amino group on chitosan (Fig. 2.5B). This double hydrogel network demonstrated glucose- and pH-responsive properties. Glucose may compete with PVA to form phenylboronate esters with CSPBA, causing the crosslinking of the developed hydrogel to loosen and the loaded insulin to be released. This glucose-responsive wound dressing not only improves wound healing but also plays an important role in controlling blood glucose levels. It provided a new promising idea in combining disease healing with wound healing [53]. Consuming local glucose at the wound site is an alternative method that aids diabetic wound healing. Bhadauriya et al. created a glucose responsive wound dressing by immobilizing yeast extract immobilized copper nanoparticles in a carbon nanofibrous (CNF) matrix. The yeast extract can promote and facilitate respiration and the conversion of glucose to ethanol (Fig. 2.5C). It can also help with cell differentiation and the formation of granular tissue. Meanwhile, copper nanoparticles (Cu NPs) were used as an anti-pathogenic agent in this wound dressing and showed promising results in reducing microbial colonies in in vitro studies. When applied to diabetic wounds in rats, this Cu-CNF-YE wound dressing significantly increased neovascular numbers while decreasing inflammations. This developed wound dressing can consume the glucose present in the wound site to maintain proper wound healing conditions. The main concern with this developed wound dressing is the ethanol produced by glucose consumption, which has been shown to be toxic to native tissue.

2.11 Self-healing wound dressing for motional wound

The motional wound is a type of wound that has specific wound dressing requirements. Wounds at stretchable parts are a good example of motional wounds. For example, wounds on the joints or the neck, any tension or other body movements that result in stretch stress. These wound dressings can be easily separated from the wound site. To keep the wound dressing on the wound surface, researchers developed self-healing dressings for motional wounds. Qu et al. [55] have been developed hydrogel with antimicrobial potency as wound dressing for motional wounds. In this study, Quaternized chitosan was crosslinked with benzaldehyde-terminated pluronic F127 using the Schiff-base reaction. Because PF127-CHO is an amphipathic molecule, it can form micelles in water and encapsulate various hydrophilic drugs. Because of the synergistic effects of the mobilized Schiff-base and PF127 micelles crosslinking interaction, this fabricated wound dressing is suitable for wounds that have experienced more stretch stress. This developed hydrogel also

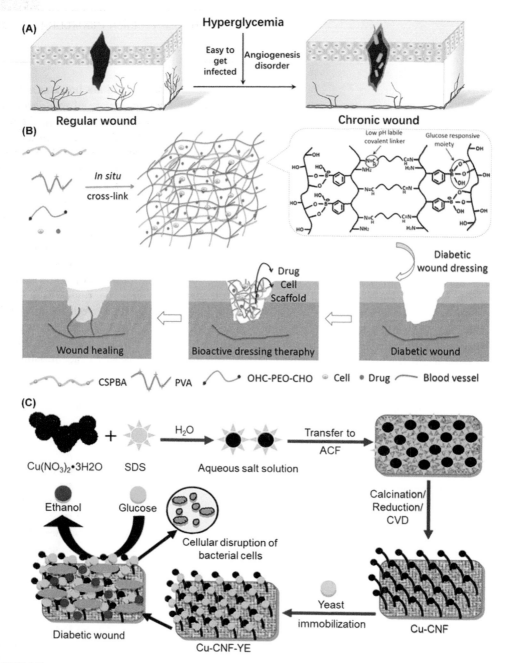

FIGURE 2.5

Glucose-responsive wound dressing: (A) hyperglycemia results in chronic wounds by disordering angiogenesis and increasing risks of infection, (B) wound dressing that was fabricated via Schiff-base and phenylboronate ester, and (C) yeast extraction was immobilized on the copper nanoparticles of the Cu-CNF wound dressing.

Reproduced with permission from Elsevier. (A), (B) [54] https://s100.copyright.com/AppDispatchServlet; (C) [40] https://s100. copyright.com/AppDispatchServlet.

revealed excellent adhesion efficacy and when employed on the elbow, it did not break or split off when the human elbow bent to a 120 °C (Fig. 2.6A). Another important feature of this developed wound dressing is that it is biodegradable, which means that it does not need to be changed during its use period, thereby avoiding the pain associated with changing it. Li et al. [56] utilized the poly (3,4-ethylenedioxythiophene) and GS (cationic guar slime): poly(styrene sulfonate) (PEDOT:PSS) for developing a self-healing conductive hydrogel applied for wounds in stretchable parts. GS discovered a cationic side chain that interacts with PEDOT:PSS occurs as a result of electrostatic interaction in which the hydroxyl groups in the GS can hydrate and form hydrogen bonds with one another (Fig. 2.6B). These two main interactions resulted in the formation of the hydrogel. Because the GS has a large number of ipsilateral hydroxyl groups, when external forces break part of the hydrogen bonds, it immediately forms new hydrogen bonds, resulting in its fast self-healing property. After self-healing, the developed hydrogel could still be stretched to 200%. Using this developed hydrogel on rat neck wounds significantly increased wound healing potency and rate when compared to GS-treated and -untreated groups. This newly developed hydrogel promotes the formation of granular tissue while also increasing collagen deposition. Despite the fact that many self-healing wound dressings have been designed and developed, only a few wound dressings for motional wounds have been developed. Mechanical properties and adhesion capacity, in addition to good self-healing ability, are required for the development and design of wound dressings for motional wounds.

2.12 **Wound dressing for infection monitoring**

Pathogenic infections retard the wound healing process, resulting in generation of chronic wound. He et al., [57] and collogues have been developed and established a wound dressing can directly monitor the microbial infection. Methylene blue was embedded into the wound dressing in this study to achieve the microbial-infection monitoring potency. Methylene blue is a cationic dye that lost its blue color when it came into contact with microbes. This developed wound dressing was double-layered, with electrospun sodium carboxymethylcellulose serving as the surface mesh and PVA foam containing an antimicrobial drug and methylene blue dye serving as the second layer. If a microbial infection occurs after this wound dressing is applied to the wound site, the color of the methylene blue will be discharged. The disadvantage of this technique is that it can only show the microbial infection qualitatively and not the level of infection. The microbial infection may cause pH and temperature changes at the wound site. Wound dressings that detect changes in pH and temperature can help to monitor infection at the wound site and allow for early treatment of the infection. The pH monitoring can be classified into two different ways; electrochemical and colorimetric methods [58,59]. Mirani et al. [60] designed a multifunction wound dressing called GelDerm which can give the pH of the wound site in a colorimetric manner. In this study, they created a simple wound dressing by incorporating drug-eluting scaffolds and color-changing pH sensors into an alginate-based dressing. pH-responsive dyes were mixed into mesoporous resin beads, which were then embedded in alginate for 3D printing of the porous sensor array. Mobile apparatus can detect changes in the color of pH sensors, and the application can detect the pH value as demonstrated in Fig. 2.7A. The dressings provided patients with a traditional and simple method of

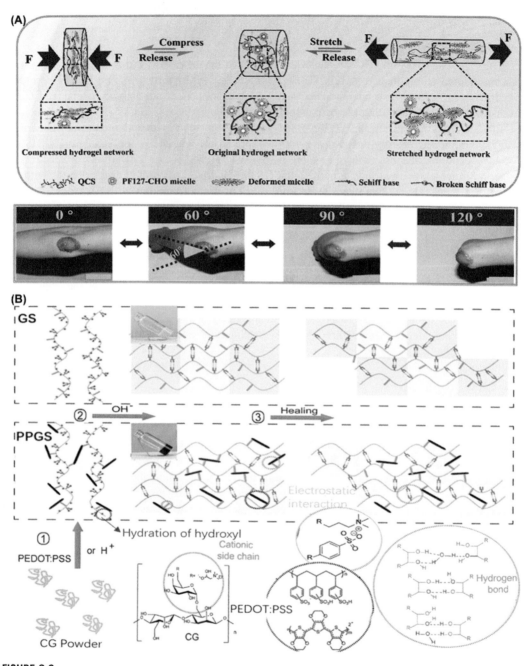

FIGURE 2.6

Self-healing dressings for motion-related wounds: (A) self-healing wound dressing composed of QCS and PF127-CHO, (B) GS/PEDOT:PSS hydrogel that was fabricated via multiple physical crosslinking [55,56].

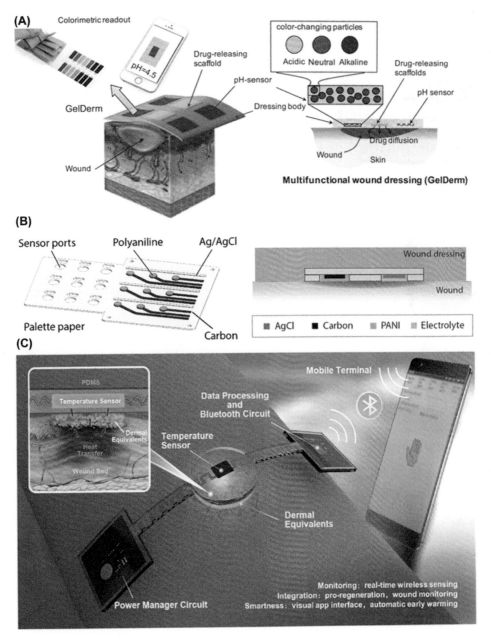

FIGURE 2.7

Wound dressing that utilizes pH and temperature changes to identify infection: (A) Schematic image of the pH-monitoring wound dressing via the colorimetric method, (B) The electrochemical method to detect pH changing via the sensor fabricated on a commercial paper, and (C) Temperature monitoring wound dressing with temperature sensor, power manager circuit and data processing and Bluetooth circuit.

detecting wound conditions at home. The electrochemical method is another method for detecting pH changes. Rahimi et al., [61] developed a pH sensor supported by two screen-printed electrodes (Ag/AgCl used as reference electrode and polyaniline coated carbon electrode) was produced on a commercial paper [62] (Fig. 2.7B, C). The sensing capability of this apparatus is due to polyaniline deprotonation and protonation in alkaline and acidic solutions, respectively. In an acidic solution, polyaniline will be doped with H^+ and transfer to the emeraldine salt status, resulting in good conductivity. The charge of the polymer would be reduced when this sensor was used in alkaline solution due to the neutralization of the H^+ ions. When the pH was changed from 4 to 10, the resulting voltage exhibited a linear relationship with the pH, and the sensitivity is approximately -50 mV/pH. This newly developed monitoring wound dressing was extremely sensitive to pH changes. Furthermore, the dressing itself is inexpensive; however, changes in pH should be detected by an external device, making it difficult to use, particularly for personal use. In this regard, new wireless pH sensitive dressings have been designed and developed by interfering the Bluetooth and power materials in the pH changing detected and monitored by the dressing, which can be sent directly to a mobile device, allowing for in-home personal usage.

2.13 Conclusion and prospects

The skin is the body's largest organ, and wounds are the most common skin problems. Microbial infection is the most common complication that develops from wounds, which can be fatal if not treated properly. Wound dressings are frequently used to prevent wound infections caused by microorganisms such as *Staphylococcus aureus* and *Escherichia coli*. Because biomaterial composition and manufacturing methods directly affect the rate of drug release and healing period, it is critical to deliver the appropriate antimicrobial agents at a specific rate during the healing period. This chapter highlighted the various wound dressings used, ranging from passive to smart dressings, as well as the specifications for the best wound dressings. Furthermore, the various types of conventional (traditional) wound dressings are discussed, as well as their disadvantages. In addition, various biomechanical wound dressing techniques were investigated. The current chapter examined and highlighted a number of smart wound dressing approaches. However, in order to be commercialize the efficient wound dressing materials in the global market, wound dressing based on natural polymers and advanced techniques should be extensively studied and evaluated.

References

[1] S.A. Eming, P. Martin, M. Tomic-Canic, Wound repair and regeneration: mechanisms, signaling, and translation, Sci. Transl. Med. 6 (2014) 265sr6.

[2] S. Elliot, T.C. Wikramanayake, I. Jozic, M. Tomic-Canic, A modeling conundrum: murine models for cutaneous wound healing, J. Investig. Dermatol. 138 (2018) 736—740.

[3] S.A. Eming, T.A. Wynn, P. Martin, Inflammation and metabolism in tissue repair and regeneration, Science 356 (2017) 1026—1030.

[4] C. Lindholm, R. Searle, Wound management for the 21st century: combining effectiveness and efficiency, Int. Wound J. 13 (2016) 5–15.

[5] J.G. Powers, C. Higham, K. Broussard, T.J. Phillips, Wound healing and treating wounds: chronic wound care and management, J. Am. Acad. Dermatol. 74 (2016) 607–625.

[6] B. Rowson, S.M. Duma, Annals of biomedical engineering 2019 year in review, Ann. Biomed. Eng. 48 (2020) 1587–1589.

[7] D.N. Moholkar, P.S. Sadalage, D. Peixoto, A.C. Paiva-Santos, K.D. Pawar, Recent advances in biopolymer-based formulations for wound healing applications, Eur. Polym. J. 160 (2021) 110784.

[8] Z. Gao, C. Su, C. Wang, Y. Zhang, C. Wang, H. Yan, et al., Antibacterial and hemostatic bilayered electrospun nanofibrous wound dressings based on quaternized silicone and quaternized chitosan for wound healing, Eur. Polym. J. 159 (2021) 110733.

[9] X. Zhang, M. Qin, M. Xu, F. Miao, C. Merzougui, X. Zhang, et al., The fabrication of antibacterial hydrogels for wound healing, Eur. Polym. J. 146 (2021) 110268.

[10] F. Fiorentini, G. Suarato, P. Grisoli, A. Zych, R. Bertorelli, A. Athanassiou, Plant-based biocomposite films as potential antibacterial patches for skin wound healing, Eur. Polym. J. 150 (2021) 110414.

[11] E. Rezvani Ghomi, S. Khalili, S. Nouri Khorasani, R. Esmaeely Neisiany, S. Ramakrishna, Wound dressings: current advances and future directions, J. Appl. Polym. Sci. 136 (2019) 47738.

[12] D.Z. Zmejkoski, N.M. Zdravković, D.D. Trišić, M.D. Budimir, Z.M. Marković, N.O. Kozyrovska, et al., Chronic wound dressings—Pathogenic bacteria anti-biofilm treatment with bacterial cellulose-chitosan polymer or bacterial cellulose-chitosan dots composite hydrogels, Int. J. Biol. Macromol. 191 (2021) 315–323.

[13] A.I. Raafat, N.M. El-Sawy, N.A. Badawy, E.A. Mousa, A.M. Mohamed, Radiation fabrication of Xanthan-based wound dressing hydrogels embedded ZnO nanoparticles: in vitro evaluation, Int. J. Biol. Macromol. 118 (2018) 1892–1902.

[14] H. Fang, D. Li, L. Xu, Y. Wang, X. Fei, J. Tian, et al., A reusable ionic liquid-grafted antibacterial cotton gauze wound dressing, J. Mater. Sci. 56 (2021) 7598–7612.

[15] J. Borges-Vilches, J. Poblete, F. Gajardo, C. Aguayo, K. Fernández, Graphene oxide/polyethylene glycol aerogel reinforced with grape seed extracts as wound dressing, J. Mater. Sci. 56 (2021) 16082–16096.

[16] Y.-F. Goh, I. Shakir, R. Hussain, Electrospun fibers for tissue engineering, drug delivery, and wound dressing, J. Mater. Sci. 48 (2013) 3027–3054.

[17] E.M. Tottoli, R. Dorati, I. Genta, E. Chiesa, S. Pisani, B. Conti, Skin wound healing process and new emerging technologies for skin wound care and regeneration, Pharmaceutics 12 (2020) 735.

[18] M. Li, Y. Liang, J. He, H. Zhang, B. Guo, Two-pronged strategy of biomechanically active and biochemically multifunctional hydrogel wound dressing to accelerate wound closure and wound healing, Chem. Mater. 32 (2020) 9937–9953.

[19] X. Yuan, M.I. Setyawati, D.T. Leong, J. Xie, Ultrasmall Ag + -rich nanoclusters as highly efficient nanoreservoirs for bacterial killing. Nano, Research 7 (2014) 301–307.

[20] Y. Liang, J. He, B. Guo, Functional hydrogels as wound dressing to enhance wound healing, ACS Nano 15 (2021) 12687–12722.

[21] H. Naeimi, M. Golestanzadeh, Z. Zahraie, Synthesis of potential antioxidants by synergy of ultrasound and acidic graphene nanosheets as catalyst in water, Int. J. Biol. Macromol. 83 (2016) 345–357.

[22] Z. Ouyang, S. Cui, H. Yu, D. Xu, C. Wang, D. Tang, et al., Versatile sensing devices for self-driven designated therapy based on robust breathable composite films, Nano Res. 15 (2022) 1027–1038.

[23] H. Qiu, F. Pu, Z. Liu, X. Liu, K. Dong, C. Liu, et al., Hydrogel-based artificial enzyme for combating bacteria and accelerating wound healing, Nano Res. 13 (2020) 496–502.

[24] A. Farazin, M. Mohammadimehr, A. Ghorbanpour-Arani, Simulation of different carbon structures on significant mechanical and physical properties based on MDs method, Struct. Eng. Mech. An. Int J. 78 (2021) 691–702.

[25] A. Farazin, M. Mohammadimehr, Nano research for investigating the effect of SWCNTs dimensions on the properties of the simulated nanocomposites: a molecular dynamics simulation, Adv. Nano Res. 9 (2020) 83–90.

[26] E.A. Kamoun, S.A. Loutfy, Y. Hussein, E.-R.S. Kenawy, Recent advances in PVA-polysaccharide based hydrogels and electrospun nanofibers in biomedical applications: a review, Int. J. Biol. Macromol. 187 (2021) 755–768.

[27] K. Nuutila, E. Eriksson, Moist wound healing with commonly available dressings, Adv. Wound Care 10 (2021) 685–698.

[28] Y. Zhong, H. Xiao, F. Seidi, Y. Jin, Natural polymer-based antimicrobial hydrogels without synthetic antibiotics as wound dressings, Biomacromolecules 21 (2020) 2983–3006.

[29] M. Naseri-Nosar, Z.M. Ziora, Wound dressings from naturally-occurring polymers: a review on homopolysaccharide-based composites, Carbohydr. Polym. 189 (2018) 379–398.

[30] J.M. Souza, M. Henriques, P. Teixeira, M.M. Fernandes, R. Fangueiro, A. Zille, Comfort and infection control of chitosan-impregnated cotton gauze as wound dressing, Fibers Polym. 20 (2019) 922–932.

[31] A.S. Montaser, M. Rehan, W.M. El-Senousy, S. Zaghloul, Designing strategy for coating cotton gauze fabrics and its application in wound healing, Carbohydr. Polym. 244 (2020) 116479.

[32] M. Rehan, S. Zaghloul, F.A. Mahmoud, A.S. Montaser, A. Hebeish, Design of multi-functional cotton gauze with antimicrobial and drug delivery properties, Mater. Sci. Eng. C. 80 (2017) 29–37.

[33] Y. Ding, Z. Sun, R. Shi, H. Cui, Y. Liu, H. Mao, et al., Integrated endotoxin adsorption and antibacterial properties of cationic polyurethane foams for wound healing, ACS Appl. Mater. Interfaces 11 (2018) 2860–2869.

[34] X. Zhao, Y. Liang, B. Guo, Z. Yin, D. Zhu, Y. Han, Injectable dry cryogels with excellent blood-sucking expansion and blood clotting to cease hemorrhage for lethal deep-wounds, coagulopathy and tissue regeneration, Chem. Eng. J. 403 (2021) 126329.

[35] R.C. Op't Veld, X.F. Walboomers, J.A. Jansen, F.A. Wagener, Design considerations for hydrogel wound dressings: strategic and molecular advances, Tissue Eng. Part. B: Rev. 26 (2020) 230–248.

[36] Q. Wei, J. Duan, G. Ma, W. Zhang, Q. Wang, Z. Hu, Enzymatic crosslinking to fabricate antioxidant peptide-based supramolecular hydrogel for improving cutaneous wound healing, J. Mater. Chem. B 7 (2019) 2220–2225.

[37] J. Zhu, F. Li, X. Wang, J. Yu, D. Wu, Hyaluronic acid and polyethylene glycol hybrid hydrogel encapsulating nanogel with hemostasis and sustainable antibacterial property for wound healing, ACS Appl. Mater. Interfaces 10 (2018) 13304–13316.

[38] S.A. Connery, J. Yankowitz, L. Odibo, O. Raitano, D. Nikolic-Dorschel, J.M. Louis, Effect of using silver nylon dressings to prevent superficial surgical site infection after cesarean delivery: a randomized clinical trial, Am. J. Obstet. Gynecol. 221 (2019) 57. e1.

[39] E. Hahnel, M. El Genedy, T. Tomova-Simitchieva, A. Hauß, A. Stroux, A. Lechner, et al., The effectiveness of two silicone dressings for sacral and heel pressure ulcer prevention compared with no dressings in high-risk intensive care unit patients: a randomized controlled parallel-group trial, Br. J. Dermatol. 183 (2020) 256–264.

[40] P. Bhadauriya, H. Mamtani, M. Ashfaq, A. Raghav, A.K. Teotia, A. Kumar, et al., Synthesis of yeast-immobilized and copper nanoparticle-dispersed carbon nanofiber-based diabetic wound dressing material: simultaneous control of glucose and bacterial infections, ACS Appl. Bio Mater. 1 (2018) 246–258.

[41] Y. Zhu, J. Zhang, J. Song, J. Yang, Z. Du, W. Zhao, et al., A multifunctional pro-healing zwitterionic hydrogel for simultaneous optical monitoring of pH and glucose in diabetic wound treatment, Adv. Funct. Mater. 30 (2020) 1905493.

[42] P. Martin, J. Lewis, Actin cables and epidermal movement in embryonic wound healing, Nature 360 (1992) 179–183.

[43] S.O. Blacklow, J. Li, B.R. Freedman, M. Zeidi, C. Chen, D.J. Mooney, Bioinspired mechanically active adhesive dressings to accelerate wound closure, Sci. Adv. 5 (2019) 3963 eaaw3963.

[44] M. Li, Y.P. Liang, J.H. He, H.L. Zhang, B.L. Guo, Perspective on theoretical methods and modeling relating to electro-catalysis processes, Chem. Commun. 56 (2020) 9937−9949 (Camb.).

[45] M. Li, J. Chen, M.T. Shi, H.L. Zhang, P.X. Ma, B.L. Guo, Electroactive anti-oxidant polyurethane elastomers with shape memory property as non-adherent wound dressing to enhance wound healing, Chem. Eng. J. 375 (2019) 121999.

[46] G. Li, Y. Wang, S. Wang, Z. Liu, Z. Liu, J. Jiang, A thermo-and moisture-responsive zwitterionic shape memory polymer for novel self-healable wound dressing applications, Macromol. Mater. Eng. 304 (2019) 1800603.

[47] N. Ninan, A. Forget, V.P. Shastri, N.H. Voelcker, A. Blencowe, Antibacterial and anti-inflammatory pH-responsive tannic acid-carboxylated agarose composite hydrogels for wound healing, ACS Appl. Mater. Interfaces 8 (2016) 28511−28521.

[48] J. He, Z. Zhang, Y. Yang, F. Ren, J. Li, S. Zhu, et al., Injectable self-healing adhesive pH-responsive hydrogels accelerate gastric hemostasis and wound healing, Nano-micro Lett. 13 (2021) 1−17.

[49] G. Kiaee, P. Mostafalu, M. Samandari, S. Sonkusale, A pH-mediated electronic wound dressing for controlled drug delivery, Adv. Healthc. Mater. 7 (2018) 1800396.

[50] J. Wang, X.-Y. Chen, Y. Zhao, Y. Yang, W. Wang, C. Wu, et al., pH-switchable antimicrobial nanofiber networks of hydrogel eradicate biofilm and rescue stalled healing in chronic wounds, ACS Nano 13 (2019) 11686−11697.

[51] A.S. Montaser, M. Rehan, M.E. El-Naggar, pH-Thermosensitive hydrogel based on polyvinyl alcohol/ sodium alginate/N-isopropyl acrylamide composite for treating re-infected wounds, Int. J. Biol. Macromol. 124 (2019) 1016−1024.

[52] X. Lin, X. Guan, Y. Wu, S. Zhuang, Y. Wu, L. Du, et al., An alginate/poly (N-isopropylacrylamide)-based composite hydrogel dressing with stepwise delivery of drug and growth factor for wound repair, Mater. Sci. Eng. C. 115 (2020) 111123.

[53] H. Cho, M.R. Blatchley, E.J. Duh, S. Gerecht, Acellular and cellular approaches to improve diabetic wound healing, Adv. Drug. Deliv. Rev. 146 (2019) 267−288.

[54] L. Zhao, L.J. Niu, H.Z. Liang, H. Tan, C.Z. Liu, F.Y. Zhu, pH and glucose dual responsive injectable hydrogels with insulin and fibroblasts as bioactive dressings for diabetic wound healing, ACS Appl. Mater. Interfaces 9 (2017) 37563−37574.

[55] J. Qu, X. Zhao, Y. Liang, T. Zhang, P.X. Ma, B. Guo, Antibacterial adhesive injectable hydrogels with rapid self-healing, extensibility and compressibility as wound dressing for joints skin wound healing, Biomaterials 183 (2018) 185−199.

[56] S. Li, L. Wang, W. Zheng, G. Yang, X. Jiang, Rapid fabrication of self-healing, conductive, and injectable gel as dressings for healing wounds in stretchable parts of the body, Adv. Funct. Mater. 30 (2020) 2002370.

[57] H. He, F. An, Q. Huang, Y. Kong, D. He, L. Chen, et al., Metabolic effect of AOS-iron in rats with iron deficiency anemia using LC-MS/MS based metabolomics, Food Res. Int. 130 (2020) 108913.

[58] V. Sridhar, K. Takahata, A hydrogel-based passive wireless sensor using a flex-circuit inductive transducer, Sens. Actuators A: Phys. 155 (2009) 58−65.

[59] R. Rahimi, U. Brener, S. Chittiboyina, T. Soleimani, D.A. Detwiler, S.A. Lelievre, et al., Laser-enabled fabrication of flexible and transparent pH sensor with near-field communication for in-situ monitoring of wound infection, Sens. Actuators B: Chem. 267 (2018) 198−207.

[60] B. Mirani, E. Pagan, B. Currie, M.A. Siddiqui, R. Hosseinzadeh, P. Mostafalu, et al., An advanced multifunctional hydrogel-based dressing for wound monitoring and drug delivery, Adv. Healthc. Mater. 6 (2017) 1700718.

[61] M.T. Rahimi, S. Sarvi, M. Sharif, S. Abediankenari, E. Ahmadpour, R. Valadan, et al., Immunological evaluation of a DNA cocktail vaccine with co-delivery of calcium phosphate nanoparticles (CaPNs) against the Toxoplasma gondii RH strain in BALB/c mice, Parasitol. Res. 116 (2017) 609−616.

[62] D. Lou, Q. Pang, X.C. Pei, S.R. Dong, S. J. Li, W.Q. Tan, L. Ma, Flexible wound healing system for pro-regeneration, temperature monitoring and infection early warning, Biosens. Bioelectron. 162 (2020) 112275.

Nanoparticles as potential antimicrobial agents for enzyme immobilization in antimicrobial wound dressings

Lakshmi Kanth Kotarkonda[1], Tej Prakash Sinha[1], Sanjeev Bhoi[1], Amit Tyagi[2], Akshay Kumar[1], Vijay Pal Singh[3] and Subhashini Bharathala[1]

[1]*Department of Emergency Medicine, All India Institute of Medical Sciences, New Delhi, New Delhi, India* [2]*Institute of Nuclear Medicine and Allied Sciences, Defence Research & Development Organisation, Timarpur, New Delhi, New Delhi, India* [3]*CSIR-Institute of Genomics and Integrative Biology, New Delhi, New Delhi, India*

3.1 Introduction

Infectious diseases caused by pathogenic microbes are the leading cause of death and have a negative impact on human health. Microbes such as bacteria, protozoa, viruses, and fungi cause a global burden by causing various health issues. Antibiotics are the first-line treatment for many microbial infections, but they are becoming less effective due to bacterial resistance [1,2].

Antibiotic overuse has resulted in the emergence of multidrug resistance (MDR) bacteria, which is associated with rising morbidity, mortality, and healthcare costs. The majority of antibiotics work by inhibiting transcription and translation mechanisms and disrupting cell membranes, or by competing with enzymes involved in cell wall synthesis. However, microorganisms have evolved a plethora of mechanisms to combat the effects of antibiotics [3,4]. Nanoparticle-based treatment is a promising approach for combating microbial resistance. Nanoparticles have been shown to be effective antimicrobial agents against a wide range of microbes, with inherent properties that allow them to avoid the problems associated with bacterial resistance. The nanoparticles have an antibacterial effect in a variety of ways, including direct contact with the negatively charged bacterial cell wall and accessing permeability via the formation of pores on the surface of bacteria [5,6].

Furthermore, due to their ideal properties such as larger surface area, transfer resistance, and efficient enzyme loading, nanoparticles act as effective substantiate materials for enzyme immobilization. Enzyme immobilization on nanoparticles has been shown to be effective against various biofilm matrices produced by various bacterial strains. They extensively penetrate the biofilm by targeting specific protein matrices and inhibiting cell membrane permeability [7].

Traditional wound dressings containing antiseptic or antibiotic drugs are the leading cause of difficult-to-treat drug-resistant strains. Novel nanoparticle-based drug delivery technologies in wound healing applications have been investigated to mitigate the emergence of antibiotic resistance strains [8]. Nanoparticles in antimicrobial dressings reduce bacterial populations in wounds through a variety

Antimicrobial Dressings. DOI: https://doi.org/10.1016/B978-0-323-95074-9.00009-9

of methods, lowering the risk of inflammation and infection. Furthermore, advances in computational and molecular techniques have resulted in engineered nanoparticles serving as support systems for enzyme immobilization. Nanoparticles have been widely used in wound healing applications due to their high stability and cost-effective preparation methods for enhanced antibacterial activity and therapeutic efficacy by accelerating the wound healing process [9–11].

By lowering the pH of the wound, the immobilized enzymes stimulate cytokine production and cell multiplication at the wound site, assisting in wound closure. They also provide debriding effects and necrotic tissue clearance for subsequent repair. Furthermore, the altered physicochemical properties of nanoparticles affect the kinetic stability and activity of immobilized enzymes, which is dependent on the type of chemical interactions in the system [12,13].

3.2 Antimicrobial activity of nanoparticles

Microbial infections are the leading cause of infection, sepsis, and mortality, and they can be effectively treated with several classes of antibiotics due to significant outcomes and cost-effectiveness in a variety of infections. The majority of antibiotics inhibit cell wall synthesis, transcription and translation machinery, membrane depolarization, and various metabolic pathways, which is often predictable and leads to the development of resistance mechanisms through the expression of resistance enzymes such as β-lactamases and aminoglycosides that degrade the antibiotics, cell wall component modification, and expression of efflux pumps [14]. In comparison to antibiotics, the advantage of using nanoparticles as antimicrobials is that their mode of action is extraneous to other mechanisms and less prone to bacterial resistance. Several antimicrobial nanomaterials have been developed and used in a variety of applications [15].

3.2.1 Types of antimicrobial nanoparticles

Nanoparticles as antibacterial agents act as competent antibiotic substitutes and are gaining widespread interest in filling gaps in the fight against biofilm-forming multidrug-resistant bacteria. Antimicrobial nanoparticles, both inorganic and organic, have a wide range of innate and diverse physicochemical properties [16] as shown in Fig. 3.1.

3.2.1.1 Inorganic nanoparticles

Antimicrobial nanoparticles of various sizes and shapes were used in wound healing applications. Inorganic nanoparticles act as potential antimicrobials, both intrinsically and through surface functionalized attachments, to provide antimicrobial effects that last. Silver, copper, gold, zinc, iron, and aluminum are the most commonly used nanoparticles in wound dressing. Among the various metallic nanoparticles, silver offers advanced functional versatility while remaining inexpensive and having good electrical and thermal conductivity [17,18]. Both Gram-positive and Gram-negative bacteria are effectively combated by silver nanoparticles. Similarly, zinc oxide nanoparticles have piqued researchers' interest due to their innovative and promising antimicrobial applications. Magnetic nanoparticles' exceptional properties, such as mobility, high mass structure, and large surface-to-volume ratio, earned them a special place as antimicrobials and versatile carriers for enzyme support. They

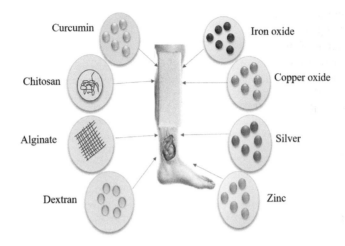

FIGURE 3.1

Types of nanoparticles used in antimicrobial dressings.

can be easily separated and recovered by using an external magnetic system. Magnetic nanoparticles also extend the lifetime of enzymes by varying recovery cycles, pH, and temperature ranges, allowing for the effective immobilization of a wide range of enzymes and biomolecules with improved properties [19–21]. Gold nanoparticles were thought to be safe and nontoxic, and they were widely used as antimicrobial agents. Their antibacterial properties are linked to dispersibility, particle size, and surface modifications, which can be altered by adjusting the reaction conditions [22].

Aluminum nanoparticles are thermodynamically stable, and their antimicrobial property has been attributed to reactive oxygen species generation by disrupting the cell wall and causing cell death. They also act as free radical scavengers, causing a slight decrease in the bacterium's extracellular protein content [23]. The different types of nanoparticles and their mechanism of action are listed in Table 3.1.

3.2.1.2 Organic nanoparticles

To combat multidrug-resistant bacteria, a wide range of organic nanoparticles with intrinsic antimicrobial properties or in conjugation with other metals have been widely used in wound healing. Organic nanoparticles' high biocompatibility and low toxicity have made them useful as potential antimicrobials in wound healing, primarily dextran, alginate, and chitosan. Because bacterial cell walls are negatively charged, positively charged organic polymers with quaternary ammonium groups interact most effectively with the cell wall [28,47]. Some synthetic biodegradable polymers with quaternary ammonium moieties and aromatic or heterocyclic structures, such as polystyrene, poly-lactic acid, and polyvinyl pyridine, are also being studied as antimicrobials for wound healing. However, disadvantages of using synthetic polymers include lower biocompatibility and wound healing ability. As a result, enzyme immobilization has been incorporated within the polymeric network to provide desired mechanical and bioactive properties for accelerating wound healing [48].

Table 3.1 Antimicrobial activity of organic and inorganic nanoparticles.

Nanoparticles	Size	Organism	Mechanism of action	References
Silver	5,15,55 nm	Escherichia *coli*, Actinomycetes, Streptococcus *aureus*, Human immunodeficiency virus	Membrane destabilization and altering the permeability, reactive oxygen species generation, DNA damage	[8,24,25]
Copper	15−100 nm	Vibrio *vulnificus*, Vibrio *harveyi*, and Vibrio *parahaemolyticus*, Staphylococcus *aureus*, Escherichia *coli*	Cell membrane dysfunction and enzyme malfunction, destruction of reactive oxygen species balance	[22,26,27]
Zinc oxide	20−40 nm	Escherichia *coli*, Staphylococcus *aureus*, Fusarium *oxysporum*, Candida *albicans*, Aspergillus *niger*	Reactive oxygen species, Inhibits the synthesis of mycotoxins and mitochondrial damage, inhibit protein synthesis	[28−31]
Iron oxide	10−100 nm	Staphylococcus *epidermidis*, Klebsiella *pneumoniae*, Pseudomonas *aeruginosa*, Mucor *piriformis*, Aspergillus *niger*	Reactive oxygen species generation, inhibits spore generation. DNA damage	[32−34]
Gold	5−15 nm	Candida *albicans*, Staphylococcus *aureus*, E. *coli*, Pseudomonas *flourescens*	Membrane disruption, reactive oxygen species generation	[35,36]
Selenium	20−100 nm	Staphylococcus *aureus*, Lactobacillus *acidophilus* Trichophyton *rubrum*, Candida *albicans*	Membrane depolarization, reactive oxygen species generation	[10,37,38]
Nickel	3−100 nm	Pseudomonas *aeruginosa*, Staphylococcus *aureus*, Bacillus *subtilis*, Micrococcus *luteus*, Escherichia *coli*, Aspergillus *niger*, Candida *albicans*	Alter membrane permeability, Oxidative damage	[17,39]
Chitosan	10−200 nm	Staphylococcus *aureus*, Escherichia *coli*, Pseudomonas *aeruginosa*, Candida *albicans*, Klebsiella *pneumoniae*	Disruption of membrane, Mitochondrial damage, Inhibit protein synthesis machinery	[2,3,40−42]
Dextran	20−80 nm	Bacillus *subtilis*, Staphylococcus *aureus*, Streptococcus *pyogenes*, Escherichia *coli*, Pseudomonas *aeruginosa*, Klebsiella *pneumoniae*, Proteus *vulgaris*	Morphological damage, Cell wall disruption, loss of cytosolic contents	[18,43,44]
Alginate	50−100 nm	Staphylococcus *aureus*, Pseudomonas *aeruginosa*, Escherichia *coli*, Propionibacterium *acnes*	Cell membrane disruption, Prevention of biofilm	[13,45,46]

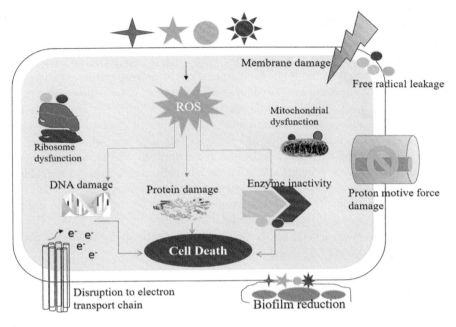

FIGURE 3.2

Mechanisms of antimicrobial activity of nanoparticles.

3.2.2 **Mechanisms of antimicrobial activity of nanoparticles**

Nanoparticles exhibit antimicrobial activity through a variety of mechanisms, ranging from physical disruption of the cell membrane to changes in cell chemical composition. The antecedent step in the microbial cell inhibition process is cellular penetration. As the primary mechanism of penetration, they were adsorbed or diffused at the cell surface. Adsorption occurs through the attachment of negatively charged functional groups, which causes cell membrane interaction, ion channel disruption, reactive oxygen species generation, mitochondrial dysfunction, and uptake of free ions, all of which harm microbial cells. Small-dimensional nanoparticles interact with bacterial cell walls via receptor—ligand interactions, Van der Waals forces, and hydrophobic interactions [49,50]. The antimicrobial activity of nanoparticles is illustrated in Fig. 3.2.

3.3 **Pathophysiology of wound healing**

Wound healing's primary function is to repair the damaged epithelial barrier, which is a complex process divided into three stages. The loss of first-line infection defense leaves the body vulnerable to external pathogens, resulting in fluid loss. The progression of phases results in tissue reconstitution, which includes the hemostasis, inflammation, and proliferation phases, followed by the healing phase, which includes the formation of mature scar tissue [51]. The progressive use of nanoparticles has accompanied a new perception of wound regeneration by improving skin retention and augmentation of biological and synthetic molecules for controlled drug release, bacterial load eradication or reduction, and re-epithelialization advancement [52]. The stages of wound healing are described in Fig. 3.3. After bleeding, the initial hemostasis phase appears, which is controlled by vascular constriction, thrombus formation, coagulation propagation, clot termination, and ultimately removed by fibrinolysis. The damaged Endothelial layer exposes the basal lamina, allowing the wound to heal. The bleeding is restricted and damaged cells are removed during the second stage of inflammation. Thrombin activated platelets, which released several growth factors to draw attention to WBC, growth factors, and nutrients that accelerate wound healing and protect the skin from infection. The wound is refurbished in the third stage of proliferation by the release of proangiogenic factors from platelets and inflammatory cells, which leads to fibroblast proliferation and angiogenesis. Later, fibroblast differentiation to myofibroblasts causes wound contraction by gripping the wound edges. The wound is completely closed by collagen fibers during the final maturation stage. Apoptosis will remove the cellular debris used for wound repair. The use of debriding enzymes at this stage speeds up the wound healing process by removing necrotic tissue [53,54].

3.3.1 Immobilization of enzymes on nanoparticles

Enzymes are biocatalysts that are found in plants and animals and are used in a variety of biochemical and chemical reactions. Enzymes' properties such as ease of production, substrate specificity,

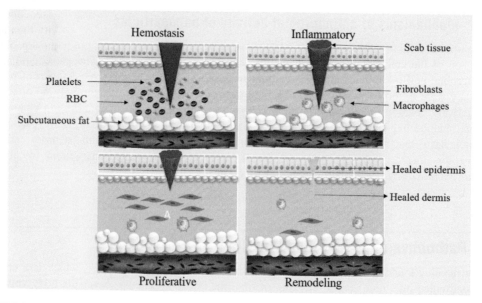

FIGURE 3.3

Overview of stages in wound healing.

and green chemistry synthesis have led to their widespread use in various sectors of the food, healthcare, and pharmaceutical industries [43].

Because of their proteinaceous nature, enzymes are very sensitive to physical and chemical changes and degrade rapidly, resulting in less stability and loss of activity by altering tertiary structures. The advancement of high-throughput nanotechnology has made the use of enzymes on solid support more cost-effective while preserving enzyme activity and allowing for easy recovery from the reaction medium. The immobilization also improves the enzyme's stability and activity [55]. Because of their ideal properties such as specific surface area, mass transfer resistance, and effective enzyme loading, nanoparticles are functionally used for enzyme immobilization. Nanoparticles immobilized with enzymes have the potential to balance the key factors that determine biocatalyst efficiency, such as surface area and unique chemical, mechanical, optical, and electric properties [56]. Immobilized enzymes by physical adsorption or conjugation have improved stability, high enzyme loading, and easy separation from products. The excellent biocompatibility provided by nanoparticles for enzyme immobilization has several advantages, including cost-effective synthesis, improved moisture retention, infection control, debriding action, and pain reduction [40]. The applications of immobilized enzymes on nanoparticles are illustrated in Fig. 3.4.

The application of site-specific enzyme immobilization has seen remarkable advances in organic chemistry and nanotechnology, allowing for the incorporation of a wide range of enzymes onto the support. The stability at specific conditions, cost, and variety of functional groups present on the surface are all factors to consider when selecting support systems for enzyme immobilization.

Furthermore, enzyme-bound nanoparticles dispersed in aqueous solutions exhibit Brownian movement, indicating enhanced activities such as pH tolerance, increased structural rigidity, selectivity, heat, and functional stability, as well as improved performance in organic solvents. Immobilized enzymes with altered structural changes can be used repeatedly in subsequent reactions by shortening the extraction and purification procedures and having higher activity than native enzymes [26,57]. In general, native enzymes are more active in aqueous media than organic solvents, and immobilized enzymes have properties that can be used in non-aqueous reactions with altered catalytic properties [58,59].

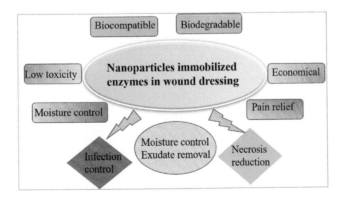

FIGURE 3.4

Applications of immobilized enzymes on nanoparticles.

3.3.2 Types of enzymes used in antimicrobial wound dressings

Wound healing is a significant process in which the living body sustains tissue damage while reinforcing anatomical impartiality by restoring functionality to the injured parts. The presence of dead tissue in the wound serves as a reservoir for bacterial colonization, promoting high levels of inflammation and reducing cellular migration, both of which are required for wound regeneration [25]. Debridement with appropriate enzymes was used to aid in the wound healing process. Among the various debridement methods, enzymatic debridement is a highly effective strategy that uses proteolytic enzymes to eliminate necrotic tissue while also providing moist wound healing and serves as an adjunct to the autolytic debridement process [60]. The various debriding enzymes and their applications is shown in Table 3.2. Further, wound heterogeneity needs an advanced understanding of the cellular cascades and intrinsic mechanisms for personalized wound care treatment.

3.3.2.1 Collagenase

Collagenase, a water-soluble proteinase obtained from bacteria such as Clostridium histolyticum, is the most commonly used enzyme in wound debridement. It has specific activity against non-doable cells, such as healthy cells, making it suitable for wound management. It has a low fibrinolytic activity and selective collagen degradation, so it does not increase the risk of bleeding in wounds. Collagenase activity has been demonstrated against various types of collagen, including type I and type III collagen. MMP-1, 8, and 13 were the collagenases with the ability to break down the collagen triple helix in mammals [42,65].

3.3.2.2 Elastase

Elastases belongs to the serine protease family, which primarily hydrolyzes peptide bonds. They aid in wound healing by disrupting the integrity of the vessel wall, causing lymphatic system distortion, and evoking an important skin defense mechanism. Furthermore, they are involved in the dissolution of tissue elastin to make room for new collagen, resulting in wound matrix formation [66].

Table 3.2 Types of enzymes and support material used in wound healing applications.

Enzyme	Immobilized support	Method of immobilization	Application	References
Papain	Sodium alginate and Magnetic iron oxide, Chitosan, and zinc oxidePolycaprolactone	Adsorption, Covalent bondCross-linkingPrecipitation	Wound healing, biomedical applications	[48,50,61,62]
Trypsin	Chitosan nanofibers, silver nanoparticles, gold nanoparticles	Adsorption, AdsorptionCross-linking and ionic interactions	Wounds and burns, wound healing, biomedical applications	[47,63,64]
Collagenase	Chitosan nanofibers	Covalent bond	Wound healing	[16]
Serratiopeptidase	Chitosan/magnetic nanoparticles, Liposomes, magnetic iron oxide	Covalent bonding, covalent encapsulation	Anti-inflammatory, Antimicrobial	[6,51,60]

3.3.2.3 Papain

Papain is a proteolytic enzyme derived from the papaya fruit (*Carica papaya*). It has a nonspecific proteolytic function. By stimulating its activity, the enzyme with sulfhydryl groups breaks down fibrinous components in necrotic tissue. According to some studies, combining urea and papain was more effective in influencing the biological activity of recombinant human platelet-derived growth factors [31]. It also has anti-inflammatory, bactericidal, and bacteriostatic properties. The enzyme also stimulates cytokine production, which promotes local cell multiplication by altering the pH of the wound, and it also aids in wound edge consolidation by reducing scars. In nature, free papain is very sensitive and deactivates in harsh climatic conditions, which can be overcome by immobilization on polymeric support to improve wound healing [27].

3.3.2.4 Papain-urea-chlorophyllin copper enzyme composite

The composite enzyme complex is made up of chlorophyllin and anti-agglutinin, which prevents erythrocyte agglutination. The combination of copper and papain-urea allows for faster wound healing while also reducing odor. It also removes necrotic tissue debris and promotes proper circulation [64].

3.3.2.5 Streptokinase

Streptokinase is a bacterial and human serine protease that acts as a debriding enzyme by converting plasminogen to plasmin, allowing fibrin degradation and preventing bleeding [67].

3.3.3 Methods of enzyme immobilization

Several methods were utilized to accomplish enzyme immobilization, having both advantages and limitations. The most common methods are physical adsorption, encapsulation covalent bonding, and cross-linking that are shown in Fig. 3.5. The actual deciding part of the immobilization process is the selection of a relevant immobilization method, as it plays a significant role in concluding the enzyme activity [44]. The catalyst specifications, overall enzymatic activity, enzyme deactivation reactivation characteristics, and toxicity of immobilization reagents are the major factors to consider for immobilized enzymes [37].

3.3.3.1 Immobilization by physical adsorption

The salt linkages and hydrophobic interactions formed between the enzyme and the support, which can dry on electrode surfaces, prevented enzyme immobilization by physical adsorption. Several weak forces govern physical adsorption, including hydrophobic interaction, van der Walls, electrostatic forces, and hydrogen bonds. Surface area, material type, and electrostatic interactions between the carrier and enzyme all play important roles in enzyme adsorption. The enzymes were protected from proteolysis, aggregation, and hydrophobic interaction with the interfaces by being adsorbed [45]. The carriers used for enzyme immobilization are organic and inorganic in nature. As support for enzyme adsorption, inorganic carriers such as silica, titanium, hydroxyapatite, and biodegradable compounds of natural origin such as cellulose, chitosan, and alginate were frequently used. Eco-friendly supports with high cation exchange properties, good water-holding capacity, and high enzyme retainability, such as coconut fibers, cellulose, and kaolin, are ideal for oxidation and reduction reactions [35]. The advantages of the adsorption method are that it is simple, requires

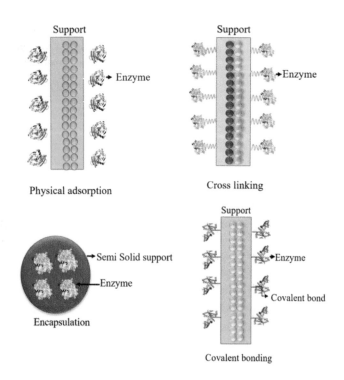

FIGURE 3.5

Methods of enzyme immobilization.

few steps for activation, and is inexpensive. Physical adsorption has several advantages, but it also has some disadvantages, such as poor operation stability and susceptibility to parameters such as ionic strength, temperature, and pH. The most significant disadvantage is desorption, which occurs when the enzyme separates from the support material as a result of minor changes in temperature, pH, and ionic strength. Particle-particle interactions, flow rate, agitation, and collisions are all physical factors that contribute to desorption. Nonspecific binding results in low enzyme activity. Furthermore, if not carefully controlled, overloading the support with an enzyme can result in low catalytic activity, and the presence of spacers between the enzymes and the support can result in steric hindrance [68,69].

3.3.3.2 Immobilization by encapsulation

The enzyme is entangled in the internal structure of the polymer material via covalent or non-covalent bonds in the encapsulation method. Entrapped enzymes are free to move in a constrained environment. Several materials such as nylon and cellulose nitrate have been utilized to encapsulate are in the range of 10−100 μm in diameter [63]. Polyacrylamide gels, cellulose acetate, gelatin, and alginate are common encapsulation matrix materials because they are water-soluble. Some biological cells, such as erythrocytes, are also used as encapsulation capsules. Encapsulated enzymes have higher operational and thermal stability in this method because encapsulation immobilization preserves enzyme mobility and increases activity [70,71]. The advantages of the encapsulation method are that it is simple to prepare, inexpensive, and causes less conformational change in the enzyme. The limitations include the possibility of microbial contamination and enzyme leakage [32].

3.3.3.3 Immobilization by covalent binding

The covalent binding immobilization method uses chemical bonds to connect the nonessential group of immobilized enzymes to the functional group of carriers. Because vigorous conditions can destroy the active conformation of the enzyme, the reaction to form covalent chemical bonds is usually carried out under mild conditions [30]. Because some carriers with no functional groups or the reaction conditions cause a reduction or loss of enzyme activity, activation of the carriers with some functional groups improves the conditions for enzyme immobilization. Chemical bonds that support covalent bonds in enzymes include amino, hydroxyl, carboxyl, thiol, imidazole, and phenol rings. As support systems, cellulose, polyacrylamide, silica, gelatin, and agarose carriers are used [72]. The covalent binding method of enzyme immobilization can intensely integrate the enzyme with the carrier and prevent enzyme shedding or leakage. The benefits of covalent bonding are straightforward: a wide range of support materials with varying functional groups, and no desorption or leakage issues. However, the limitations are that it frequently results in low recovery due to the destruction of the enzyme active conformation during the covalent reaction, steric hindrance of the enzyme, multiple attachments to the supports, and enzyme inactivation [73].

3.3.3.4 Immobilization by co-polymerization or cross-linking

Immobilization of enzymes on carriers via cross-linking involves the intermolecular cross-linking of the enzymes, either in the presence or absence of solid support. This method is based on bond formation between enzyme molecules via bi-or multi-functional reagents, resulting in three-dimensional hydrophobic cross-linked aggregates [74]. Cross-linking can occur between molecules or within molecules, and is referred to as intermolecular or intramolecular cross-linking. Chemical cross-linking, in general, refers to the formation of covalent bonds using a multifunctional reagent. Each functional reagent binds to two molecules of the enzyme to form a network in the cross-linking method [39]. Bifunctional cross-linkers are classified into two types: homobifunctional (two or more functionally identical groups) and heterobifunctional (two or more functionally varied groups). When compared to physical adsorption or encapsulation, cross-linking or covalent binding is performed under relatively harsh conditions. The benefits of this method are that it is quick and simple. The limitations are conformational changes in the enzyme caused by structural changes and the risk of denaturation when using polyfunctional agents [75,76] and is shown in Fig. 3.5.

3.3.4 Advantages of enzyme immobilization on nanoparticles

The process of immobilization significantly alters the enzyme properties due to changes in the physical and chemical nature of the carrier, which in turn changes the conformation of the enzyme as well as the nature of catalysis. The altered reactivity of various kinetic constants such as pH, substrate specificity, Michaelis constant, and substrate specificity can be caused by conformational changes in the enzyme molecule, which retains the activity for an extended period of suitable storage due to the immobilization process [77,78]. The major advantages of immobilized enzymes are described briefly.

3.3.4.1 Stability

Enzyme stability is a critical factor in the success of biocatalysts. The stability of native enzymes is largely determined by their internal structure, whereas the stability of immobilized enzymes is determined by several factors such as the type of interaction with the carrier, the type and number of bonds, the properties of the spacer, and the immobilized conditions [79]. Enzyme immobilization also provides molecular confinement and a favorable microenvironment, which can be accomplished by using the right carriers. Immobilization of enzymes has emerged as a powerful mechanism for increasing enzyme stability while retaining most enzyme properties such as selectivity, stability, and specific activity [80]. Furthermore, the porous support allows for the free dispersion of enzyme molecules without the risk of interacting with the outward interactions and stabilizes the enzymes in the solution against interaction with external molecules by preventing autolysis, aggregation, and proteolytic degradation [29]. Furthermore, it prevents enzyme interaction with soluble proteins, assuming operational stability without affecting structural stability under storage, heat, and pH conditions [81].

3.3.4.2 Recovery and reusability

The significant progress in enzyme immobilization has facilitated increased recovery and reusability, lowering the cost of enzyme use and accelerating the growth of biocatalysts. Because of the low economic burden of using immobilized enzymes, they are appealing for recast use over a longer period of time [82]. Following the termination of a chemical reaction, the immobilized enzymes were recovered from the reaction components by implementing optimal reusability protocols such as membrane centrifugation for efficient recovery of the enzymes, which is the first choice for solid supports, and also using stimuli sensitive smart carriers that can be separated more productively from the reaction mixture [34].

3.3.4.3 Enhanced efficacy and selectivity

Because of the reduced substrate inhibiting effect, many types of enzymes immobilized using various techniques have higher activity than native enzymes. Several factors, including changes in the microenvironment and diffusion effect, conformational flexibility, orientation, and binding mode, all contributed to an increase in enzyme efficacy. Similarly, the presence of diffusion constraints can influence the selectivity of two reactions that may occur concurrently in the same reaction system. The process of enzyme immobilization causes a change in temperature or pH, which affects the enzyme's activity and selectivity [83,84].

3.4 Antimicrobial wound dressings

Antimicrobial wound dressings have emerged as viable options for infection control and healing enhancement. The primary goal of wound care is to inhibit or eliminate pathogenic bacteria while also accelerating healing. Several wound dressings have been used as passive barriers to keep wounds safe from external contamination. Traditional wound dressings such as gauze, plasters, and cotton wool bandages are used as primary or secondary dressings to keep the wound clean. However, they have limitations such as being dry in nature and requiring frequent changes to protect from the infusion of healthy tissues, as well as becoming glued to the wound, making removal difficult [41,46,85].

Instead of an upright envelope, modernized wound dressings have been developed to accelerate wound healing. They promote dehydration healing based on the origin and type of wound. Wound dressings made of synthetic polymers have been classified as interactive, passive, or bioactive. Passive dressings, such as gauze and tulle, are used to cover the wound and restore its underneath function. The semi-elusive interactive dressings come in foam, film hydrogel, and hydrocolloid forms. They act as a barrier, preventing bacteria from entering the wound microenvironment [86]. With their anti-inflammatory properties, honey dressings promote epithelial restoration, granulation, and angiogenesis. Furthermore, honey's acidic pH stimulates macrophage phagocytosis and inhibits biofilm formation [87]. Dressings based on hydrogels with a polymeric network of hydrophilic groups trap wound exudates. They also allow for gas diffusion in a moist environment. Because of their excellent biocompatibility, low toxicity, and biodegradability, bioactive dressings made from biomaterials such as collagen, alginate, hyaluronic acid, chitosan alginate, and elastin play an important role in the healing process. Depending on the nature and type of wound, these polymeric materials are frequently used alone or in combination with other polymers. They are also infused with growth factors and antimicrobials to speed up the healing process [36,62].

To address the shortcomings of traditional dressings, advanced dressings containing antibiotics or other antiseptic composites were developed to replace nondrug-based dressings. Aminoglycosides, quinolones, tetracycline, and broad-spectrum cephalosporins are common antibiotics embedded in wound dressings. They work by modifying or inhibiting bacterial nucleic acid and protein synthesis, resulting in metabolic abnormalities by transmuting the integrity of the bacterial cell wall [38]. However, due to the emergence of MDR bacteria and other resistant microbes, novel nanoparticle-based antimicrobial dressings for the treatment of various types of wounds have been developed. The majority of chronic wound patients with diabetes and advanced age have complications from impaired circulation at the extremities, rendering intravenous and oral antibiotics ineffective. Because systemic antibiotics are ineffective at reducing bacterial counts in granulating wounds, the dressings may have a dual benefit in the prevention of depreciated and chronically infected wounds [88]. As a result, more localized inhibitory action is frequently beneficial in accelerating wound healing. Furthermore, several bacterial toxins and enzymes impede wound healing, which can be reduced by using antimicrobial dressings. In chronic wounds, elevated levels of inflammatory cytokines and enzymes cause tissue destruction, which can spread locally or systemically, resulting in sepsis. Furthermore, the bacteria frequently form biofilms with sessile bacterial exudates, which inhibits both infiltration and antibiotic susceptibility [12,33]. Unlike topical agents, which are used with a very limited contact exposure for prolonged inhibitory effects with reduced toxicity and adverse effects of systemic absorption of the drug, antimicrobial nano dressings drain the invasive infection risk by minimizing bacterial colonization in wounds by dispensing adequate contact time for action. They also reduce the bioburden caused by major trauma wounds and burns and speed up healing. Antimicrobial dressings release drugs at the wound surface in a controlled or sustained release manner to provide effective antimicrobial action while also keeping the wound moist for healing [24,61].

3.5 Conclusions and future prospects

Antibiotic resistance has become a critical issue as a result of excessive antibiotic use, and it is to blame for the failure of conventional antimicrobial therapies. Nanotechnology advancements in recent years have made it easier to adapt nanoparticles for a wide range of biomedical applications, including antimicrobial dressings. As a result, research into novel antimicrobials for wound treatment would continue, with improved properties. Nanoparticles with intrinsic antimicrobial activity can also be used to support enzyme immobilization and high enzyme loading. As a result, nanoparticles functionalized with enzymes can be investigated further and used as potential antimicrobials in wound healing.

A combination of different types of nanomaterials in the future makes them an appealing solution in wound healing. Despite the fact that several nanoparticles have been investigated, bio nanocomposite-based new approaches to wound healing are still in their infancy. To investigate their benefits, efficacy and toxicity studies of novel nanoparticles in appropriate animal models under a variety of conditions such as defined pathogen strains and mixed communities of pathogens are required. Furthermore, the long-term safety of nanoparticles as antimicrobials in efficient tissue repair and wound healing necessitates more extensive research to delineate detailed molecular mechanisms for effective translational studies in wound management.

References

[1] S. Dhingra, et al., Microbial resistance movements: an overview of global public health threats posed by antimicrobial resistance, and how best to counter, Front. Public. Health (2020). Available from: https://doi.org/10.3389/fpubh.2020.535668.

[2] G. Maugeri, et al., Identification and antibiotic-susceptibility profiling of infectious bacterial agents: a review of current and future trends, Biotechnol. J (2019). Available from: https://doi.org/10.1002/biot.201700750.

[3] B. Aslam, et al., Antibiotic resistance: a rundown of a global crisis, Infect. Drug. Resist. (2018) 1645—1658. Available from: https://doi.org/10.2147/IDR.S173867.

[4] S. Bharathala, R. Singh, P. Sharma, Controlled release and enhanced biological activity of chitosan-fabricated carbenoxolone nanoparticles, Int. J. Biol. Macromol. 164 (2020) 45—52. Available from: https://doi.org/10.1016/j.ijbiomac.2020.07.086.

[5] R. Singh, M.S. Smitha, S.P. Singh, The role of nanotechnology in combating multi-drug resistant bacteria, J. Nanosci. Nanotechnol. (2014) 4745—4756. Available from: https://doi.org/10.1166/jnn.2014.9527.

[6] L. Wang, C. Hu, L. Shao, The antimicrobial activity of nanoparticles: present situation and prospects for the future, Int. J. Nanomed. (2017) 1227—1249. Available from: https://doi.org/10.2147/IJN.S121956.

[7] N. Fattahian Kalhor, et al., Interaction, cytotoxicity and sustained release assessment of a novel antitumor agent using bovine serum albumin nanocarrier, J. Biomol. Struct. Dyn. 38 (9) (2020) 2546—2558. Available from: https://doi.org/10.1080/07391102.2019.1638303.

[8] W. Gao, et al., Nanoparticle-based local antimicrobial drug delivery, Adv. Drug. Deliv. Rev. (2018) 46—57. Available from: https://doi.org/10.1016/j.addr.2017.09.015.

[9] S. Bharathala, P. Sharma, Biomedical applications of nanoparticles', in *nanotechnology in modern*, Anim. Biotechnol. Concepts Appl. (2019) 113—132. Available from: https://doi.org/10.1016/B978-0-12-818823-1.00008-9.

[10] S.S.D. Kumar, et al., Recent advances on silver nanoparticle and biopolymer-based biomaterials for wound healing applications, Int. J. Biol. Macromol. (2018) 165−175. Available from: https://doi.org/10.1016/j.ijbiomac.2018.04.003.

[11] F. Paladini, M. Pollini, Antimicrobial silver nanoparticles for wound healing application: progress and future trends, Materials (2019). Available from: https://doi.org/10.3390/ma12162540.

[12] S. El-Ashram, et al., Naturally-derived targeted therapy for wound healing: beyond classical strategies, Pharmacol. Res (2021). Available from: https://doi.org/10.1016/j.phrs.2021.105749.

[13] A. Naskar, K.S. Kim, Recent advances in nanomaterial-based wound-healing therapeutics, Pharmaceutics (2020). Available from: https://doi.org/10.3390/pharmaceutics12060499.

[14] O. Pacios, et al., Strategies to combat multidrug-resistant and persistent infectious diseases, Antibiotics (2020). Available from: https://doi.org/10.3390/antibiotics9020065.

[15] E. Sánchez-López, et al., Metal-based nanoparticles as antimicrobial agents: an overview, Nanomaterials (2020). Available from: https://doi.org/10.3390/nano10020292.

[16] M. Terreni, M. Taccani, M. Pregnolato, New antibiotics for multidrug-resistant bacterial strains: latest research developments and future perspectives, Molecules (2021). Available from: https://doi.org/10.3390/molecules26092671.

[17] D. Lahiri, et al., Microbiologically-synthesized nanoparticles and their role in silencing the biofilm signaling cascade, Front. Microbiol. (2021). Available from: https://doi.org/10.3389/fmicb.2021.636588.

[18] K. McNamara, S.A.M. Tofail, Nanoparticles in biomedical applications, Adv. Phys.: X (2017) 54−88. Available from: https://doi.org/10.1080/23746149.2016.1254570.

[19] M. Noh, et al., Magnetic nanoparticle-embedded hydrogel sheet with a groove pattern for wound healing application, ACS Biomater. Sci. Eng. 5 (8) (2019) 3909−3921. Available from: https://doi.org/10.1021/acsbiomaterials.8b01307.

[20] M.N.F. Shaheen, D.E. El-hadedy, Z.I. Ali, Medical and microbial applications of controlled shape of silver nanoparticles prepared by ionizing radiation, BioNanoScience (2019). Available from: https://doi.org/10.1007/s12668-019-00622-2.

[21] K.S. Siddiqi, et al., Properties of zinc oxide nanoparticles and their activity against microbes, Nanoscale Res. Lett (2018). Available from: https://doi.org/10.1186/s11671-018-2532-3.

[22] X. Gu, et al., Preparation and antibacterial properties of gold nanoparticles: a review, Environ. Chem. Lett. (2021) 167−187. Available from: https://doi.org/10.1007/s10311-020-01071-0.

[23] N. Doskocz, K. Affek, M. Załęska-Radziwiłł, Effects of aluminium oxide nanoparticles on bacterial growth, in: E3S Web of Conferences, 2017. Available from: https://doi.org/10.1051/e3sconf/20171700019.

[24] R.G. Frykberg, J. Banks, Challenges in the treatment of chronic wounds, Adv. Wound Care 4 (9) (2015) 560−582. Available from: https://doi.org/10.1089/wound.2015.0635.

[25] A.C.D.O. Gonzalez, et al., Wound healing - a literature review, An. Brasileiros de. Dermatologia (2016) 614−620. Available from: https://doi.org/10.1590/abd1806-4841.20164741.

[26] U. Guzik, K. Hupert-Kocurek, D. Wojcieszynska, Immobilization as a strategy for improving enzyme properties- application to oxidoreductases, Molecules (2014) 8995−9018. Available from: https://doi.org/10.3390/molecules19078995.

[27] R.F. Hakim, Fakhrurrazi, Dinni, Effect of Carica papaya extract toward incised wound healing process in mice (Mus musculus) clinically and histologically, Evid. Complement. Altern. Med., 2019. Available from: https://doi.org/10.1155/2019/8306519.

[28] S. Hamdan, et al., Nanotechnology-driven therapeutic interventions in wound healing: potential uses and applications, ACS Cent. Sci. 3 (3) (2017) 163−175. Available from: https://doi.org/10.1021/acscentsci.6b00371.

[29] P. Han, X. Zhou, C. You, Efficient multi-enzymes immobilized on porous microspheres for producing inositol from starch, Front. Bioeng. Biotechnol. (2020) 8. Available from: https://doi.org/10.3389/fbioe.2020.00380.

[30] M.E. Hassan et al. Impact of immobilization technology in industrial and pharmaceutical applications, 3 Biotech, 2019. Available from: https://doi.org/10.1007/s13205-019-1969-0.

[31] W. Heitzmann, P.C. Fuchs, J.L. Schiefer, Historical perspectives on the development of current standards of care for enzymatic debridement, Medicina (Lithuania) 56 (12) (2020) 1−8. Available from: https://doi.org/10.3390/medicina56120706.

[32] A.A. Homaei, et al., Enzyme immobilization: an update, J. Chem. Biol. (2013) 185−205. Available from: https://doi.org/10.1007/s12154-013-0102-9.

[33] S. Homaeigohar, A.R. Boccaccini, Antibacterial biohybrid nanofibers for wound dressings, Acta Biomater. (2020) 25−49. Available from: https://doi.org/10.1016/j.actbio.2020.02.022.

[34] H.T. Imam, P.C. Marr, A.C. Marr, Enzyme entrapment, biocatalyst immobilization without covalent attachment, Green. Chem. (2021) 4980−5005. Available from: https://doi.org/10.1039/d1gc01852c.

[35] T. Jesionowski, J. Zdarta, B. Krajewska, Enzyme immobilization by adsorption: a review, Adsorption (2014) 801−821. Available from: https://doi.org/10.1007/s10450-014-9623-y.

[36] E.A. Kamoun, E.R.S. Kenawy, X. Chen, A review on polymeric hydrogel membranes for wound dressing applications: PVA-based hydrogel dressings, J. Adv. Res. (2017) 217−233. Available from: https://doi.org/10.1016/j.jare.2017.01.005.

[37] M.R. Khan, Immobilized enzymes: a comprehensive review, Bull. Natl Res. Cent. 45 (1) (2021). Available from: https://doi.org/10.1186/s42269-021-00649-0.

[38] D. Kowalczuk, et al., Characterization of ciprofloxacin-bismuth-loaded antibacterial wound dressing, Molecules (Basel, Switz.) 25 (21) (2020). Available from: https://doi.org/10.3390/molecules25215096.

[39] Z. Li, Z. Lin, Recent advances in polysaccharide-based hydrogels for synthesis and applications, Aggregate 2 (2) (2021). Available from: https://doi.org/10.1002/agt2.21.

[40] X. Lyu, et al., Immobilization of enzymes by polymeric materials, Catalysts (2021). Available from: https://doi.org/10.3390/catal11101211.

[41] T. Maheswary, A.A. Nurul, M.B. Fauzi, The insights of microbes' roles in wound healing: a comprehensive review, Pharmaceutics 13 (7) (2021). Available from: https://doi.org/10.3390/pharmaceutics13070981.

[42] S.K. McCallon, D. Weir, J.C. Lantis, Optimizing wound bed preparation with collagenase enzymatic debridement, J. Am. Coll. Clin. Wound Specialists (2014) 14−23. Available from: https://doi.org/10.1016/j.jccw.2015.08.003.

[43] G.K. Meghwanshi, et al., Enzymes for pharmaceutical and therapeutic applications, Biotechnol. Appl. Biochem. (2020) 586−601. Available from: https://doi.org/10.1002/bab.1919.

[44] N.R. Mohamad, et al., An overview of technologies for immobilization of enzymes and surface analysis techniques for immobilized enzymes, Biotechnol. Biotechnol. Equip. (2015) 205−220. Available from: https://doi.org/10.1080/13102818.2015.1008192.

[45] N.F. Mokhtar, et al., The immobilization of lipases on porous support by adsorption and hydrophobic interaction method, Catalysts (2020) 1−17. Available from: https://doi.org/10.3390/catal10070744.

[46] I. Negut, V. Grumezescu, A.M. Grumezescu, Treatment strategies for infected wounds, Molecules (2018). Available from: https://doi.org/10.3390/molecules23092392.

[47] G. Suarato, R. Bertorelli, A. Athanassiou, Borrowing from nature: biopolymers and biocomposites as smart wound care materials, Front. Bioeng. Biotechnol (2018). Available from: https://doi.org/10.3389/fbioe.2018.00137.

[48] R. Song, et al., Current development of biodegradable polymeric materials for biomedical applications, Drug. Design, Dev. Ther. (2018) 3117−3145. Available from: https://doi.org/10.2147/DDDT.S165440.

[49] N. Beyth, et al., Alternative antimicrobial approach: nano-antimicrobial materials, Evid.-based Complement. Altern. Med (2015). Available from: https://doi.org/10.1155/2015/246012.

[50] Y.N. Slavin, et al., Metal nanoparticles: understanding the mechanisms behind antibacterial activity, J. Nanobiotechnol (2017). Available from: https://doi.org/10.1186/s12951-017-0308-z.

[51] H.A. Wallace, P.M. Zito, Wound healing phases, StatPearls. Available from: http://www.ncbi.nlm.nih.gov/pubmed/29262065, 2019.

[52] E. Bellu, et al., Nanomaterials in skin regeneration and rejuvenation, Int. J. Mol. Sci. 22 (13) (2021). Available from: https://doi.org/10.3390/ijms22137095.

[53] K. Banerjee, et al., Nanoceutical adjuvants as wound healing material: precepts and prospects, Int. J. Mol. Sci (2021). Available from: https://doi.org/10.3390/ijms22094748.

[54] A. Barroso, et al., Nanomaterials in wound healing: from material sciences to wound healing applications, Nano Sel. 1 (5) (2020) 443−460. Available from: https://doi.org/10.1002/nano.202000055.

[55] R. Chapman, M.H. Stenzel, All wrapped up: stabilization of enzymes within single enzyme nanoparticles, J. Am. Chem. Soc (2019). Available from: https://doi.org/10.1021/jacs.8b10338.

[56] M. Seema, A.S. Kumar, Enzyme immobilization by nanoparticles, Res. J. Biotechnol. (2021) 206−211.

[57] M. Sharifi, et al., Enzyme immobilization onto the nanomaterials: application in enzyme stability and prodrug-activated cancer therapy, Int. J. Biol. Macromol. (2020) 665−676. Available from: https://doi.org/10.1016/j.ijbiomac.2019.12.064.

[58] J. An et al., Recent advances in enzyme-nanostructure biocatalysts with enhanced activity, Catalysts, 2020. Available from: https://doi.org/10.3390/catal10030338.

[59] K. Xu, et al., Immobilization of multi-enzymes on support materials for efficient biocatalysis, Front. Bioeng. Biotechnol (2020). Available from: https://doi.org/10.3389/fbioe.2020.00660.

[60] D.C. Thomas, et al., The role of debridement in wound bed preparation in chronic wound: a narrative review, Ann. Med. Surg (2021). Available from: https://doi.org/10.1016/j.amsu.2021.102876.

[61] R. Smith, et al., Antibiotic delivery strategies to treat skin infections when innate antimicrobial defense fails, Antibiotics (2020). Available from: https://doi.org/10.3390/antibiotics9020056.

[62] J. Su, et al., Hydrogel preparation methods and biomaterials for wound dressing, Life (2021). Available from: https://doi.org/10.3390/life11101016.

[63] T. Tamer, A. Omer, M. Hassan, Methods of enzyme immobilization, Int. J. Curr. Pharm. Rev. Res. 7 (6) (2016) 385−392.

[64] D. Telgenhoff, et al., Influence of papain urea copper chlorophyllin on wound matrix remodeling, Wound Repair. Regen. 15 (5) (2007) 727−735. Available from: https://doi.org/10.1111/j.1524-475X.2007.00279.x.

[65] R. Di Pasquale, et al., Collagenase-assisted wound bed preparation: an in vitro comparison between Vibrio alginolyticus and Clostridium histolyticum collagenases on substrate specificity, Int. Wound J. 16 (4) (2019) 1013−1023. Available from: https://doi.org/10.1111/iwj.13148.

[66] S. Zhu, et al., The emerging roles of neutrophil extracellular traps in wound healing, Cell Death Dis (2021). Available from: https://doi.org/10.1038/s41419-021-04294-3.

[67] D. Diwan, et al., Thrombolytic enzymes of microbial origin: a review', Int. J. Mol. Sci. 22 (19) (2021). Available from: https://doi.org/10.3390/ijms221910468.

[68] H. Essa, et al., Influence of pH and ionic strength on the adsorption, leaching and activity of myoglobin immobilized onto ordered mesoporous silicates, J. Mol. Catal. B: Enzymatic 49 (1−4) (2007) 61−68. Available from: https://doi.org/10.1016/j.molcatb.2007.07.005.

[69] H.H. Nguyen, M. Kim, An overview of techniques in enzyme immobilization, Appl. Sci. Converg. Technol. 26 (6) (2017) 157−163. Available from: https://doi.org/10.5757/asct.2017.26.6.157.

[70] M.L. Cacicedo, et al., Immobilized enzymes and their applications', *Biomass*, Biofuels, Biochem.: Adv. Enzyme Technol. (2019) 169−200. Available from: https://doi.org/10.1016/B978-0-444-64114-4.00007-8.

[71] Y.A.H. Samarasinghe, et al., Recombinant enzymes used in fruit and vegetable juice industry, Microb. Enzyme Technol. Food Appl. (2017) 375−395. Available from: https://doi.org/10.1201/9781315368405.

[72] P. Zucca, E. Sanjust, Inorganic materials as supports for covalent enzyme immobilization: methods and mechanisms, Molecules (2014) 14139−14194. Available from: https://doi.org/10.3390/molecules 190914139.

[73] Z. Zhao, M.C. Zhou, R.L. Liu, Recent developments in carriers and non-aqueous solvents for enzyme immobilization, Catalysts (2019). Available from: https://doi.org/10.3390/catal9080647.

[74] H. Yamaguchi, Y. Kiyota, M. Miyazaki, Techniques for preparation of cross-linked enzyme aggregates and their applications in bioconversions, Catalysts (2018). Available from: https://doi.org/10.3390/catal8050174.

[75] M. Fernández, J. Orozco, Advances in functionalized photosensitive polymeric nanocarriers, Polymers, 2021. Available from: https://doi.org/10.3390/polym13152464.

[76] R. Parhi, Cross-linked hydrogel for pharmaceutical applications: a review', Adv. Pharm. Bull. (2017) 515−530. Available from: https://doi.org/10.15171/apb.2017.064.

[77] R. Datta, et al., How enzymes are adsorbed on soil solid phase and factors limiting its activity: a review, Int. Agrophys. (2017) 287−302. Available from: https://doi.org/10.1515/intag-2016-0049.

[78] S.B. Sigurdardóttir, et al., Enzyme immobilization on inorganic surfaces for membrane reactor applications: mass transfer challenges, enzyme leakage and reuse of materials, Adv. Synth. Catal. (2018) 2578−2607. Available from: https://doi.org/10.1002/adsc.201800307.

[79] A. Basso, S. Serban, Industrial applications of immobilized enzymes—a review, Mol. Catal (2019). Available from: https://doi.org/10.1016/j.mcat.2019.110607.

[80] R.C. Rodrigues, et al., Stabilization of enzymes via immobilization: multipoint covalent attachment and other stabilization strategies, Biotechnol. Adv (2021). Available from: https://doi.org/10.1016/j.biotechadv.2021.107821.

[81] C. Silva, et al., Practical insights on enzyme stabilization, Crit. Rev. Biotechnol. (2018) 335−350. Available from: https://doi.org/10.1080/07388551.2017.1355294.

[82] M. Bilal, et al., Multi-point enzyme immobilization, surface chemistry, and novel platforms: a paradigm shift in biocatalyst design, Crit. Rev. Biotechnol. (2019) 202−219. Available from: https://doi.org/10.1080/07388551.2018.1531822.

[83] S. Ali, et al., Enzymes immobilization: an overview of techniques, support materials and its applications, Int. J. Sci. Technol. Res. 6 (07) (2017) 64−72. Available at: http://www.ijstr.org.

[84] M. Araki, et al., The effect of conformational flexibility on binding free energy estimation between kinases and their inhibitors, J. Chem. Inf. Model. 56 (12) (2016) 2445−2456. Available from: https://doi.org/10.1021/acs.jcim.6b00398.

[85] M. Rodrigues, et al., Wound healing: a cellular perspective, Physiol. Rev. 99 (1) (2019) 665−706. Available from: https://doi.org/10.1152/physrev.00067.2017.

[86] V. Brumberg, et al., Modern wound dressings: hydrogel dressings, Biomedicines (2021). Available from: https://doi.org/10.3390/biomedicines9091235.

[87] H. Scepankova, et al., Role of honey in advanced wound care, Molecules. (2021). Available from: https://doi.org/10.3390/molecules26164784.

[88] Z.B. Nqakala, et al., Advances in nanotechnology towards development of silver nanoparticle-based wound-healing agents, Int. J. Mol. Sci (2021). Available from: https://doi.org/10.3390/ijms222011272.

Polyelectrolyte assembly with nanoparticle-immobilized enzymes

4

Pratik Kolhe, Maitri Shah and Sonu Gandhi

DBT-National Institute of Animal Biotechnology (NIAB), Hyderabad, Telangana, India

4.1 Introduction

Enzymes are bio-catalysts that are used in all anabolic and catabolic reactions in the body at both the macroscopic and microscopic levels. Because of the site-specific nature of enzymes, they have been used in a variety of other industrial applications, including food, chemicals, pharmaceuticals, biofuel, textiles, and many others as a green synthetic approach. Enzymatic reactions with substrate specificity do not require high reaction conditions and can react at low temperatures with high efficacy. Because enzymes are proteinaceous, the main issue that arises when working with enzymes is the stability of their structure. Long-term storage affects the stability of quaternary structure, which in turn affects the activity of an enzyme gold [1]. One of the most difficult challenges is the recovery of soluble enzymes for subsequent use. To address such issues, the enzyme immobilization technique has gained popularity in recent years. Immobilization refers to the physical trapping, insolubilization, or attachment of enzyme to a solid matrix.

The immobilized enzyme has been studied for a very long time and it can be seen from an inclining the interest in this field from the recent research and reviews. As the immobilized enzymes are highly efficient, the commercial use has [2−4]. Immobilized enzymes are much more stable and flexible to handle than native or free enzymes. Because of the immobilization of these enzymes, the rate of enzymatic reaction can also be accountable and manageable, with the added benefit that the end of the reaction is not harmed by side products [5]. Immobilization strengthens enzyme properties such as enzyme activity in organic solvent, pH sustainability, heat tolerance, and immobilization adduct in protein structural ruggedness. The immobilized enzymes can regenerate and are ready to enter the next reaction. There is no need for additional processes such as extraction and purification. The alteration occurs as the immobilization process results into structural changes in the enzyme and the developed microenvironment for enzyme activity [6−8], which has been peculiar than bulk mixture. The primary goal of enzyme immobilization is to keep enzymes out of the environment and to maximize the benefits of enzyme catalysis, which is made possible by using a low-cost synthetic support with a high binding capacity [9].

The stability of immobilized enzyme molecules is determined by several factors, including the nature and properties of the carrier to which the enzyme is attached (chemical and physical),

Antimicrobial Dressings. DOI: https://doi.org/10.1016/B978-0-323-95074-9.00006-3

its interaction with the carrier, the nature of the microenvironment, the presence of a spacer between the enzyme and the carrier molecule to be immobilized, the nature of the spacer and linker, and the methods and conditions of immobilization used. Along with the previously mentioned conditions, additional overall experimental conditions such as pH, temperature, and storage conditions influence enzyme activity [10]. When epoxy-hydrolase adsorbed on DEAE (Diethyl aminoethyl)-cellulose resin with ionic bonding, its activity increases twice than the natal enzyme [11]. Because immobilization conditions such as enzyme loading, pH, carrier, and binding chemistry are frequently required to retain high activity, extensive screening for immobilization conditions such as enzyme loading, pH, carrier, and binding chemistry is frequently required.

There has been a lot of progress in the development and implementation of site-specific protein or enzyme immobilization in recent years. Organic chemistry and molecular biology advancements have resulted in the development of a number of highly effective, efficient, site-specific, and significant applications of localized proteins and enzymes onto supports [12,13].

Polyelectrolyte assembly is used to create thin films with multifunctional properties. Polyelectrolyte assembly is also used to create ionic layers one on top of the other, resulting in multiple ionic layers and a thin ionic film. Enzymes are encapsulated or immobilized within the film's 2D or 3D framework. Alternatively charged molecules form large layers as a controllable 3D ionic film framework. Physical forces such as Van-der-Walls forces, weak dispersive forces, electrostatic forces, and capillary forces influence the self-assembled polyelectrolyte system. Polyanion and polycations share an electrostatic attractive force in polyelectrolyte systems, allowing them to form complex structures. They also form hydrogen bonds, covalent bonds, and host−guest interactions between the reactive species. The Langmuir−Blodgett method, covalent-binding method, and spontaneous adsorption method can all be used to explain how such catalytic or multifunctional films are made.

4.2 Enzyme immobilization methods

Enzyme immobilization can be accomplished in a number of ways, the most common of which are physical and chemical. Chemical techniques result in a covalent bond between the support and the enzyme, whereas physical techniques result in weak contacts between the matrix and the enzyme. Enzyme immobilization techniques include physical adsorption, encapsulation, chemical crosslinking, electrochemical polymerization, and layer-by-layer (LBL) assembly.

4.2.1 Physical techniques for enzyme immobilization

Nanofibers, nanocomposites, nanorods, nanoparticles, nanosheets, and other nanostructures have been reported for nano-immobilization and the use of nanocatalysts [14,15]. For immobilization on the surface of nanoparticles, various physical methods have been used. Methods are classified as adsorption or encapsulation based on the binding interaction between the enzyme and the nano-surface [16]. Furthermore, one can use a combination of two methods to compensate for the shortcomings of one technique.

4.2.2 Chemical methods for enzyme immobilization

When polyanion and polycation solutions are mixed, the electrostatic force of attraction produces a polyion complex that is exceptionally stable for dissociation. Because the polyion complex has multiple binding sites, dissociation of one or more ions has no effect on overall activity. In general, such versatile membrane solutions were created by combining weak poly-acid and weak poly-base solutions, which are similar to weak acid and bases or somsetimes chemical modification in functional groups like silanation, hydroxylation. It does not, however, completely dissociate at a specific pH and is not fully charged. As a result of this property, it produces buffer-like characteristics and can be molded with desired characteristics. As polyanion solutions, polyacids such as poly-maleic acid (PMA), poly-acrylic acid (PAA), poly-L-lysine (PLL), and poly-L-arginine (PLA) are used. Polyelectrolyte complex formation works by LBL or Polyion Complex (PIC) mechanism. PIC method involves mixing of polyanion and polycation altogether to form a complex membranous structure, possesses heteromorphous nature [17] for example property of surrounding like pH weak polyacids shows its activity at specific pH range by changing its range protonation property of acid arises due to its ionizable groups also varies and cause change in overall charge of layer or complex. Similarly a reasonable ionic strength of solution reduces its repulsiveness among the charged molecules and offers electrostatic assembly [18−22] (Fig. 4.1).

4.3 Recent advancement in enzyme immobilization

Nanoparticles are also thought to be an effective carrier material for enzyme immobilization due to properties such as efficient biocatalysis, a high surface-to-volume ratio, and resistance to mass transfer [23−26]. According to scientific evidence, immobilized enzyme reduces protein unraveling and increases enzyme stability and activity. Several reviews of enzyme immobilization on various types of nanoparticles have been published, including metal nanoparticles, metal oxide nanoparticles, magnetic nanoparticles, porous and polymeric nanoparticles [27,28].

Inorganic molecules such as alkali earth elements (Ca, Mg), transition metal and metal-oxides, mixed metal oxides (TiO_2, Fe-Cd), organic molecules such as carbonous material such as graphene, graphite, graphene oxide, carbon nanotubes, organic polymer, hydrogel, and so on can be used to create nanoparticles [29]. Because of their quantum properties and high surface-to-volume ratio, nanoparticles such as magnetic nanoparticles have been used for immobilization. Furthermore, magnetic nanoparticles can be easily separated by applying an external magnetic field [30−32]. The following are the primary criteria for nano-biocatalysis: (1) Enzyme should cover the surface area of the matrix or solid support, ensuring that the biomolecule content per unit mass of particles is high. (2) Reduces mass transfer strain on the substrate [33,34]. The use of nanomaterials avoids the problem of steric hindrance caused by bulky support between biomolecules. Nanomaterials exhibit Brownian motion in aqueous solution, resulting in higher enzymatic activity than free enzymes. Enzyme activity increases after binding with nanoparticles due to the interaction of free substrate with immobilized enzyme. Weak attachment of nanoparticles to substrates results in multiple binding possibilities to substrate molecules, eventually leading to maximum concentration of substrate surrounding nanoparticles [35−37].

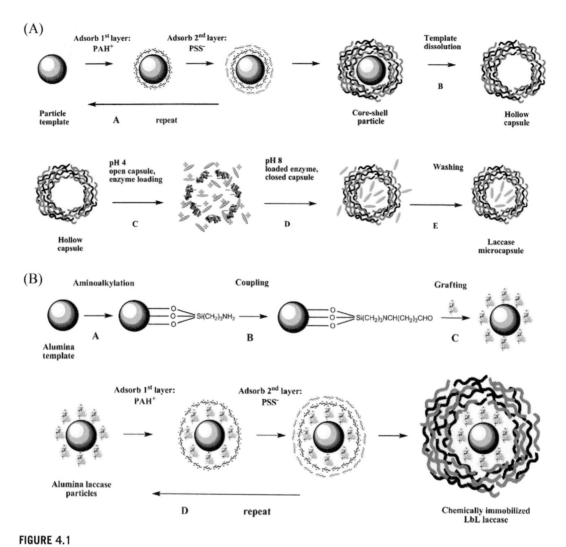

FIGURE 4.1

Schematic diagram of (A) physical and (B) chemical adsorption of layer-by-layer assembly [22].

Reprinted with the permission for figure obtained from Elsevier (C. Crestini, R. Perazzini, R. Saladino, Oxidative functionalisation of lignin by layer-by-layer immobilised laccases and laccase microcapsules, Appl. Catal. A. Gen. 372 (2) (2010) 115–123).

As nanomaterials are an emerging trend in the field of research and technology by the virtue of its nano-dimensional size [38,39]. The route of nanomaterial synthesis influences properties such as size, surface to charge ratio, and physiochemical properties. As a result, it is critical to select an appropriate method for the desired application [40,41]. Fig. 4.2 depicts various methods for synthesis of nanomaterials [42,43].

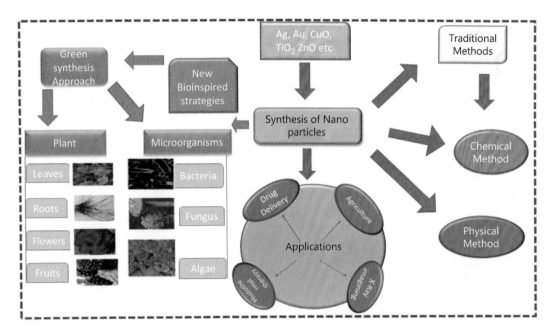

FIGURE 4.2

Schematic representation of various methods for synthesis of nanoparticles [43].

Reprinted with the permission for figure obtained from Elsevier (G.A. Naikoo, M. Mustaqeem, I.U. Hassan, T. Awan, F. Arshad, H. Salim, Qurashi, et al. Bioinspired and green synthesis of nanoparticles from plant extract with antiviral and antimicrobial properties: A critical review. J. Saudi Chem. Soc. 25(9) (2021) 101304. https://doi.org/10.1016/j.jscs.2021.101304.

At least one dimension of nanoparticles is in the nanometer range. There are various types of nanoparticles based on their composition, such as polymer nanoparticles, liposomes, metallic nanoparticles, carbon-based nanoparticles, core-shell nanoparticles, and so on [44]. However, there are several challenges faced during characterization of these nanoparticles following:

1. Variably size
2. Not high degree of synthesis at laboratory
3. Undiscoverable methods for pre-sampling and pre-process
4. Unavailability of calibration purpose during analytical measurement
5. Difficulties in data analysis and interpretation
6. Difficulty measuring the exact concentration of nanomaterial at during and after reaction

Despite such difficulties in characterization of nanoparticles based on the specific parameter to be studied, various techniques can be used to determine size, charge present on the surface, atomic configuration, surface study, optical properties, magnetic properties, and so on [45,46]. Based on these properties, nanomaterials can be further used for applications in [47,48], diagnostics [49,50], theragnostic, etc. with the help of further modifications such as enzyme immobilization. Following are some of the nano-enzymes explained here:

4.3.1 Carbon based nano-catalyst

Graphene [51], graphene oxide (GO) [52−57], reduced graphene oxide (rGO) [58−61], carbon nanotubes (CNT) [62,63], single-walled carbon nanotubes, multiwalled carbon nanotubes MWCNT [64,65], GO-nanosheets [66], etc. are some of the examples of carbon based nanomaterials, which are used for the immobilization of biomolecules like enzymes [67] and proteaceous molecules. Because of the conjugated structure and multidimensional shapes, slight chemical or functional modification can be used for the immobilization of the enzyme and other biomolecules. Carbon-based nanomaterials have a wide range of applications in biomedicine, including sensing, drug delivery, targeting, and diagnostics [59]. GO [68,69] has continuous conjugated planner structure with double sided geometry, CNT (carbon nanotube) has rolled to form tubular structure from graphene, having two dimensional geometry [70]. Surface functionalization of these nanomaterials not only improves enzyme binding but also increases the effect of enzyme activity. There have been reports of functionalization increasing the effect of immobilization; for example, the hydrolytic activity of lipase from *Candida rugosa* increased by 55% when immobilized on the GO with amine functionalized lipase [71]. Yan and co-worker observed in their study that immobilized enzyme had more thermostability than free enzyme, for that they used polyethelyeneglycol-deposited GO [72] nanosheets and checked the activity of trypsin at elevated temperature.

4.3.2 Metal/metal-oxide polymeric nanomaterials

Inorganic elements are also used as a platform for enzyme immobilization. These nano-enzymes has diverse application mostly in the area of sensing and detection along with drug deliver [73−75], regulation in blood coagulation [76]. Gold (Au) [77−82] and Silver (Ag) has numerous applications in the biomedical field as low bio-toxicity of both metals [83]. Immobilization on gold nanoparticles (AuNPs) through enormous interaction of analyte with thiol, carboxylate, amines [84,85]. Hung shows that when a silver nano-catalyst is coated with lipase for the production of biodiesel, the production yield of biodiesel is increased by up to 95% compared to free enzyme, which yields only 86% [86−88]. Researchers demonstrated the activity of immobilized protein on poly-dopamine (PDA) functionalized magnetic nanoparticles (Fe_3O_4@PDA). Trypsin molecules that bind to the magnetic surface were easily accessible to the target, with increasing binding capacity towards the substrate [89]. The extent to which nanoparticles are immobilized on their surfaces is also determined by the morphology and shape of the nanoplates, nanofibers, nanoflower, nano-cubes, nanoflakes, nano-multipods, and nano-disks. The impact of nanospheres, different shapes and sizes of nanomaterials, varies [90,91]. Zhang and colleagues investigated the effect of morphology and enzyme activity in which they immobilized horseradish peroxidase (HRP) on ZnO nanoparticles in various shapes such as nanospheres, nano-disks, and nano-multipods. The experiment demonstrates how morphological characteristics influence the activity of an immobilized enzyme. In their study, they discovered that multipods self-polymerized and amine conjugated with HRP had the best catalytic performance compared to others [92] (Fig. 4.3). PDA was used as an

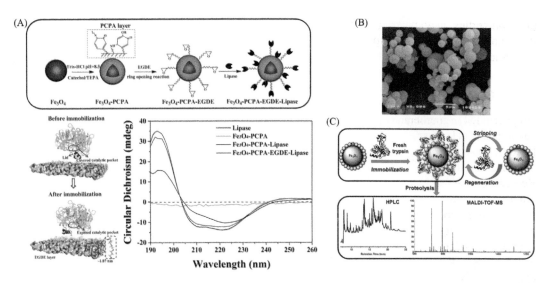

FIGURE 4.3

Iron oxide (Fe$_3$O$_4$) nanoparticles are used for immobilization of different enzymes. (A) Systemic mechanism of CPA-modified iron oxide with ester linker, its structural simulation. (B) SEM image of formed microcapsule multilayer mechanism. (C) Reversible immobilization of trypsin on reusable gold-coated iron oxide.

Reprinted with the permission for figure obtained from Elsevier [93–95].

adherent constituent which eases enzyme immobilization, but PDA is not economic for regular use and recovered enzyme activity also not good, to overcome these issues, new low-cost polymer system has been employed like catechol/tetraethylene pentamine (CPA) binary system. Other metal framework intermediate substrates could serve as a ligand for binding of enzyme [93–95].

4.3.3 Luminescence based nanozymes

Another type of nanomaterial is luminescent nanomaterials, which consist of luminescent semiconductor nanostructures, primarily quantum dots (QD) [96,97]. Because of their physical, electrical, optical, and chemical properties such as photoluminescence, photo-blanching, and manageability, these materials have been extensively used in research [98–101]. Because of the luminescent property, it is simple to track the enzyme's delivery and site specificity. Morphology and the size of such materials has impact on the enzyme activity, a group of researchers studied the effect of variable size of QD with enzymatic activity, where CdSe-Zn nano shells of size 5 and 10 nm were used to check enzymatic activity of cellulase. They observed the activity of QD-enzyme conjugate of size 10 nm diameter showing increased rate of hydrolysis than QD-enzyme conjugate of size 5 nm diameter. Smaller QDs with higher surface curvature promote natural enzyme assembly and have fewer enzyme-to-enzyme neighbor contacts than larger QDs. The greatest improvement in enzyme performance was observed when fewer enzymes were immobilized on the QDs, supporting the previously stated hypothesis about size and morphology [102].

4.4 Polyelectrolyte assembly and fabrication

The mechanism of LBL deposition of polyelectrolytes on the surface and membrane formation is discussed here. Other biomolecules can be easily encapsulated or bonded to the layer due to the presence of an ion-layer and charged molecules on the layer. Instead of polyelectrolytes, scientists can now use biopolymers such as proteins to create protein-LBL membranes that have more properties than polyion membranes [103–105]. The method is similar in that the anionic substrate is first treated with the polyanion solution, and then the mixture of protein and polycation is applied to it. More suitable surfaces, such as polyelectrolyte brushes and dendrimers, have been chosen for improved immobilization. Brushes made of polyelectrolytes are dense, macromolecular, and have a charged surface. Because of the charge on their surfaces, polyelectrolyte brushes can also immobilize nanomaterials, nanocomposites, and nanoconjugates. These immobilized small molecules can be used to create electrochromic devices capable of representing data in electrical and color differential forms [106,107] (Fig. 4.4).

A surface with some surface characteristics for biological applications is required to form a massive multilayered complex structure. The surface should have a good response to biological

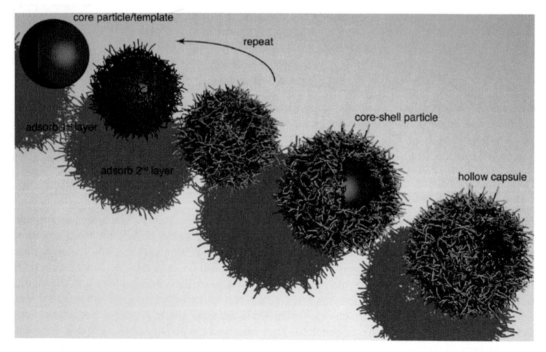

FIGURE 4.4

Layer-by-layer assembly of polyelectrolyte on core particle or a surface [107].

Reprinted with the permission for figure obtained from Elsevier (A.P.R. Johnston, C. Cortez, A.S. Angelatos, F. Caruso, Layer-by-layer engineered capsules and their applications, Curr. Opin. Colloid. Interface Sci. 11 (4) (2006) 203–209).

systems while also not being harmful to biological systems. The surface of a polyelectrolyte membrane (PEM) can be modified to improve its activity. When the surface is neutral, such as graphite electrodes, it is treated with a concentrated acid solution, such as nitric acid, sulfuric acid (H_2SO_4), or concentrated chromic acid, to generate a negative charge [17]. The charged anionic surface can then be used further to fabricate PEM.

4.4.1 Features of polyelectrolyte membrane membranes

4.4.1.1 Surface having physio-chemically adherent nature

Materials such as chitosan and alginate derived from chitinous material, which are mechanically adhesive, biodegradable, non-toxic, anti-bacterial, anti-fungal, and hemostatic, have been used. This material is naturally presenting functional groups that can be used for crosslinking. Another method is to synthesize such adherent substances in their chemical anomeric forms. Catechol biomimetic polymer that is a modified analog of Levodopa (L-DOPA) [108,109]. Hyaluronic acid and poly (ethylenimine) have also been used as catechol modified polymers that have been adsorbed on the silicon or gold surface. The hydroxyl group or any hydroxyl derivative present in the compound is responsible for polymer adhesion.

Because of the involvement of a medical adherent compound, the LBL assembly for the PEM multilayer has acquittance for wound healing. However, there is still a need for research into good adherent molecules that do not get removed due to the presence of biological fluids and the use of non-natural, non-biodegradable polymers that create toxicity and impede the tissue formation process.

4.4.1.2 Dynamic reactive surface

We discussed modified surfaces with modifiers, but these surfaces are static in nature, which means they do not react to external factors present for the modification. Dynamic reactive surfaces, on the other hand, are defined as stimuli reactive surfaces that undergo conformational changes in response to external factors such as temperature, humidity, stressors such as chemically developed or mechanically applied stressors, or even light [110]. J. C. Rodríguez-Cabellz and group worked on multilayer film made up of chitosan and some Elastic like Recombiner (ELR) polymer which is a genetically mutated polypeptide. These polypeptide chains change phase in a temperature-dependent manner, with free random peptides gaining spiral-coiled form as they are exposed to heat. The dependable transitions of ELRs were measured using parameters such as pH and ionic strength in the subsequent experiment. They found that change in the wettability or hydrophilicity of the surface above 11 pH, 12 M ionic strength of solution at temperature 50ₒC. Furthermore, the genetically derived ELRs aid in cell adhesion. Okano and colleagues created a multilayered thin film of calcium-binding protein alginate (ALG)-chitosan that was modified with poly(N-isopropylacrylamide) (PNIPAAm) and was capable of transplanting and culture cells [111]. They demonstrated in their experiment the possibility of culturing human osteoblast cells at $\pm 4°C$ temperature than the conventional temperature as a result of increase in hydrophilicity of the cell surface due to PNIPAAn chains. Sharp changes in parameters like pH and ionic strength can be harmful to cells and biologically active species because they are unable to withstand hypotonic, hypertonic, severe acid, and severe alkaline environments. Cells, on the other hand, handled temperature variations in the system better for temperatures up to 37°C, as demonstrated by cell-sheet technology used to test the response of the cell on multilayer for growth and adhesion [112].

Another factor that has received attention is light, which cannot be easily controlled in the case of implants. Most light-sensitive molecules absorb in the UV range, causing biological cells to be damaged. The introduction of certain light responsive or radicle groups, on the other hand, can shift absorption towards the visible region. Light is non-invasive and can be more effective because of its penetration power. The Ahmad, Saqib, and Barret azobenzene-functionalized film demonstrated distinct hydrophilic/hydrophobic character upon UV light application [113].

4.5 Layer-by-layer assembly for immobilization of enzyme

Surface modification by chemical or plasma treatment can result in polyelectrolyte assembly or LBL deposition over the surface, activating the surface to attract oppositely charged layers, one above the other. Adsorption of the polyanion on the modified surface is followed by polycation treatment for further activation. Unbound ions are expelled or removed by intermediate washing during each step of adsorption. These steps were repeated several times to create a framework of polyanion and polycation. The thickness of the framework is proportional to the amount of deposition in this case. Because of its simple mechanism, this technique makes it simple to architect complex 3D films [114].

In the LBL self-assembly process, a charged substrate is brought into contact with an oppositely charged polyelectrolyte polymer, as shown in the schematic. This mechanism facilitates overcompensation and surface charge reversal, resulting in a negatively charged surface on the substrate. The sample is then washed in water and placed in contact with a charged polyelectrolyte polymer, where the same thing occurs, except the substrate now has a net surface charge of the electrolyte used. Adsorption of polyelectrolytes results in a polymer brush with randomly ordered coils and loops that can be changed with additional adsorbed polyelectrolytes [115]. The charges of the outermost polyelectrolyte layer were designed to match the charge of a specific biomolecule. The process of surface neutralization and charge reversal can now be repeated to produce an alternating polymer and biomolecule film. This method can be applied to any surface-charged system, and the remaining components can be swapped out to achieve the desired film composition [116]. The adsorption of cationic and anionic amphiphiles on solids as a function of concentration, pH, ionic strength, and temperature [117], that can be easily changed to influence the response of the LBL synthesis. Haller reported the reversible adsorption of a single monolayer of sodium dodecyl sulfate on a silicon single crystal surface with a positive surface charge due to amino-propyl-silanization [118,119].

Polyelectrolytes have been modified to become programmable nanocomponents. The multilayer structure in the electrolyte solution remains electroneutral due to the exchange of ions and solvent with the external electrolyte, avoiding the influence of excess potential. The nano-assembly can be functionalized to create a multifunctional surface with infinite control over the interface features. It is worth noting that multilayer structures based on electrostatic interactions can be stable within a typical pH range, but can also be destroyed in extreme situations such as high pH or high salt solution concentration [115] (Fig. 4.5).

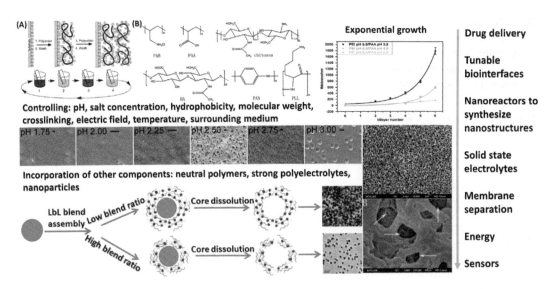

Controlling: pH, salt concentration, hydrophobicity, molecular weight, crosslinking, electric field, temperature, surrounding medium

Incorporation of other components: neutral polymers, strong polyelectrolytes, nanoparticles

FIGURE 4.5

Layer-by-layer assembly of polyelectrolytes. Schematic of synthesis of layer-by-layer assembly, chemical structure of various polyelectrolytes, and their applications. (A) Procedure for layer-by-layer assembly formation. (B) Chemical constituents of weak poly acid [119].

Reprinted with the permission for figure obtained from Elsevier (W. Yuan, G.M. Weng, J. Lipton, C.M. Li, P.R. Van Tassel, A.D. Taylor, Weak polyelectrolyte-based multilayers via layer-by-layer assembly: approaches, properties, and applications Adv. Colloid Interface Sci. (2021) 102200).

4.5.1 Limitations to Layer-by-Layer assembly

Multilayer assembly is primarily determined by the intrinsic properties of the polymer materials, such as polymer molar mass, inorganic ion size, and shape. Several extrinsic properties, such as substrate amount or size, temperature, pH, and viscosity, either facilitate or hinder polyelectrolyte multilayer formation. When working with weak polyacids, if the pH of the solution is not kept within the range of its dissociation constant (pKa), the charge density of the solution and the electrostatic interaction between the layers will be affected [110]. Adsorption, electrochemical, encapsulation, and covalent binding localization of enzyme on the surface of nanoparticles can all be performed after optimizing and screening all of the physical favorable conditions for multilayer formation. Nanomaterials such as nanoparticles, nanospheres, nanotubes, dendrimers, nanoplates, nanorods, and so on can be used as immobilization surfaces. Because the structural properties of these materials differ, the final multilayer may have varying characteristics [120,121].

4.6 Further applications of layer-by-layer assembly

A solid support or implant is required to assemble the biomolecule immobilized multilayers. The implant should provide a bio-interface between the biological site and the multilayer. To ensure that the implant does not adsorb nonspecific proteins, it should be coated with hydrophilic groups or materials with viscous properties [122,123]. Furthermore, the solid support preserves the biomolecules' activity and structural features. The supporting material or film should sufficiently cover the entire surface of the tissue when layering or implanting over it. The coating material determines the subsequent formation of multilayers over one another. The number of adsorption cycles required to produce PEM is proportional to the film thickness. PEM fabrication control can maintain mechanical strength, surface properties, and thin film durability. Because natural polymer is non-toxic, biodegradable, and bioavailable, it is used as a solid support for the LBL-assembly. Natural polymer like gelatin, protein like albumin [124], polypeptides (poly-lysine, poly-L-aspartic acid), polysaccharide (chitosan, chondroitin, dextran) are some well-known examples employed in LBL assembly. While polymers like poly (ethylenimine) (PEI), poly (dimethyl diallyl ammonium chloride) etc., are used as polycation solution and, PAA, poly (vinyl sulfate), poly (styrene sulfonate) (PSS) as polyanion solution.

We talked about how polyelectrolyte systems work and how multilayers form throughout this chapter. LBL polyelectrolyte systems build multilayered structures of poly-ions by repeating and alternately changing the electrolyte solution of polyanion and polycation. In the following section, we will go over the major applications of PEM assemblies.

4.6.1 Permeable mediator

Polyelectrolyte membrane complexes are built using either the LBL or PIC methods, and they have distinct structural features such as multilayers formed by ion staking. PEM has a nano-porous structure embedded with nanoparticles, which increases the membrane's porosity. PEM membranes, in which polyanionic cellulose (PAC) and PAA are utilized as weak polyelectrolytes, form porous structures with sizes ranging from 500 to 20 nm, that can be managed by varying the quantity of salt used depending on the permeability required [125]. The nano-porous property of the membrane changes as the membrane's components change. For example, if the number of polycations in the membrane increases. Negative ions will penetrate the membrane more easily than cations. The species' penetration is determined by the particle's molecular size and the cation-anion ratio. Using the concept of ion penetration and size, one can create a PEM complex with variable micro and nano-pore size. The multifunctional PEM system has been applied previously to monitor the gaseous exchange, movement of ions like ion balance sodium (Na^+)/ calcium (Ca^{2+}) [126], chloride (Cl^-)/fluoride (F^-) [127]. Permeability of the membrane can be controlled by managing the experimental parameters like ionic strength of solution, thickness of the membrane, changing the cation/anion ratio, polyelectrolyte solution which eventually affects the pore size, molecular size and separation [128].

4.6.2 As a surface coating

Polyelectrolyte membrane can be used as a membrane, but it can also be used as a surface coating. Regardless of the nature of the surface, polyelectrolyte complexes can be used for coating to alter the physio-chemical properties of the surface. These modified surfaces can be used to either create an appropriate environment for cell culture or to adsorb protein on the surface. The degree of protein adsorption on the surface can be determined using the electrostatic interaction between the protein molecule and the PEM. The method can also be used for immunological testing [107,129,130] (Fig. 4.6).

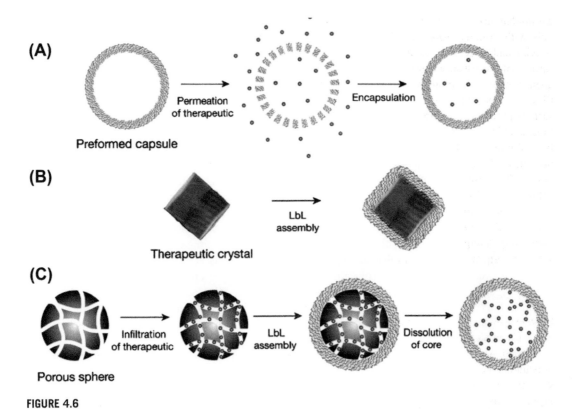

FIGURE 4.6

Encapsulation methods for the formation of layer-by-layer assemblies. (A) Capsule formation and release of therapeutics. (B) Encapsulation of therapeutics over the crystalline material. (C) Mechanism of therapeutic encapsulation on porous material and systemic release [107].

Reprinted with the permission for figure obtained from Elsevier (A.P.R. Johnston, C. Cortez, A.S. Angelatos, F. Caruso, Layer-by-layer engineered capsules and their applications, Curr. Opin. Colloid. Interface Sci. 11 (4) (2006) 203–209).

4.6.3 Biomineralization activator

Bone regeneration, like wound healing, is a major research interest in the field of biomedical sciences. The application of implants, patches, and multilayers that can be biomineralized or generate osteo-tissue. These fibrous polymers, which are formed by encapsulating fibers, promote bone growth and stimulate osteoblast proliferation and maturation [131,132].

In recent research, for bone regeneration scientists have developed micro-balance system composed of chitosan and chondroitin sulfate [133]. This system act as an ion reservoir, permeable for calcium (Ca^{2+}) and phosphate (PO_4^{3-}) and is able to precipitate calcium phosphate over multilayer which in turn triggers biomineralization [134]. It is difficult to immobilize such particles inside the multilayer, and it is also difficult to penetrate the multilayer with nanoparticles. Because the size of the particles within the multilayer is comparable to that of nanoparticles and immobilized particles, the use of such materials may prove to be more osteoconductive in nature than free particles. One of the approaches used for biomineralization was the titanium alloy disk as a surface upon which gelatinized dopamine and its analogs are used as surface modifiers. A different modified surface developed by Hu and co-workers [135], BMP-2 (Bone Morphogenetic Protein) and fibronectin were deposited on chitosan-Gelatin surface, used for multilayer fabrication [102].

4.6.4 Immobilized biomolecule

Polyionic solution is mixed in proportion with the required amount of biomolecules to be immobilized on the surface of the membrane or the solution of nanomaterial coated with the biomolecule during the formation of PEM of desired dimension and structural features [136]. This results in the entrapment of biomolecules in the layer, and after mixing the components and drying, we obtain the bio-immobilized layer. When biomolecules are mixed with polycation solutions, unreacted or unbounded residues can cross-link with the biomolecule. Protein multilayers formed by a weak polyelectrolyte can be more soluble and stable than proteins formed by a strong polyion. Strong polyion with protein is insoluble at physiological pH, whereas activity is close to its isoelectric point with weak poly electrolyte solubility. Biomolecules can be crosslinked to PEM with the help of crosslinking agents to increase the stability of the fabricated bio multilayer [137,138]. Electrolyte should be added in a linear fashion to act as an electrostatic polyion glue for effective protein polyelectrolyte multilayer assembly. Polyion glue can be used to bind protein arrays in nano-assembly.

Biomolecular multilayers or thin films are deposition on Silicon-rubber without any pre-modification using the LBL or PEM assembly method. Silicon is used as a support material in a variety of biomedical applications due to its inert nature. Multilayers of polyion can be coated on silicon rubber and treated with plasma to make it hydrophilic. As a biocompatible coating agent on silicon rubber, poly electrolytes such as poly-lysin, PSS, PEI, gelatin, fibronectin, and hyaluronic acid can be used [139]. It is very beneficial to use 3D material to grow cells as it provides a 3D surface for proliferation, maturation, and to study cell behavior [140].

Because polycation layers are negatively charged, DNA molecules or polynucleotide molecules can easily adsorb or form bonds with them. Polyion film coated with DNA was used in the medical implant process to perform local gene therapy [141,142]. Decher performed adsorption experiment on PEI and polyallylamine hydrochloride (PAH) multilayer at slight acidic pH 5.2 with DNA conc 0.1 mg/mL. Adsorption of DNA molecule was confirmed by FT-IR analysis. Fig. 4.7 shows a general apporch for encapsulation of DNA molecules using core-shell nanoparticles [143].

Enzymes like lysosomes [144,145], HRP [146,147], globular protein like albumin [148], immunoglobulins [149,150], secretary protein which helps in digestive and respiration process like glucoamylase [151], glucose oxidase [152−154], and glucose isomerase [155] are used for the multilayer assembly.

Because of their rapid nature, substrate selectivity, and reusability, nanocatalysts are a new area of research. Because of their unique properties such as specificity, energy efficacy, tunability to different sizes, surface characteristics, porous nature, and acid-base properties, nanoparticles are widely used in biocatalysis. Furthermore, the specificity of enzymes can be altered chemically [37,156,157].

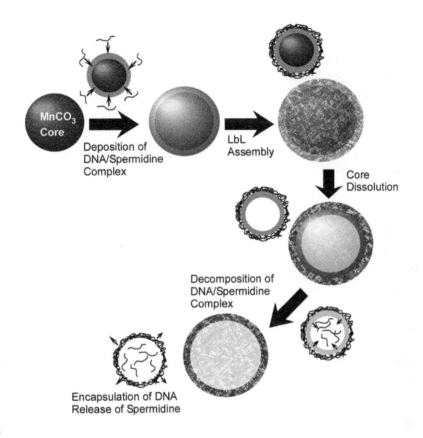

FIGURE 4.7

Encapsulation of DNA molecules with the help of layer-by-layer assembly [143].

Reprinted with the permission for figure obtained from Elsevier (K. Ariga, Y.M. Lvov, K. Kawakami, Q. Ji, J.P. Hill, Layer-by-layer self-assembled shells for drug delivery, Adv. Drug. Deliv. Rev. 63 (9) (2011) 762–771).

4.7 Conclusion

Enzymes are active catalysts that catalyze a wide range of biological reactions and are frequently used in catalysis-related applications. The ability of nanoparticles to improve the efficacy of immobilized enzymes has piqued the interest of enzyme-nanoparticle system researchers. The immobilization procedure has been used to improve enzyme activity and stability in both aqueous and non-aqueous conditions. The support matrix in enzyme immobilization must be carefully selected and designed. Nanoparticles have recently emerged as a useful technique for developing effective supports for enzyme stabilization due to their small size and large surface area. Nanoparticles have a significant influence on the mechanical properties of the material, such as stiffness and elasticity. They also provide biocompatible environments for enzyme immobilization.

Furthermore, improved enzymatic activity on enzyme-nanoparticle conjugates is aided by five primary physicochemical mechanisms: higher enzyme density; increased incident substrate mass transport due to both substrate attraction to the nanoparticle surface and movement of the enzyme-nanoparticle bioconjugate; nanoparticle morphology; nanoparticle surface chemistry leading to more active enzymes; and favorable enzyme orientation for the increased enzyme-substrate in By providing a better understanding of how enzymes are immobilized on nanoparticles, these mechanisms can be used to improve the design of novel enzyme-nanoparticle conjugate systems. As many examples in this book chapter demonstrate, the enhancement is still not completely predictable, and more research is needed to develop a framework for understanding the enzyme-nanoparticle systems. Polyelectrolyte complexes, which are composed of polyanions and polycations, are highly efficient and reactive when it comes to immobilizing biomolecules. The complexes have been used in a variety of applications outside of biological research due to their remarkable properties such as permittivity, selectivity, and the ability to encapsulate tiny molecules or biomolecules. Polyelectrolyte membranes are excellent materials for immobilizing biomolecules because they allow for the preservation or enhancement of the biomolecules' features, capabilities, and functions. As a result, polyelectrolyte complexes may be used in a variety of applications, including biosensor [158−161] development, bioreactor and biofuel cell development.

Convoluted three-dimensional PEM configurations can be generated in multidimensional scales using the polyelectrolyte multilayer approach. As wound dressings, free-standing membranes could be used. Porous scaffolds could be used in tissue engineering as new cell expansion supports, and tubular structures could be used as blood vessel mimics. Furthermore, using cells as multilayer elements may allow for the creation of hybrid structures that function and look like tissues. Among all 3D devices studied, spherical capsules have received the most attention. Diffusion, cell targeting, degradation, and cargo release are some of the processes that can be used for drug delivery and biosensing. Finally, the LBL method can be used to create structures that resemble biological cells or tissues in the form of compartmentalized systems with varying size scales of organization.

Acknowledgments

The authors acknowledge the Department of Biotechnology (DBT), New Delhi, for financing (Grant number BT/PR34216/AAQ/1/765/2019), the intensification of research in high priority area (IRHPA) program from Science and Engineering Research Board (SERB), New Delhi (Grant Number IPA/2020/000069, CRG/2020/003014, WEA/2020/000036) and the institutional grant from National Institute of Animal Biotechnology (DBT-NIAB), Hyderabad (Grant Number C0038).

Authors' contributions

Conceptualization, S.G., P.K., M.S.; investigation, S.G., P.K., M.S.; writing—original draft preparation, P.K. and M.S.; writing-review and editing M.S., and P.K.; supervision, S.G.

Conflict of Interest

The authors declare no conflict of interest.

References

[1] C.K. Lee, A.N. Au-Duong, Enzyme immobilization on nanoparticles: recent applications, Emerg. Areas. Bioeng. (2017). Available from: https://doi.org/10.1002/9783527803293.ch4.

[2] R. DiCosimo, J. McAuliffe, A.J. Poulose, G. Bohlmann, Industrial use of immobilized enzymes, Chem. Soc. Rev. 42 (15) (2013) 6437−6474. Available from: https://doi.org/10.1039/C3CS35506C.

[3] S. Cantone, V. Ferrario, L. Corici, C. Ebert, D. Fattor, P. Spizzo, et al., Efficient immobilisation of industrial biocatalysts: criteria and constraints for the selection of organic polymeric carriers and immobilisation methods, Chem. Soc. Rev. 42 (15) (2013) 6262−6276. Available from: https://doi.org/10.1039/C3CS35464D.

[4] R. Ahmad, M. Sardar, Enzyme immobilization: an overview on nanoparticles as immobilization matrix, Biochem. Anal. Biochem 4 (2) (2015) 1000178. Available from: https://doi.org/10.4172/2161-1009.1000178.

[5] G. Massolini, E. Calleri, Immobilized trypsin systems coupled on-line to separation methods: recent developments and analytical applications, J. Sep. Sci. 28 (1) (2005) 7−21. Available from: https://doi.org/10.1002/JSSC.200401941.

[6] U. Guzik, K. Hupert-Kocurek, D. Wojcieszyńska, Immobilization as a strategy for improving enzyme properties-application to oxidoreductases, Molecules 19 (7) (2014) 8995−9018. Available from: https://doi.org/10.3390/molecules19078995.

[7] R.C. Rodrigues, C. Ortiz, Á. Berenguer-Murcia, R. Torres, R. Fernández-Lafuente, Modifying enzyme activity and selectivity by immobilization, Chem. Soc. Rev. 42 (15) (2013) 6290−6307. Available from: https://doi.org/10.1039/C2CS35231A.

[8] R.A. Sheldon, S. Van Pelt, Enzyme immobilisation in biocatalysis: why, what and how, Chem. Soc. Rev. 42 (15) (2013) 6223−6235. Available from: https://doi.org/10.1039/c3cs60075k.

[9] A. Liese, L. Hilterhaus, Evaluation of immobilized enzymes for industrial applications, Chem. Soc. Rev. 42 (15) (2013) 6236−6249. Available from: https://doi.org/10.1039/C3CS35511J.

[10] C. Mateo, J.M. Palomo, G. Fernandez-Lorente, J.M. Guisan, R. Fernandez-Lafuente, Improvement of enzyme activity, stability and selectivity via immobilization techniques, Enzyme Microb. Technol. 40 (6) (2007) 1451−1463. Available from: https://doi.org/10.1016/j.enzmictec.2007.01.018.

[11] A. Ursini, P. Maragni, C. Bismara, B. Tamburini, Enzymatic method of preparation of optically active trans-2-amino cyclohexanol derivatives, Synth. Commun. 29 (8) (1999) 1369−1377. Available from: https://doi.org/10.1080/00397919908086112.

[12] C. Garcia-Galan, Á. Berenguer-Murcia, R. Fernandez-Lafuente, R.C. Rodrigues, Potential of different enzyme immobilization strategies to improve enzyme performance, Adv. Synth. Catal. 353 (16) (2011) 2885−2904. Available from: https://doi.org/10.1002/ADSC.201100534.

[13] R.C. Rodrigues, Á. Berenguer-Murcia, R. Fernandez-Lafuente, Coupling chemical modification and immobilization to improve the catalytic performance of enzymes, Adv. Synth. Catal. 353 (13) (2011) 2216−2238. Available from: https://doi/full/10.1002/adsc.201100163.

[14] M. Misson, H. Zhang, B. Jin, Nanobiocatalyst advancements and bioprocessing applications, J. R. Soc. Interface 12 (102) (2015). Available from: https://doi/abs/10.1098/rsif.2014.0891.

[15] M. Bilal, H.M.N. Iqbal, Chemical, physical, and biological coordination: an interplay between materials and enzymes as potential platforms for immobilization, Coord. Chem. Rev. 388 (2019) 1−23. Available from: https://doi.org/10.1016/J.CCR.2019.02.024.

[16] S. Datta, L.R. Christena, Y.R.S. Rajaram, Enzyme immobilization: an overview on techniques and support materials, 3 Biotech. 3 (1) (2013) 1−9. Available from: https://doi.org/10.1007/S13205-012-0071-7.

[17] S. Yakubi, Polyelectrolyte complex membranes for immobilizing biomolecules, and their applications to bio-analysis, Anal. Sci. 27 (7) (2011) 695. Available from: https://doi.org/10.2116/analsci.27.695.

[18] J. Fu, J. Ji, W. Yuan, J. Shen, Construction of anti-adhesive and antibacterial multilayer films via layer-by-layer assembly of heparin and chitosan, Biomaterials 26 (33) (2005) 6684−6692. Available from: https://doi.org/10.1016/J.BIOMATERIALS.2005.04.034.

[19] M. Salomäki, I.A. Vinokurov, J. Kankare, Effect of temperature on the buildup of polyelectrolyte multilayers, Langmuir 21 (24) (2005) 11232−11240. Available from: https://doi.org/10.1021/LA051600K.

[20] S.S. Shiratori, M.F. Rubner, pH-Dependent thickness behavior of sequentially adsorbed layers of weak polyelectrolytes, Macromolecules 33 (11) (2000) 4213−4219. Available from: https://doi.org/10.1021/MA991645Q.

[21] D. Yoo, S.S. Shiratori, M.F. Rubner, Controlling bilayer composition and surface wettability of sequentially adsorbed multilayers of weak polyelectrolytes, Macromolecules 31 (13) (1998) 4309−4318. Available from: https://doi.org/10.1021/ma9800360.

[22] C. Crestini, R. Perazzini, R. Saladino, Oxidative functionalisation of lignin by layer-by-layer immobilised laccases and laccase microcapsules, Appl. Catal. A. Gen. 372 (2) (2010) 115−123. Available from: doi.org/10.1016/J.APCATA.2009.10.012.

[23] W. Feng, P. Ji, Enzymes immobilized on carbon nanotubes, Biotechnol. Adv. 29 (6) (2011) 889−895. Available from: https://doi.org/10.1016/J.BIOTECHADV.2011.07.007.

[24] M.N. Gupta, M. Kaloti, M. Kapoor, K. Solanki, Nanomaterials as matrices for enzyme immobilization. 39 (2) (2011) 98−109. https://doi/abs/10.3109/10731199.2010.516259

[25] M.L. Verma, C.J. Barrow, M. Puri, Nanobiotechnology as a novel paradigm for enzyme immobilisation and stabilisation with potential applications in biodiesel production, Appl. Microbiol. Biotechnol. 97 (1) (2012) 23−39. Available from: https://doi.org/10.1007/S00253-012-4535-9.

[26] S.A. Ansari, Q. Husain, Potential applications of enzymes immobilized on/in nano materials: a review, Biotechnol. Adv. 30 (3) (2012) 512−523. Available from: https://doi.org/10.1016/J.BIOTECHADV.2011.09.005.

[27] K. Min, Y.J. Yoo, Recent progress in nanobiocatalysis for enzyme immobilization and its application, Biotechnol. Bioprocess. Eng. 19 (4) (2014) 553−567. Available from: https://doi.org/10.1007/S12257-014-0173-7.

[28] E.P. Cipolatti, M.J.A. Silva, M. Klein, V. Feddern, M.M.C. Feltes, J.V. Oliveira, et al., Current status and trends in enzymatic nanoimmobilization, J. Mol. Catal. B. Enzym. 99 (2014) 56−67. Available from: https://doi.org/10.1016/J.MOLCATB.2013.10.019.

[29] N. Singh, B.S. Dhanya, M.L. Verma, Nano-immobilized biocatalysts and their potential biotechnological applications in bioenergy production, Mater. Sci. Energy Technol. 3 (2020) 808−824. Available from: https://doi.org/10.1016/J.MSET.2020.09.006.

[30] S.K. Wang, A.R. Stiles, C. Guo, C.Z. Liu, Harvesting microalgae by magnetic separation: a review, Algal. Res. 9 (2015) 178−185. Available from: https://doi.org/10.1016/J.ALGAL.2015.03.005.

[31] L. Lei, X. Liu, Y. Li, Y. Cui, Y. Yang, G. Qin, et al., Study on synthesis of poly(GMA)-grafted Fe_3O_4/SiOX magnetic nanoparticles using atom transfer radical polymerization and their application for lipase immobilization, Mater. Chem. Phys. 125 (3) (2011) 866−871. Available from: https://doi.org/10.1016/J.MATCHEMPHYS.2010.09.031.

[32] R.S. Chouhan, M. Horvat, J. Ahmed, N. Alhokbany, S.M. Alshehri, S. Gandhi, Magnetic nanoparticles—a multifunctional potential agent for diagnosis and therapy, Cancers 13 (9) (2021) 2213. Available from: https://doi.org/10.3390/CANCERS13092213.

[33] S.V. Mohan, G. Mohanakrishna, S.S. Reddy, B.D. Raju, K.S.R. Rao, P.N. Sarma, et al., Self-immobilization of acidogenic mixed consortia on mesoporous material (SBA-15) and activated carbon to enhance fermentative hydrogen production, Int. J. Hydrog. Energy 33 (21) (2008) 6133−6142. Available from: https://doi.org/10.1016/J.IJHYDENE.2008.07.096.

[34] A. Basso, S. Serban, Industrial applications of immobilized enzymes—a review, Mol. Catal. 479 (2019) 110607. Available from: https://doi.org/10.1016/J.MCAT.2019.110607.

[35] S. Ding, A.A. Cargill, I.L. Medintz, J.C. Claussen, Increasing the activity of immobilized enzymes with nanoparticle conjugation, Curr. Opin. Biotechnol. 34 (2015) 242−250. Available from: https://doi.org/10.1016/j.copbio.2015.04.005.

[36] A.E. Prigodich, A.H. Alhasan, C.A. Mirkin, Selective enhancement of nucleases by polyvalent DNA-functionalized gold nanoparticles, J. Am. Chem. Soc. 133 (7) (2011) 2120−2123. Available from: https://doi.org/10.1021/ja110833r.

[37] G.A. Somorjai, H. Frei, J.Y. Park, Advancing the frontiers in nanocatalysis, biointerfaces, and renewable energy conversion by innovations of surface techniques, J. Am. Chem. Soc. 131 (46) (2009) 16589−16605. Available from: https://doi.org/10.1021/JA9061954.

[38] R.G. Kerry, G.P. Mahapatra, G.K. Maurya, S. Patra, S. Mahari, G. Das, Molecular prospect of type-2 diabetes: nanotechnology based diagnostics and therapeutic intervention, Rev. Endocr. Metab. Disord. 22 (2020) 424−451. Available from: https://doi.org/10.1007/S11154-020-09606-0.

[39] S. Sasidharan, S. Raj, S. Sonawane, S. Sonawane, D. Pinjari, A.B. Pandit, et al., Nanomaterial synthesis: chemical and biological route and applications, Nanomater. Synth. Des. Fabr. Appl. (2019) 27−51. Available from: https://doi.org/10.1016/B978-0-12-815751-0.00002-X.

[40] P.T. Sekoai, C.N.M. Ouma, S.P. du Preez, P. Modisha, N. Engelbrecht, D.G. Bessarabov, et al., Application of nanoparticles in biofuels: an overview, Fuel 237 (2019) 380−397. Available from: https://doi.org/10.1016/J.FUEL.2018.10.030.

[41] P.G. Jamkhande, N.W. Ghule, A.H. Bamer, M.G. Kalaskar, Metal nanoparticles synthesis: an overview on methods of preparation, advantages and disadvantages, and applications, J. Drug. Deliv. Sci. Technol. 53 (2019) 101174. Available from: https://doi.org/10.1016/J.JDDST.2019.101174.

[42] A. García-Quintero, M. Palencia, A critical analysis of environmental sustainability metrics applied to green synthesis of nanomaterials and the assessment of environmental risks associated with the nano-technology, Sci. Total. Environ. (2021) 793. Available from: https://doi.org/10.1016/j.scitotenv.2021.148524.

[43] G.A. Naikoo, M. Mustaqeem, I.U. Hassan, T. Awan, F. Arshad, H. Salim, et al., Bioinspired and green synthesis of nanoparticles from plant extracts with antiviral and antimicrobial properties: a critical review, J. Saudi. Chem. Soc. 25 (9) (2021) 101304. Available from: https://doi.org/10.1016/j.jscs.2021.101304.

[44] I. Khan, K. Saeed, I. Khan, Nanoparticles: properties, applications and toxicities, Arab. J. Chem. 12 (7) (2019) 908−931. Available from: https://doi.org/10.1016/J.ARABJC.2017.05.011.

[45] S. Mourdikoudis, R.M. Pallares, N.T.K. Thanh, Characterization techniques for nanoparticles: comparison and complementarity upon studying nanoparticle properties, Nanoscale 10 (27) (2018) 12871−12934. Available from: https://doi.org/10.1039/C8NR02278J.

[46] P. Khanna, A. Kaur, D. Goyal, Algae-based metallic nanoparticles: synthesis, characterization and applications, J. Microbiol. Methods. 163 (2019) 105656. Available from: https://doi.org/10.1016/J.MIMET.2019.105656.

[47] S. Mahari, A. Roberts, S. Gandhi, Probe-free nanosensor for the detection of Salmonella using gold nanorods as an electroactive modulator, Food Chem. 390 (2022) 133219. Available from: https://doi.org/10.1016/j.foodchem.2022.133219.

[48] D. Shahdeo, A. Roberts, G.J. Archana, N.S. Shrikrishna, S. Mahari, K. Nagamani, et al., Label free detection of SARS CoV-2 receptor binding domain (RBD) protein by fabrication of gold nanorods deposited on electrochemical immunosensor (GDEI, Biosens. Bioelectron. 212 (2022) 114406. Available from: https://doi.org/10.1016/j.bios.2022.114406.

[49] A. Roberts, R.S. Chouhan, D. Shahdeo, N.S. Shrikrishna, V. Kesarwani, M. Horvat, et al., A recent update on advanced molecular diagnostic techniques for COVID-19 pandemic: an overview, Front. Immunol. (2021) 12. Available from: https://doi.org/10.3389/fimmu.2021.732756.

[50] S. Narlawar, S. Coudhury, S. Gandhi, Magnetic properties-based biosensors for early detection of cancer, Biosens. Based Adv. Cancer Diagnostics (2022) 165−178. Available from: https://doi.org/10.1016/B978-0-12-823424-2.00010-7.

[51] A. Roberts, P.P. Tripathi, S. Gandhi, Graphene nanosheets as an electric mediator for ultrafast sensing of urokinase plasminogen activator receptor-A biomarker of cancer, Biosens. Bioelectron. 141 (15) (2019) 111398. Available from: https://doi.org/10.1016/j.bios.2019.111398.

[52] J. Zhang, F. Zhang, H. Yang, X. Huang, H. Liu, J. Zhang, et al., Graphene oxide as a matrix for enzyme immobilization, Langmuir 26 (9) (2010) 6083−6085. Available from: https://doi.org/10.1021/la904014z.

[53] L. Jin, K. Yang, K. Yao, S. Zhang, H. Tao, S.T. Lee, et al., Functionalized graphene oxide in enzyme engineering: a selective modulator for enzyme activity and thermostability, ACS Nano 6 (6) (2012) 4864−4875. Available from: https://doi/abs/10.1021/nn300217z.

[54] T. Kuila, S. Bose, P. Khanra, A.K. Mishra, N.H. Kim, J.H. Lee, et al., Recent advances in graphene-based biosensors, Biosens. Bioelectron. 26 (12) (2011) 4637−4648. Available from: https://doi.org/10.1016/J.BIOS.2011.05.039.

[55] L. Feng, S. Zhang, Z. Liu, Graphene based gene transfection, Nanoscale 3 (3) (2011) 1252−1257. Available from: https://doi.org/10.1039/C0NR00680G.

[56] D. Shahdeo, A. Roberts, N. Abbineni, S. Gandhi, Graphene based sensors, Compr. Anal. Chem. 91 (2020) 175−199. Available from: https://doi.org/10.1016/BS.COAC.2020.08.007.

[57] A. Roberts, N. Chauhan, S. Islam, S. Mahari, B. Ghawri, R.K. Gandham, Graphene functionalized field-effect transistors for ultrasensitive detection of Japanese encephalitis and Avian influenza virus, Sci. Reports. 10 (1) (2020) 1−12. Available from: https://doi.org/10.1038/s41598-020-71591-w.

[58] A. Roberts, V. Kesarwani, R. Gupta, S. Gandhi, Electroactive reduced graphene oxide for highly sensitive detection of secretory non-structural 1 protein: a potential diagnostic biomarker for Japanese encephalitis virus, Biosens. Bioelectron. 198 (2022) 113837. Available from: https://doi.org/10.1016/j.bios.2021.113837.

[59] S. Mahari, S. Gandhi, Electrochemical immunosensor for detection of avian Salmonellosis based on electroactive reduced graphene oxide (rGO) modified electrode, Bioelectrochemistry 144 (2022) 108036. Available from: https://doi.org/10.1016/j.bioelechem.2021.108036.

[60] J. Dey, A. Roberts, S. Mahari, S. Gandhi, P.P. Tripathi, Electrochemical detection of Alzheimer's disease biomarker, β-secretase enzyme (BACE1), with one-step synthesized reduced graphene oxide, Front. Bioeng. Biotechnol. (2022) 10. Available from: https://doi.org/10.3389/fbioe.2022.873811.

[61] S. Shukla, Y. Haldorai, I. Khan, S.M. Kang, C.H. Kwak, S. Gandhi, et al., Bioreceptor-free, sensitive and rapid electrochemical detection of patulin fungal toxin, using a reduced graphene oxide@SnO_2 nanocomposite, Mater. Sci. Eng. C. 113 (2020) 110916. Available from: https://doi.org/10.1016/j.msec.2020.110916.

[62] R. Ahmad, S.K. Khare, Immobilization of Aspergillus niger cellulase on multiwall carbon nanotubes for cellulose hydrolysis, Bioresour. Technol. 252 (2018) 72−75. Available from: https://doi.org/10.1016/J.BIORTECH.2017.12.082.

[63] S.G. Mhaisalkar, J.N. Tey, S. Gandhi, I.P.M. Wijaya, A. Palaniappan, J. Wei, et al., Direct detection of heroin metabolites using a competitive immunoassay based on a carbon-nanotube liquid-gated field-effect transistor, Small 6 (9) (2010) 993−998. https://doi/full/10.1002/smll.200902139.

[64] L. Wang, R. Xu, Y. Chen, R. Jiang, Activity and stability comparison of immobilized NADH oxidase on multi-walled carbon nanotubes, carbon nanospheres, and single-walled carbon nanotubes, J. Mol. Catal. B. Enzym. 69 (3−4) (2011) 120−126. Available from: https://doi.org/10.1016/J. MOLCATB.2011.01.005.

[65] S.S. Karajanagi, A.A. Vertegel, R.S. Kane, J.S. Dordick, Structure and function of enzymes adsorbed onto single-walled carbon nanotubes, Langmuir 20 (26) (2004) 11594−11599. Available from: https:// doi.org/10.1021/la047994h.

[66] I.V. Pavlidis, T. Vorhaben, D. Gournis, G.K. Papadopoulos, U.T. Bornscheuer, H. Stamatis, Regulation of catalytic behaviour of hydrolases through interactions with functionalized carbon-based nanomaterials, J. Nanoparticle. Res. 14 (2012) 842. Available from: https://doi.org/10.1007/S11051-012-0842-4.

[67] F. Zhang, B. Zheng, J. Zhang, X. Huang, H. Liu, S. Guo, et al., Horseradish peroxidase immobilized on graphene oxide: physical properties and applications in phenolic compound removal, J. Phys. Chem. C. 114 (18) (2010) 8469−8473. Available from: https://doi.org/10.1021/jp101073b.

[68] J. Farmakes, I. Schuster, A. Overby, L. Alhalhooly, M. Lenertz, Q. Li, et al., Enzyme immobilization on graphite oxide (GO) surface via one-pot synthesis of GO/metal-organic framework composites for large-substrate biocatalysis, ACS Appl. Mater. Interfaces 12 (20) (2020) 23119−23126. Available from: https://doi/abs/10.1021/acsami.0c04101.

[69] K. Chaudhary, N. Yadav, P. Venkatesu, D.T. Masram, Evaluation of utilizing functionalized graphene oxide nanoribbons as compatible biomaterial for lysozyme, ACS Appl. Bio. Mater. 4 (8) (2021) 6112−6124. Available from: https://doi.org/10.1021/ACSABM.1C00450.

[70] M. Ma, B. Liu, L. Meng, Applications of carbon-based nanomaterials in biofuel, Cell. (2017) 39−58. Available from: https://doi.org/10.1007/978-3-319-45459-7_3.

[71] Z.B. Qu, L.F. Lu, M. Zhang, G. Shi, Colorimetric detection of carcinogenic aromatic amine using layer-by-layer graphene oxide/cytochrome c composite, ACS Appl. Mater. Interfaces. 10 (13) (2018) 11350−11360. Available from: https://doi.org/10.1021/acsami.8b01176.

[72] A.M. Díez-Pascual, A.L. Díez-Vicente, Poly(propylene fumarate)/polyethylene glycol-modified graphene oxide nanocomposites for tissue engineering, ACS Appl. Mater. Interfaces. 8 (28) (2016) 17902−17914. Available from: https://doi.org/10.1021/ACSAMI.6B05635.

[73] K. Dhara, J. Stanley, T. Ramachandran, B.G. Nair, S.B. Satheesh, Pt-CuO nanoparticles decorated reduced graphene oxide for the fabrication of highly sensitive non-enzymatic disposable glucose sensor, Sens. Actuators B Chem. 195 (2014) 197−205. Available from: https://doi.org/10.1016/J.SNB.2014.01.044.

[74] W. Xie, N. Ma, Enzymatic transesterification of soybean oil by using immobilized lipase on magnetic nano-particles, Biomass Bioenergy 34 (6) (2010) 890−896. Available from: https://doi.org/10.1016/J. BIOMBIOE.2010.01.034.

[75] P. Mishra, T. Munjal, S. Gandhi, Nanoparticles for detection, imaging, and diagnostic applications in animals, Nanosci. Sustain. Agric. (2019) 437−477. Available from: https://doi.org/10.1007/978-3-319-97852-9_19.

[76] E. Sanfins, C. Augustsson, B. Dahlbäck, S. Linse, T. Cedervall, Size-dependent effects of nanoparticles on enzymes in the blood coagulation cascade, Nano. Lett. 14 (8) (2014) 4736−4744. Available from: https://doi.org/10.1021/nl501863u.

[77] F. Hao, F. Geng, X. Zhao, R. Liu, Q.S. Liu, Q. Zhou, Chirality of gold nanocluster affects its interaction with coagulation factor XII, NanoImpact 22 (2021) 100321. Available from: https://doi.org/10.1016/J. IMPACT.2021.100321.

[78] D. Lan, B. Li, Z. Zhang, Chemiluminescence flow biosensor for glucose based on gold nanoparticle-enhanced activities of glucose oxidase and horseradish peroxidase, Biosens. Bioelectron. 24 (4) (2008) 934−938. Available from: https://doi.org/10.1016/j.bios.2008.07.064.

[79] H. Yazdani, S.E. Hooshmand, R.S. Varma, Gold nanoparticle-catalyzed multicomponent reactions, ACS Sustain. Chem. Eng. 9 (49) (2021) 16556−16569. Available from: https://doi/abs/10.1021/acssuschemeng. 1c04361.

[80] D. Shahdeo, A.B. Chandra, S. Gandhi, Urokinase plasminogen activator receptor-mediated targeting of a stable nanocomplex coupled with specific peptides for imaging of cancer, Anal. Chem. 93 (34) (2021) 11868−11877. Available from: https://doi/abs/10.1021/acs.analchem.1c02697.

[81] A. Roberts, S. Mahari, D. Shahdeo, S. Gandhi, Label-free detection of SARS-CoV-2 spike S1 antigen triggered by electroactive gold nanoparticles on antibody coated fluorine-doped tin oxide (FTO) electrode, Anal. Chim. Acta. 1188 (2021) 339207. Available from: https://doi.org/10.1016/J.ACA.2021. 339207.

[82] D. Shahdeo, V. Kesarwani, D. Suhag, J. Ahmed, S.M. Alshehri, S. Gandhi, et al., Self-assembled chitosan polymer intercalating peptide functionalized gold nanoparticles as nanoprobe for efficient imaging of urokinase plasminogen activator receptor in cancer diagnostics, Carbohydr. Polym. 266 (2021) 118138. Available from: https://doi.org/10.1016/j.carbpol.2021.118138.

[83] A. Mishra, M. Sardar, Cellulase assisted synthesis of nano-silver and gold: application as immobilization matrix for biocatalysis, Int. J. Biol. Macromol. 77 (2015) 105−113. Available from: https://doi.org/ 10.1016/j.ijbiomac.2015.03.014.

[84] K. Saha, S.S. Agasti, C. Kim, X. Li, V.M. Rotello, Gold nanoparticles in chemical and biological sensing, Chem. Rev. 112 (5) (2012) 2739−2779. Available from: https://doi/full/10.1021/cr2001178.

[85] I. Ardao, J. Comenge, M.D. Benaiges, G. Álvaro, V.F. Puntes, Rational nanoconjugation improves biocatalytic performance of enzymes: aldol addition catalyzed by immobilized rhamnulose-1-phosphate aldolase, Langmuir. 28 (15) (2012) 6461−6467. Available from: https://doi/abs/10.1021/ la3003993.

[86] K. Dumri, D.H. Anh, Immobilization of lipase on silver nanoparticles via adhesive polydopamine for biodiesel production, Enzyme Res. 2014 (2014) 389739. Available from: https://doi.org/10.1155/2014/ 389739.

[87] S. Gandhi, H. Arami, K.M. Krishnan, Detection of cancer-specific proteases using magnetic relaxation of peptide-conjugated nanoparticles in biological environment, Nano. Lett. 16 (6) (2016) 3668−3674. Available from: https://doi/abs/10.1021/acs.nanolett.6b00867.

[88] A. Tomitaka, H. Arami, S. Gandhi, K.M. Krishnan, Lactoferrin conjugated iron oxide nanoparticles for targeting brain glioma cells in magnetic particle imaging, Nanoscale 7 (40) (2015) 16890−16898. Available from: https://doi.org/10.1039/C5NR02831K.

[89] G. Cheng, S.Y. Zheng, Construction of a high-performance magnetic enzyme nanosystem for rapid tryptic digestion, Sci. Rep. 4 (1) (2014) 1−10. Available from: https://doi.org/10.1038/srep06947.

[90] J.N. Vranish, M.G. Ancona, S.A. Walper, I.L. Medintz, Pursuing the promise of enzymatic enhancement with nanoparticle assemblies, Langmuir 34 (9) (2018) 2901−2925. Available from: https://doi.org/ 10.1021/ACS.LANGMUIR.7B02588.

[91] J.C. Breger, E. Oh, K. Susumu, W.P. Klein, S.A. Walper, M.G. Ancona, et al., Nanoparticle size influences localized enzymatic enhancement - a case study with phosphotriesterase, Bioconjug. Chem. 30 (7) (2019) 2060−2074. Available from: https://doi/abs/10.1021/acs.bioconjchem.9b00362.

[92] Y. Zhang, H. Wu, X. Huang, J. Zhang, S. Guo, Effect of substrate (ZnO) morphology on enzyme immobilization and its catalytic activity, Nanoscale Res. Lett. 6 (1) (2011) 1−7. Available from: https://doi. org/10.1186/1556-276X-6-450.

[93] V.G. Matveeva, L.M. Bronstein, Magnetic nanoparticle-containing supports as carriers of immobilized enzymes: key factors influencing the biocatalyst performance, Nanomater 11 (9) (2021) 2257. Available from: https://doi.org/10.3390/NANO11092257.

[94] W. Tang, C. Chen, W. Sun, P. Wang, D. Wei, Low-cost mussel inspired poly(Catechol/Polyamine) modified magnetic nanoparticles as a versatile platform for enhanced activity of immobilized enzyme, Int. J. Biol. Macromol. 128 (2019) 814–824. Available from: https://doi.org/10.1016/J.IJBIOMAC. 2019.01.161.

[95] Y. Cao, L. Wen, F. Svec, T. Tan, Y. Lv, Magnetic AuNP@Fe$_3$O$_4$ nanoparticles as reusable carriers for reversible enzyme immobilization, Chem. Eng. J. 286 (2016) 272–281. Available from: https://doi.org/10.1016/J.CEJ.2015.10.075.

[96] J.C. Breger, K. Susumu, G. Lasarte-Aragonés, S.A. Díaz, J. Brask, I.L. Medintz, et al., Quantum dot lipase biosensor utilizing a custom-synthesized peptidyl-ester substrate, ACS Sens. 5 (5) (2020) 1295–1304. Available from: https://doi.org/10.1021/ACSSENSORS.9B02291.

[97] G.A. Ellis, S.N. Dean, S.A. Walper, I.L. Medintz, Quantum dots and gold nanoparticles as scaffolds for enzymatic enhancement: recent advances and the influence of nanoparticle size, Catalysts 10 (1) (2020) 83. Available from: https://doi.org/10.3390/CATAL10010083.

[98] J.C. Claussen, W.R. Algar, N. Hildebrandt, K. Susumu, M.G. Ancona, I.L. Medintz, Biophotonic logic devices based on quantum dots and temporally-staggered Förster energy transfer relays, Nanoscale 5 (24) (2013) 12156–12170. Available from: https://doi.org/10.1039/C3NR03655C.

[99] J.C. Claussen, N. Hildebrandt, K. Susumu, M.G. Ancona, I.L. Medintz, Complex logic functions implemented with quantum dot bionanophotonic circuits, ACS Appl. Mater. Interfaces. 6 (6) (2014) 3771–3778. Available from: https://doi/abs/10.1021/am404659f.

[100] I.L. Medintz, H.T. Uyeda, E.R. Goldman, H. Mattoussi, Quantum dot bioconjugates for imaging, labelling and sensing, Nat. Mater. 4 (6) (2005) 435–446. Available from: https://doi.org/10.1038/nmat1390.

[101] E. Petryayeva, I.L. Medintz, W.R. Algar, Quantum dots in bioanalysis: a review of applications across various platformsfor fluorescence spectroscopy and imaging, Appl. Spectrosc. 67 (3) (2013) 215–252. Available from: https://doi.org/10.1366/12-06948.

[102] H. Lee, Y. Lee, A.R. Statz, J. Rho, T.G. Park, P.B. Messersmith, et al., Substrate-independent layer-by-layer assembly by using mussel-adhesive-inspired polymers, Adv. Mater. 20 (9) (2008) 1619–1623. Available from: https://doi.org/10.1002/ADMA.200702378.

[103] V. Smuleac, D.A. Butterfield, D. Bhattacharyya, Layer-by-layer-assembled microfiltration membranes for biomolecule immobilization and enzymatic catalysis, Langmuir. 22 (24) (2006) 10118–10124. Available from: https://doi/abs/10.1021/la061124d.

[104] T. Hoshi, T. Noguchi, J.I. Anzai, The preparation of amperometric xanthine sensors based on multi-layer thin films containing xanthine oxidase, Mater. Sci. Eng. C. 26 (1) (2006) 100–103. Available from: https://doi.org/10.1016/J.MSEC.2005.06.001.

[105] L. Yuri, A. Katsuhiko, K. Toyoki, Layer-by-layer assembly of alternate protein/polyion ultrathin films, Chem. Lett. 23 (12) (1994) 2323–2326. Available from: https://doi/abs/10.1246/cl.1994.2323.

[106] V. Jain, M. Khiterer, R. Montazami, H.M. Yochum, K.J. Shea, J.R. Heflin, et al., High-contrast solid-state electrochromic devices of viologen-bridged polysilsesquioxane nanoparticles fabricated by layer-by-layer assembly, ACS Appl. Mater. Interfaces. 1 (1) (2009) 83–89. Available from: https://doi/abs/10.1021/am8000264.

[107] A.P.R. Johnston, C. Cortez, A.S. Angelatos, F. Caruso, Layer-by-layer engineered capsules and their applications, Curr. Opin. Colloid. Interface Sci. 11 (4) (2006) 203–209. Available from: https://doi.org/10.1016/J.COCIS.2006.05.001.

[108] E. Katz, I. Willner, A biofuel cell with electrochemically switchable and tunable power output, J. Am. Chem. Soc. 125 (22) (2003) 6803−6813. Available from: https://doi/abs/10.1021/ja034008v.

[109] M. Togo, A. Takamura, T. Asai, H. Kaji, M. Nishizawa, An enzyme-based microfluidic biofuel cell using vitamin K3-mediated glucose oxidation, Electrochim. Acta. 52 (14) (2007) 4669−4674. Available from: https://doi.org/10.1016/J.ELECTACTA.2007.01.067.

[110] R.R. Costa, J.F. Mano, Polyelectrolyte multilayered assemblies in biomedical technologies, Chem. Soc. Rev. 43 (10) (2014) 3453−3479. Available from: https://doi.org/10.1016/J.ELECTACTA.2007.01.067.

[111] T. Okano, N. Yamada, H. Sakai, Y. Sakurai, A novel recovery system for cultured cells using plasma-treated polystyrene dishes grafted with poly(N-isopropylacrylamide), J. Biomed. Mater. Res. 27 (10) (1993) 1243−1251. Available from: https://doi.org/10.1002/jbm.820271005.

[112] Z. Tang, Y. Akiyama, T. Okano, Temperature-responsive polymer modified surface for cell sheet engineering, Polym. 4 (3) (2012) 1478−1498. Available from: https://doi.org/10.3390/POLYM4031478.

[113] N.M. Ahmad, M. Saqib, C.J. Barrett, Novel azobenzene-functionalized polyelectrolytes of different substituted head groups 3: control of properties of self-assembled multilayer thin films, J. Macromol. Sci. Part. A Pure Appl. Chem. 47 (6) (2010) 571−579. Available from: https://doi.org/10.1080/10601321003741966.

[114] J. Zhang, X. Huang, L. Zhang, Y. Si, S. Guo, H. Su, et al., Layer-by-layer assembly for immobilizing enzymes in enzymatic biofuel cells, Sustain. Energy Fuels. 4 (1) (2019) 68−79. Available from: https://doi.org/10.1039/C9SE00643E.

[115] G. Decher, J.D. Hong, Buildup of ultrathin multilayer films by a self-assembly process, 1 consecutive adsorption of anionic and cationic bipolar amphiphiles on charged surfaces, Makromol. Chemie. Macromol. Symp. 46 (1) (1991) 321−327. Available from: https://doi/full/10.1002/masy.19910460145.

[116] R.L. Arechederra, K. Boehm, S.D. Minteer, Mitochondrial bioelectrocatalysis for biofuel cell applications, Electrochim. Acta. 54 (28) (2009) 7268−7273. Available from: https://doi.org/10.1016/J.ELECTACTA.2009.07.043.

[117] X. Huang, L. Zhang, Z. Zhang, S. Guo, H. Shang, Y. Li, et al., Wearable biofuel cells based on the classification of enzyme for high power outputs and lifetimes, Biosens. Bioelectron. 124−125 (2019) 40−52. Available from: https://doi.org/10.1016/J.BIOS.2018.09.086.

[118] I. Haller, Covalently attached organic monolayers on semiconductor surfaces, J. Am. Chem. Soc. 100 (26) (2002) 8050−8055. Available from: https://doi/abs/10.1021/ja00494a003.

[119] W. Yuan, G.M. Weng, J. Lipton, C.M. Li, P.R. Van Tassel, A.D. Taylor, et al., Weak polyelectrolyte-based multilayers via layer-by-layer assembly: Approaches, properties, and applications, Adv. Colloid. Interface Sci. 282 (2021) 102200. Available from: https://doi.org/10.1016/j.cis.2020.102200.

[120] N. Fukao, K.H. Kyung, K. Fujimoto, S. Shiratori, Automatic spray-LBL machine based on in-situ QCM monitoring, Macromolecules 44 (8) (2011) 2964−2969. Available from: https://doi.org/10.1021/MA200024W.

[121] A. Izquierdo, S.S. Ono, J.C. Voegel, P. Schaaf, G. Decher, Dipping versus spraying: exploring the deposition conditions for speeding up layer-by-layer assembly, Langmuir 21 (16) (2005) 7558−7567. Available from: https://doi.org/10.1021/LA047407S.

[122] H. Ai, S.A. Jones, Y.M. Lvov, Biomedical applications of electrostatic layer-by-layer nano-assembly of polymers, enzymes, and nanoparticles, Cell Biochem. Biophys. 39 (1) (2003) 23−43. Available from: https://doi.org/10.1385/CBB:39:1:23.

[123] B.D. Ratner, Surface modification of polymers: chemical, biological and surface analytical challenges, Biosens. Bioelectron. 10 (9−10) (1995) 797−804. Available from: https://doi.org/10.1016/0956-5663(95)99218-A.

[124] F. Caruso, K. Niikura, D.N. Furlong, Y. Okahata, Assembly of alternating polyelectrolyte and protein multilayer films for imiminosensing, Langmuir 13 (13) (1997) 3427−3433. Available from: https://doi.org/10.1021/LA9608223.

[125] A. Fery, B. Schöler, T. Cassagneau, F. Caruso, Nanoporous thin films formed by salt-induced structural changes in multilayers of poly(acrylic acid) and poly(allylamine), Langmuir. 17 (13) (2001) 3779−3783. Available from: https://doi/abs/10.1021/la0102612.

[126] S.U. Hong, L. Ouyang, M.L. Bruening, Recovery of phosphate using multilayer polyelectrolyte nanofiltration membranes, J. Memb. Sci. 327 (1−2) (2009) 2−5. Available from: https://doi.org/10.1016/J.MEMSCI.2008.11.035.

[127] L. Ouyang, R. Malaisamy, M.L. Bruening, Multilayer polyelectrolyte films as nanofiltration membranes for separating monovalent and divalent cations, J. Memb. Sci. 310 (1−2) (2008) 76−84. Available from: https://doi.org/10.1016/J.MEMSCI.2007.10.031.

[128] S.S. Shiratori, K. Kohno, M. Yamada, High performance smell sensor using spatially controlled LB films with polymer backbone, Sens. Actuators B Chem. 64 (1−3) (2000) 70−75. Available from: https://doi.org/10.1016/S0925-4005(99)00486-4.

[129] R. Kurita, Y. Hirata, S. Yabuki, Y. Yokota, D. Kato, Y. Sato, et al., Surface modification of thin polyion complex film for surface plasmon resonance immunosensor, Sens. Actuators B Chem. 130 (1) (2008) 320−325. Available from: https://doi.org/10.1016/J.SNB.2007.08.007.

[130] G. Ladam, P. Schaaf, G. Decher, J. Voegel, F. Cuisiner, Protein adsorption onto auto-assembled polyelectrolyte films, Biomol. Eng. 19 (2−6) (2002) 273−280. Available from: https://doi.org/10.1016/S1389-0344(02)00031-X.

[131] A.R. Boccaccini, J.J. Blaker, Bioactive composite materials for tissue engineering scaffolds, Int. J. Environ. Res. Public. Health. 2 (3) (2014) 303−317. Available from: https://doi.org/10.3390/ijerph14010066.

[132] V. Karageorgiou, D. Kaplan, Porosity of 3D biomaterial scaffolds and osteogenesis, Biomaterials. 26 (27) (2005) 5474−5491. Available from: https://doi.org/10.1021/la047994h.

[133] Á.J. Leite, P. Sher, J.F. Mano, Chitosan/chondroitin sulfate multilayers as supports for calcium phosphate biomineralization, Mater. Lett. 121 (2014) 62−65. Available from: https://doi.org/10.1016/J.MATLET.2014.01.099.

[134] D.S. Couto, N.M. Alves, J.F. Mano, Nanostructured multilayer coatings combining chitosan with bioactive glass nanoparticles, J. Nanosci. Nanotechnol. 9 (3) (2009) 1741−1748. Available from: https://doi.org/10.1166/jnn.2009.389.

[135] Y. Hu, K. Cai, Z. Luo, Y. Zhang, L. Li, M. Lai, et al., Regulation of the differentiation of mesenchymal stem cells in vitro and osteogenesis in vivo by microenvironmental modification of titanium alloy surfaces, Biomaterials 33 (13) (2012) 3515−3528. Available from: https://doi.org/10.1016/J.BIOMATERIALS.2012.01.040.

[136] K. Ariga, M. Onda, Y. Lvov, T. Kunitake, Alternate layer-by-layer assembly of organic dyes and proteins is facilitated by pre-mixing with linear polyions, Chem. Lett. 26 (1) (1997). Available from: https://doi.org/10.1246/CL.1997.25.

[137] Y. Lvov, H. Möhwald, Protein architecture: interfacing molecular assemblies and immobilization biotechnology, Marcel Dekker, New York, 2000, p. 394.

[138] F. Caruso, H. Möhwald, Protein multilayer formation on colloids through a stepwise self-assembly technique, J. Am. Chem. Soc. 121 (25) (1999) 6039−6046. Available from: https://doi.org/10.1021/JA990441M.

[139] F. Mizutani, S. Yabuki, Y. Hirata, Amperometric l-lactate-sensing electrode based on a polyion complex layer containing lactate oxidase. Application to serum and milk samples, Anal. Chim. Acta 314 (3) (1995) 233−239. Available from: https://doi.org/10.1016/0003-2670(95)00278-8.

[140] F. Mizutani, S. Yabuki, T. Sawaguchi, Y. Hirata, Y. Sato, S. Iijima, et al., Use of a siloxane polymer for the preparation of amperometric sensors: O_2 and NO sensors and enzyme sensors, Sens. Actuators B Chem. 76 (1−3) (2001) 489−493. Available from: https://doi.org/10.1016/S0925-4005(01)00596-2.

[141] Y. Hao, M. Zhang, J. He, P. Ni, Magnetic DNA vector constructed from PDMAEMA polycation and PEGylated brush-type polyanion with cross-linkable shell, Langmuir 28 (15) (2012) 6448–6460. Available from: https://doi.org/10.1021/la300208n.

[142] G.B. Sukhorukov, H. Mijhwald, G. Decher, Y.M. Lvov, Assembly of polyelectrolyte multilayer films by consecutively alternating adsorption of polynucleotides and polycations, Thin Solid. Films 284–285 (1996) 220–223. Available from: https://doi.org/10.1016/S0040-6090(95)08309-X.

[143] K. Ariga, Y.M. Lvov, K. Kawakami, Q. Ji, J.P. Hill, Layer-by-layer self-assembled shells for drug delivery, Adv. Drug. Deliv. Rev. 63 (9) (2011) 762–771. Available from: https://doi.org/10.1016/J.ADDR.2011.03.016.

[144] R. Kayushina, Y. Lvov, N. Stepina, V. Belyaev, Y. Khurgin, Construction and X-ray reflectivity study of self-assembled lysozyme/polyion multilayers, Thin Solid. Films 284–285 (1996) 246–248. Available from: https://doi.org/10.1016/S0040-6090(95)08315-4.

[145] A.A. Vertegel, R.W. Siegel, J.S. Dordick, Silica nanoparticle size influences the structure and enzymatic activity of adsorbed lysozyme, Langmuir. 20 (16) (2004) 6800–6807. Available from: https://doi/abs/10.1021/la0497200.

[146] M. Onda, Y. Lvov, K. Ariga, T. Kunitake, Sequential actions of glucose oxidase and peroxidase in molecular films assembled by layer-by-layer alternate adsorption, Biotechnol. Bioeng. 51 (2) (1996) 163–167. Available from: https://doi.org/10.1002/(SICI)1097-0290(19960720)51:2 < 163::AID-BIT5 > 3.0.CO;2-H.

[147] N.C. Veitch, Horseradish peroxidase: a modern view of a classic enzyme, Phytochemistry 65 (3) (2004) 249–259. Available from: https://doi.org/10.1016/j.phytochem.2003.10.022.

[148] M. Houska, E. Brynda, A. Solovyev, A. Broučková, P. Křížová, M. Vaníčková, et al., Hemocompatible albumin-heparin coatings prepared by the layer-by-layer technique. The effect of layer ordering on thrombin inhibition and platelet adhesion, J. Biomed. Mater. Res. - Part. A 86 (3) (2008) 769–778. Available from: https://doi.org/10.1002/JBM.A.31663.

[149] H. Ai, M. Fang, S.A. Jones, Y.M. Lvov, Electrostatic layer-by-layer nanoassembly on biological microtemplates: platelets, Biomacromolecules 3 (3) (2002) 560–564. Available from: https://doi.org/10.1021/BM015659R.

[150] M.J. McShane, Y.M. Lvov, Electrostatic self-assembly: layer-by-layer. Dekker Encycl. Nanosci. Nanotechnol. (3rd Ed) (2014) 1342–1358. https://doi.org/10.1081/E-ENN3-120013616.

[151] M. Onda, K. Ariga, T. Kunitake, Activity and stability of glucose oxidase in molecular films assembled alternately with polyions, J. Biosci. Bioeng. 87 (1) (1999) 69–75. Available from: https://doi.org/10.1016/S1389-1723(99)80010-3.

[152] M. Onda, Y. Lvov, K. Ariga, T. Kunitake, Sequential reaction and product separation on molecular films of glucoamylase and glucose oxidase assembled on an ultrafilter, J. Ferment. Bioeng. 82 (5) (1996) 502–506. Available from: https://doi.org/10.1016/S0922-338X(97)86992-9.

[153] S. Tirkeş, L. Toppare, S. Alkan, U. Bakir, A. Önen, Y. Yǎci, et al., Immobilization of glucose oxidase in polypyrrole/polytetrahydrofuran graft copolymers, Int. J. Biol. Macromol. 30 (2) (2002) 81–87. Available from: https://doi.org/10.1016/S0141-8130(02)00011-9.

[154] A. Gürsel, S. Alkan, L. Toppare, Y. Yaǧci, Immobilization of invertase and glucose oxidase in conducting H-type polysiloxane/polypyrrole block copolymers, React. Funct. Polym. 57 (1) (2003) 57–65. Available from: https://doi.org/10.1016/J.REACTFUNCTPOLYM.2003.07.004.

[155] W. Kong, L.P. Wang, M.L. Gao, H. Zhou, X. Zhang, W. Li, et al., Immobilized bilayer glucose isomerase in porous trimethylamine polystyrene based on molecular deposition, J. Chem. Soc. Chem. Commun. 11 (1994) 1297–1298. Available from: https://doi.org/10.1039/C39940001297.

[156] M. Rai, S.S. da Silva, Nanotechnology for bioenergy and biofuel production, 2017. https://doi.org/10.1007/978-3-319-45459-7.

[157] H.V. Lee, J.C. Juan, 2017. Nanocatalysis for the conversion of nonedible biomass to biogasoline via deoxygenation reaction, 301−323. https://doi.org/10.1007/978-3-319-45459-7_13.

[158] S. Nandi, A. Mondal, A. Roberts, S. Gandhi, Biosensor platforms for rapid HIV detection, Adv. Clin. Chem. 98 (2020) 1−34. Available from: https://doi.org/10.1016/BS.ACC.2020.02.001.

[159] N.S. Shrikrishna, S. Mahari, N. Abbineni, S.A. Eremin, S. Gandhi, New trends in biosensor development for pesticide detection. biosensors in agriculture: recent trends and future perspectives, Concepts Strateg. Plant. Sci. (2021) 137−168. Available from: https://doi.org/10.1007/978-3-030-66165-6_8.

[160] A. Kaushik, R. Khan, P. Solanki, S. Gandhi, H. Gohel, Y.K. Mishra, et al., From nanosystems to a biosensing prototype for an efficient diagnostic: a special issue in honor of Professor Bansi D. Malhotra, Biosens 11 (10) (2021) 359. Available from: https://doi.org/10.3390/BIOS11100359.

[161] D. Shahdeo, S. Gandhi, Next generation biosensors as a cancer diagnostic tool. Biosens. Based Adv. Cancer Diagnostics, Compr. Anal. Chem. 91 (2022) 179−199. Available from: https://doi.org/10.1016/BS.COAC.2020.08.007.

Role of debridement and its biocompatibility in antimicrobial wound dressings

5

Mohit and Bodhisatwa Das

Department of Biomedical Engineering, Indian Institute of Technology Ropar, Ropar, Punjab, India

5.1 Introduction

A wound occurs when skin tissue is damaged, resulting in non-functionalities, as a result of any external factor. This wound will allow bacteria to colonize, potentially leading to infection and harsh conditions at the wound site.

The presence of devitalized and necrotic tissue at the wound attracts microbes, prolonging the inflammatory stage, mechanically inhibiting contraction, and preventing re-epithelialization. As a result, debridement is the process of removing necrotic tissue from wounds and bacteria to leave a healthy surface that heals more quickly [1]. Non-debrided tissue may also obstruct underlying fluid collections or abscesses, making wound depth assessment difficult. Various neutrophil-derived enzymes, such as collagenase, myeloperoxidase, elastase, acid hydrolase, and lysozymes, act in autolytic debridement during the early stages of wound healing. Protease synthesis by wound cells protects the wound bed from protease action and the degradation of intact tissue at the wound edge. Although debridement occurs naturally, it can be accelerated by assisted debridement [2].

5.2 Wound

5.2.1 Different kinds of wound

5.2.1.1 Acute wound

The structure of natural skin tissue is disrupted, and the tissue loses its function. Acute wounds exhibit continuous molecular events that result in the preservation of the original structure. The acute wound healing mechanism is primarily identified in four steps, namely hemostasis, inflammation, proliferation, and remodeling [3].

5.2.1.2 Chronic wound

A chronic wound occurs when acute wounds do not heal properly or become infected. Extreme trauma in some serious accidents may result in chronic wounds. All chronic wounds begin as acute wounds, and the difficulty arises only when the site of damage does not heal at the expected rate.

Antimicrobial Dressings. DOI: https://doi.org/10.1016/B978-0-323-95074-9.00011-7

Chronic wounds take more than four weeks to heal following the initial medical intervention. Its severity increases significantly if it does not heal beyond 8 weeks. Chronic wounds are more observed in an individual having a systemic disease such as diabetes mellitus and rheumatoid arthritis or a greater risk of bacterial incursion. Local factors can also lead to chronic wounds like venous hypertension, trauma, and arterial insufficiency [4]. Chronic wound sites were mostly made up of necrotic or sloughy tissue. These dead, devitalized cells obstruct the healing pathway and can promote the growth of microbes, which can contribute to an intense infection and further obstruct healing [5] (Table 5.1).

Table 5.1 Classification of wounds based on different criteria.

Category		Types of wound
Based on the extent of chronicity		Acute
		Chronic
Based on exposure to the outer environment	Open	Penetration
		Puncture
		Laceration
		Abrasion
		Incision
		Gunshot
	Closed	Blisters
		Hematoma
		Seroma
		Contusion
		Crush injury
Based on trauma and surgery		Contaminated wound
		Clean wound
		Specialized wound
Based on stress		Pressure ulcer
Based on changes in vessels		Lymphatic
		Arterial
		Venous
Miscellaneous		Calciphylaxis
		Sickle cell ulcer
		Malignant melanoma
		Anorectal fistulae
		Stoma-related wounds
		Neoplastic
		Vasculitis
		Pyoderma gangrene
		Factitious wounds
		Warfarin-induced skin necrosis

5.3 **Wound healing**

Hemostasis is the first step in the healing process after an injury. It involves a multistep cascade of events such as vasoconstriction, platelet aggregation, and fibrin formation to form a mesh-like structure that regulates blood loss. The inflammatory response is the next step in an acute wound, causing the release of growth factors and cytokines. The inflammatory response causes vasodilation, increasing blood flow to the injured area. Simultaneously, increased vascular permeability allows phagocytic cells, complement activation, and antibodies to remove microorganisms, bacterial toxins, enzymes, and foreign debris. Proliferation and remodeling are the next steps [2]. Fig. 5.1 defines the healing cascade of an acute wound. It summarizes all the steps of wound healing.

Several factors are responsible for the wound healing cascade. These factors are (1) local: oxygenation, infection, foreign body response, venous occlusion or insufficiency, (2) systemic factors—age and gender, sex hormones, (3) stress (4) ischemic condition, (5) diseases: diabetes, keloids, and scars, fibrosis, hereditary healing disorders like hemophilia, hepatic and renal disorders like jaundice, uremia, (6) obesity, (7) medications and drug abuse: glucocorticoid steroids, non-steroidal anti-inflammatory drugs, chemotherapy, (8) alcoholism and smoking, (9) immune system suppressed conditions: cancer, chemo and radiation therapy, AIDS, (10) malnutrition [6]. When we focus on chronic wound healing, it shows prolonged inflammation. Fig. 5.2 shows the comparison of healing steps followed in an acute wound and chronic wound. In a chronic wound, the inflammation phase appears to be a prolonged activity. The chronic wound frequently manifests dead tissue, necrotic tissue, and infection. In chronic wounds, debridement or tissue excision becomes necessary.

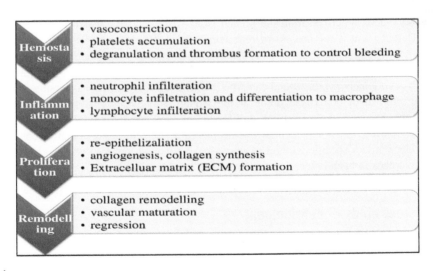

FIGURE 5.1

Acute wound-healing process.

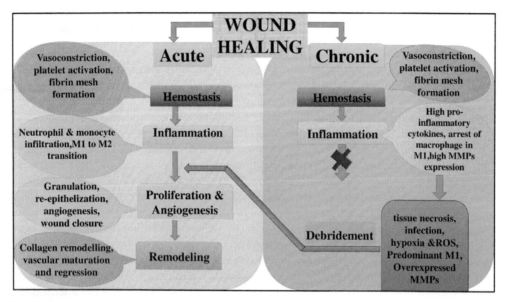

FIGURE 5.2

Differences in wound healing cascade in the acute and chronic wound: staggering of wound healing at the inflammatory phase in a chronic wound and role of debridement in resolving chronic inflammation to proliferating phase.

5.4 Debridement

The term debridement is derived from the French word "debridement," which means "to remove the constraint." Debridement is the process of removing dead and necrotic tissue from a chronic wound site, causing the wound to heal faster than usual. Eschar, hematomas, pus, foreign bodies, necrotic tissue, devitalized and infected tissue, slough, hyperkeratosis, bone fragments, and debris can all be removed with debridement [7]. In wound care, the inclusion of debridement in dressing is an intriguing concept. The removal of non-viable tissue at the wound site reveals healthy, well-perfused tissue. Furthermore, these wound bed tissues can proliferate and populate the wound bed via epithelial cell migration [8]. Debridement is an essential and irreplaceable step in the treatment of chronic wounds. Indeed, debridement is a specific method of preparing the wound bed for dressing. It can reduce the presence of microbes in open tissue, thereby shortening the healing cascade.

5.4.1 Different kinds of debridement

Different kinds of debridement are mentioned below. Fig. 5.3 depicts the chronic wound surface along with required debridement techniques.

- Autolytic
- Surgical

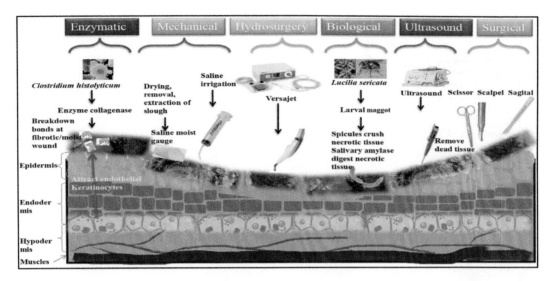

FIGURE 5.3

Wound debridement methods. Skin surface shows tissue condition in case of the wound at different stages and denotes the specified debridement techniques used for treatment.

- Enzymatic
- Mechanical
- Hydrosurgery
- Biological
- Ultrasound

5.4.1.1 Autolytic debridement

A body's natural defense involves the breakdown or lysis of damaged tissue at the wound site with the assistance of various enzymes. These enzymes can digest specific body cell components or components such as collagen, proteins, and fibrin. Dressings such as hydrocolloids and hydrogels can help to improve the debridement process. Thin films can also be used to maintain optimal temperature for the body's enzymes and moisture levels [9]. Importantly, autolytic debridement is a technique that removes non-viable, slough tissue while leaving healthy tissue alone. Pain has been linked to the maintenance of a moist local wound environment (to encourage autolytic debridement), which is a critical issue in wound care, particularly for patients [10]. This procedure is painless and safe, but it takes more time to complete and requires daily dressing changes, so it is costly. Autolytic debridement is a conservative approach used for at-home wound care [11].

5.4.1.2 Surgical debridement

To reduce the risk of infection and facilitate normal wound healing, conventional surgical debridement entails the removal of devitalized tissue that prevents normal tissue from growing. The best time and frequency for surgical debridement are unknown because they will likely vary greatly

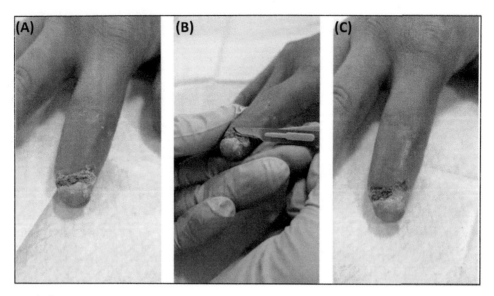

FIGURE 5.4

Sharp debridement of digital ulcers (A) image taken before (B) during (C) after, digital ulcer debridement in a patient with systemic sclerosis. The ulcer was not fully removed in one go due to the painful effect. Subsequent debridement will be done to fully expose the wound bed.

Reproduced from M. Hughes et al., Digital ulcers: should debridement be a standard of care in systemic sclerosis? Lancet Rheumatol. 2 (5) (2020), e302–e307 with editor's permission.

depending on the type of wound being treated. However, the appropriate time and interval play an obvious role in wound care [8]. Open osteocapsular debridement, arthroscopic osteocapsular debridement, and surgical debridement are all viable therapeutic options for people with symptomatic primary elbow osteoarthritis (PEOA). They can consistently improve flexion, extension, and functional outcomes while having a low complication rate [12]. In digital ulcers, surgical debridement is quite an essential method of debridement. Fig. 5.4 shows that subsequent dead tissue excision is required in a patient with systemic sclerosis having a digital ulcer [7].

5.4.1.3 Enzymatic debridement

The enzymatic method, which is recommended for fibrotic/moist wounds, is directly based on the breakage of peptide bonds. It is preferred for patients who are unsuitable for surgery. Enzymatic debriding treatments are effective for removing necrotic material from pressure ulcers, leg ulcers, and partial-thickness wounds [13–15]. Collagenase is an enzymatic debriding agent that is commonly used to treat wounds in many countries. Collagenase ointment (250 units/gm) is commonly isolated and cultivated from *Clostridium histolyticum*. If surgery is not an option, this is one of the best methods for dry wounds with fibrous slough and no granulation tissue. In vivo studies show that collagenase promotes endothelial cell and keratinocyte migration [16]. The remarks of the group Italian consensus panel are a "ready to use: set of guidelines for enzymatic debridement in burn surgery that both draw on and compete with the available scientific literature on the subject" [17].

5.4.1.4 Mechanical debridement

Surgical debridement is not always appropriate for all patients, whether they are at medical risk or their wounds do not require extensive debridement. Other types of debridement, such as mechanical debridement, are also available in this situation. These are further divided into:

Saline moistened gauge: It is a gauze (saline-moistened) is applied to the wound, which is kept there for drying, and finally, the gauze in dried condition is removed, resulting in debridement. This method can easily remove fibrin. On the other hand, while removing dried gauze, this type of debridement may cause pain and damage to the newly formed epithelium.

Saline irrigation: Here, pressurized saline irrigation such as pulsed lavage will be done. Saline-filled in a 30 mL syringe used to flush a wound will apply nearly 15 psi pressure to extract loose dead tissue present on the wound, over an 18 gauge catheter. This technique is painless, quite effective, low risk, and easily performed by a nurse [18]. In practice, antimicrobial therapy is used to supplement mechanical debridement; however, in a recent clinical trial, photodynamic therapy is being investigated as an adjunct mechanical debridement in the treatment of severe peri-implantitis with abscess. It is extremely effective at alleviating severe peri-implant symptoms [19]. Mechanical debridement can reduce the bacterial burden of venous ulcers without the use of antiseptics. This method is both safe and affordable [20].

5.4.1.5 Hydrosurgery

It is a sophisticated debridement technique in which sterile water is delivered to the injury site at high pressure. This aids in the removal of dead tissue and slough from the wound [21]. Water under high pressure is sufficient to cut the necrotic tissue. The faster the water flows, the more effective the cutting technique. This method can be combined with sharp debridement in some cases to achieve better results [22]. Currently, "Versajet," a device manufactured by Smith and Nephew, is used for this therapy. However, it is costly, requires specialized training, and can only be administered in a hospital setting. It reduces the bacterial burden in a wound by removing existing tissue [23].

Here we can clarify that, in mechanical irrigation, saline water irrigation of wound is done by syringe with a limited pressure applied but in the case of hydrosurgery, sterile water is delivered at high pressure on the wound site via a device "VERSAJET."

5.4.1.6 Biological debridement

Maggot therapy, as well as autolytic and enzymatic debridement methods, are examples of biological debridement. This type of biological debridement is also known as maggot debridement therapy (MDT) or biosurgery. Larval debridement is used to treat chronic wounds of various etiologies [24]. Biological agents such as larvae maggots are used to remove dead tissue from the wound site quickly and efficiently. These maggots' salivary enzymes digest and ingest dead cells in the wound [24]. The maggot larvae are from the green bottle blowfly (*Lucilia sericata*) and are raised in the lab. This method is highly specific and destroys only dead cells while leaving healthy cells alone. Maggot therapy's primary therapeutic actions are debridement, disinfection, and increased skin growth [25−29]. The MDT mechanism can be investigated in stages. For starters, the bodies of maggots contain spicules that remove necrotic tissue. Second, larval secretion contains proteolytic digestive enzymes that degrade necrotic tissue. The final product of this extracorporeal processing is then consumed by the larvae. It was assumed that some of the particles present in secretion have antimicrobial properties and can promote growth [30,31].

5.4.1.7 Ultrasound

Low-dose, low-frequency sonography can breakdown dead tissue. This technique is used to remove unwanted tissue (dead cells, slough) from the wound surface. This method is painless and reduces the bacterial burden. Currently, two products have been developed in this area: SONOCA by Soring and MIST therapy by Cellaration. Ultrasound debridement therapies are widely available, frequently administered in a hospital setting, and require specialized training to administer [21] (Table 5.2).

Table 5.2 Various Debridement methods, their advantages, and limitations.

S. No	Debridement type	Peculiarities	Drawbacks	References
1	Autolytic	Body's natural defense preserves healthy tissue, painless and safe, low risk of side effects	Takes more time, mostly dressing changes daily so it is costly, risk of an allergic reaction, may lead to excessive production of exudate	[9,32]
2	Surgical	Fast, urgent/emergent wound decompression is required or deeper structure, efficient in the wound with a solid layer of necrotic tissue	Risk of hemorrhage, Cost may be high if requires operating theater, Requires skilled staff, over/under excision	[7,9,32]
3	Enzymatic	Negligible trauma to healthy tissue, use in long-term care facilities	Maceration of the wound bed, allergic reaction	[13,15]
4	Mechanical	Maybe fast, removing fibrin, no special expertise needed	Does not remove devitalized tissue only, possible infection risks where water baths are used, gauge remnant act as foreign bodies, damage to newly formed epithelial	[9,33]
5	Hydrosurgery	Minimal splash, vaporization, and aerosol effect, significantly reduce bacterial	Over excision, aerosolization bleeding	[32]
6	Biological	Disinfectant, encourage skin growth, eradicate only dead tissue, cost-effective	Patient anxiety, pain in some patients	[32,34]
7	Ultrasound	Widely available, Ranging from destruction to dislocation and physical moderation	Lack of evidence supporting its use, expensive	[21,35]

5.5 Debridement in different wounds

5.5.1 Why debridement is required

The presence of necrotic tissue and slough on the surface of chronic wounds promotes the growth of microorganisms [36]. Furthermore, the presence of microbes at the wound site contributes to the ongoing inflammatory cascade, which contributes to host injury. Prostaglandin E2 and thromboxane, two inflammatory mediators, are still synthesized. This promotes the gradual infiltration of neutrophils, which are responsible for the synthesis of reactive oxygen species (ROS) and cytolytic enzymes such as collagenases, gelatinases, and stromelysins, also known as matrix metalloproteinases (MMPs). Thrombosis occurs at the target site, and the release of vaso-constricting molecules causes tissue hypoxia and degradation, which is conducive to more bacterial growth [2].

Chronic ulcers are distinguished by insufficient ECM remodeling, a failure to re-epithelialize, and persistent inflammation [37]. The epidermis fails to migrate over the wound tissue. Hyper-proliferation at wound margins disrupts normal cellular exodus over the wound bed, which may be due to apoptosis inhibition within fibroblast and keratinocyte cells. In one case, fibroblasts collected from chronic ulcers exhibit a negligible response to growth factors such as Platelet-derived growth factor-beta (PDGF-β) and Transformation growth factor-b (TGF-b), which is most likely due to senescence.

Because cells accumulating at chronic wound sites respond insignificantly to healing signals, topical growth factor treatment is unlikely to result in wound closure unless neighboring cells capable of reacting to growth factors migrate into the wound. Chronic wounds should be treated similarly to acute wounds by focusing on re-establishing the levels of cytokines, growth factors, proteases, and their natural inhibitors. This is accomplished by concentrating on necrotic tissue, bacterial load, and an abundance of exudate. Debridement is the solution to these three problems. Patients' health status and systemic flaws can be examined further for correction.

Debridement is one of the cumulative actions performed during wound bed preparation. The goals of wound bed preparation are as follows: (1) focus on all components such as debridement, bacterial balance, and exudate management. (2) Recognize the patient's health status and its impact on the wound healing process. Ultimately, the goal of wound bed preparation, whether done naturally or with skin products or grafting procedures, is to ensure the formation of good quality granulation tissue, which leads to proper wound closure.

Platelets are translocated into the wound site when it is debrided, which controls hemorrhage. Platelets are also involved in blood clotting. Inadequate platelet count or platelet aggregation suppressed by low-dose aspirin therapy can result in significant bleeding. Platelets are responsible for continuing the inflammatory phase, the second stage of wound healing, in addition to initiating the clotting cascade. In the first 48 hours, platelets are responsible for the release of growth factors, after injury. Following that, the monocytes come into contact with the wound, differentiate into macrophages, and are then in charge of secreting inflammatory cytokines. [18]. It has been shown that bEVs (bacterial extracellular vesicles) exhibit chronicity due to the persistence of microbes such as *Staphylococcus aureus* (*S. aureus*) and *Pseudomonas aeruginosa* (*P. aeruginosa*) and that the host initiates a pro-inflammatory cascade. Bacterial lipopolysaccharides (BLPS) can also activate macrophages. bEVs appear to be important in securing nutrients for the originating microbial community, either by successfully competing with surrounding microbial species or by directly

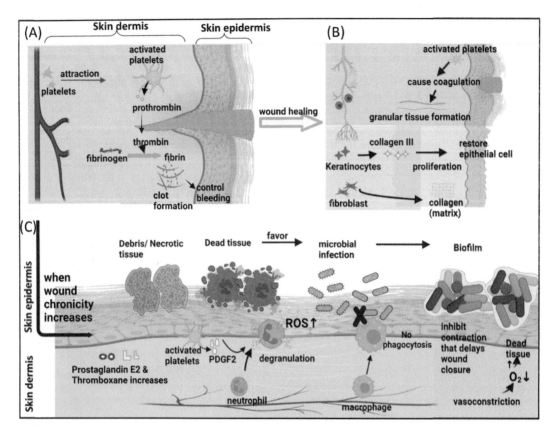

FIGURE 5.5

Wound healing mechanism. (A) Shows quick action of healing after an injury occur. (B) Normal wound healing cascade. (C) When a wound shows chronicity and does not heal in a specified time. The wound surface shows dead tissue which favors microbes' growth and further formation of biofilms, while the dermis depicts a staggering healing cascade at the cellular level.

collecting nutrients from the host [38]. Fig. 5.5 shows anatomically the cellular activity during healing. Part A and part B depict the acute wound healing cascade whereas the consequences of chronicity in the wound are elaborated in part C of Fig. 5.5 (Table 5.3).

5.5.2 Debridement procedure

The selection of a convenient debridement method for any wound should be conditional on many parameters like type of bioburden covering the wound bed, type of tissue [36,39,40], exudates, and amount of pain from the procedure, state of wound edges and skin. Additionally, the patient's environment, skills resources of caregiver, age, quality of life, regulations, and guidelines [11].

The treatment of a critical wound on a local level depends upon the:

Table 5.3 Selection criteria to choose a specific Debridement technique.

	Autolytic	Surgical	Enzymatic	Mechanical	Hydrosurgery	Biological	Ultrasound
Quickness	Poor	High	Good	Adequate	Adequate	High	High
Selectivity	Adequate	Good	High	Poor	Good	High	Poor
Pain	High	High	Good	Adequate	Good	Poor	Poor
Exudates	Adequate	High	Poor	Good	Adequate	Good	Adequate
Infection	Poor	High	Adequate	Good	Poor	Poor	Poor
Cost	High	Poor	Good	Adequate	Adequate	Adequate	High

Source: *Reproduced with permission from. S. Mancini, R. Cuomo, M. Poggialini, C. D'Aniello, and G. Botta, Autolytic debridement and management of bacterial load with an occlusive hydroactive deressing impregnated with polyhexamethylene biguanide, Acta Biomed., 88, 4, pp. 409–413, 2017, https://doi.org/10.23750/abm.v88i4.5802.*

An ongoing debridement phase: Effective debridement is a critical step in the treatment of both acute and chronic wounds. In chronic wounds, continuous debridement at regular intervals is required rather than a single time. Chronic wounds have underlying pathogenic problems that cause necrotic tissue to grow over time. Consistent removal of necrotic tissue reduces its burden on the wound and promotes healthy granulation tissue. Another advantage of debridement is that it reduces wound contamination and tissue destruction. The decomposing tissue may contribute to a favorable environment for bacterial growth. Most clinicians are concerned about wound infection.

The debridement procedure should be as gentle as possible while still being as aggressive as necessary to be effective. Mechanical debridement is usually the first step in the treatment of chronic wounds. Mechanical debridement can be used to remove loosely adherent coatings such as fibrin. It entails the application of sterile compresses. A monofilament fiber product, which causes less pain, is one therapeutic option for painful wounds. Surgical removal of adherent coatings and slough regions is common. Physical debridement with laser, plasma, or ultrasound is another option for situational wounds, as is biological debridement with medicinal larvae. Autolytic debridement includes hydrogels and proteolytic enzymes, which are used in outpatient care. It is critical to plan and discuss the necessary pain therapy with the patient ahead of time for effective debridement [41].

5.5.2.1 Debridement in the chronic wound

Liu et al. (2021) proposed a "3D principle" in the debridement of chronic wounds, which is a combination of the surgical debridement principle and clinical practices for chronic wounds infection.

3D principle includes:

- Drainage—development of unrestricted drainage,
- Disruption—adjustments of the wound environment, and
- Division—isolates the source of infection [42].

All chronic wounds contain microorganisms, and it has been proposed that a certain amount of bacteria can stimulate healing. Bacteria produce proteolytic enzymes such as hyaluronidase, which aid in wound debridement and increase neutrophil protease release.

Chronic wound infection is caused by both endogenous and exogenous factors. Diabetes, blood disease, immune disease, impaired blood circulation, and other endogenous factors are examples of endogenous factors. Exogenous factors include trauma, pressure injury, iatrogenic implants, and so on. If catastrophic consequences occur, appropriate treatment should be administered as soon as possible, and patients at high risk of non-healing wounds should avoid skin injury as much as possible. After determining the source of the infection, debridement combined with anti-infection treatment should be performed for wound care [42].

5.5.2.2 Debridement in ulcer

If we distinguish between a wound and an ulcer, the ulcer is the primary tissue breakdown caused by an internal factor, whereas a lesion is caused by an underlying disease, and the tissue disturbance is gradual. The breakdown of tissue in a wound is caused by external forces, and the disturbance of the tissue is acute. Debridement is commonly used in the treatment of ulcers to stimulate wound healing [7]. To keep blowfly larvae from escaping, apply them directly to an ulcer and cover it with a bandage. The dressing should be of a type that allows maggots to breathe and stay moist. Maggot therapy is not recommended for high-risk bleeding patients with nearby large blood vessels. [34,43,44]. Many clinicians assumed that a moist wound bed is a suitable environment for cellular and molecular processes involved in wound healing (such as autolytic debridement), and wound dressing industries followed suit. Pressure ulcers can be properly debrided using significant debriding techniques that maintain a moist and moderate wound environment [45].

Various steps for treatment of ulcer:

Tissue management: Debridement is the first step in the treatment of ulcers. More surveillance was performed on a bed, edges, and peri-lesional skin in this study. In Systemic sclerosis, digital ulcers typically contain necrotic tissue or eschar, which is dead and eventually slows healing (SSc). In a study conducted by Amanzi and a coworker, when compared to necrotic digital ulcers without fibrin, the presence of fibrin in a necrotic digital ulcer was associated with significantly prolonged healing time. Slough is another type of damaged tissue that could serve as a reservoir for pathogenic organisms. It can appear dry and adherent to moisture or loosely attached [46]. Debridement promotes tissue granulation in devitalized tissue. Aside from that, biofilms and bioburden (e.g., the diversity and virulence of the microbial load) form over the wound, acting as a barrier in the digital ulcer healing process [47,48].

Infection and inflammation: Inflammation is a non-specific response that promotes healthy wound healing. Excessive inflammation, on the other hand, can defile the tissue. *S. aureus* is the most common cause of infection in digital ulcers.

Moisture control: Maintaining the wound's moisture content is critical for ulcer healing. Excess moisture, on the other hand, has inhibitory activity due to the presence of pro-inflammatory cytokines in the wound bed.

MDT potential: Opletalova conducted a clinical trial and discovered that ulcers treated with MDT had 54.5% slough in the wound on the eighth day of treatment, while conventionally treated wounds had 66.5% slough on the same day [49].

MDT's healing potential: In the same study, they discovered that maggots cause the formation of granulation tissue and the closure of the wound. However, the healing potential is comparable to that of hydrogel dressing.

MDT antimicrobial property: Malekian and colleagues studied diabetic foot ulcers in 50 patients, some of whom had *S. aureus* and others had *P. aeruginosa* infections. These patients were treated with both MDT and the conventional method. Only after 48 hours, of MDT did the treated patient show the negligible presence of *S. aureus* on the wound. And observation after 96 hours shows a complete absence of *S. aureus* and a significant reduction in *P. aeruginosa* [50].

Cost-effectiveness of MDT: When many therapies are compared in terms of wound dressing and nursing costs, the final median cost of MDT is approximately ($293), while the cost of hydrogel is approximately ($490), which is nearly double. Furthermore, MDT and hydrogel have healing times of 9 and 28 weeks, respectively, indicating a significant cost difference between these two treatment therapies [51].

MDT and PAIN: A study compared pain in patients with and without diabetes who were treated with MDT. One can conclude that diabetic patients experience the same pain before and after using MDT. However, in the other group, patients experience more pain during MDT than before [52].

5.5.2.3 Debridement against bacteria in the wound

Infection and associated pathological inflammation are major impediments to chronic wound healing. Even if chronic wounds are not always infected, they may be colonized by a distinct microbiome that can cause infection or impede wound healing [53]. Bacterial presence at the wound site is extremely harmful in a variety of ways. Bacteria in a wound are difficult to see. Wound sampling techniques are used to tolerate the presence of bacteria, but it takes many days [54], is prone to inaccurate results [55] and is expensive. These are the reasons why many clinicians refuse to diagnose the majority of wounds. The fluorescence imaging technique developed thus far can detect the presence of bacteria. This imaging technique provides clinicians with information about the actual microbe burden at the wound. Imaging can, in fact, influence treatment decisions [56]. Bacterial fluorescence imaging is only pinned down by understanding the following points:

- Various tissues (wound tissue, tendon, and bone) and fluids (blood, pus) fluorescence signals
- Signals are given by microbes (bacteria red or cyan)
- Reasons for differing hues of tissue and bacterial fluorescence
- Artifacts in the image that can arise
- Signals by non-biological components that could be confusing (fluorescent cleansing solutions) [57].

Furthermore, Farhan and colleagues' research could pave the way for bacterial fluorescence imaging to be included in the standard diagnostic algorithm for pediatric burn patients [58]. The other obstacle in the healing cascade is the formation of biofilm [59]. Bacteria in chronic wounds do not exist in free-living planktonic states, but rather in communities known as biofilms. These could form on the wound's moist surface. As a result, biofilms are complex communities of aggregated bacteria clustered in a self-secreted extracellular polysaccharide matrix (EPS). They form when bacteria attach to a solid surface, followed by the formation of a microcolony. Mature biofilms form a protected microenvironment within the polysaccharide, which forms a complex structure with water channels for the transfer of nutrients and waste products. The glycocalyx covering around bacteria protects it from the host's defense mechanism. Phagocytic cells are unable to remove this covering and can only penetrate biofilms imperfectly. This is because, in addition to the mechanical EPS barrier, bacteria within the film release some products. In this context, debridement is an important step in dressing that reduces bacteria bioburden as well as their released toxins, as well as carries away debris and devitalized tissue, reducing the source of nutrients for remaining bacteria [60].

Furthermore, biofilm growth can be influenced by biofilm-forming potential, necessitating the consideration of both synergistic and antagonistic relationships. More research has confirmed the presence of biofilms in chronic wounds. Native bacteria may develop resistance to antimicrobial therapy and an immune response as a result of biofilm formation. As a result of this increase in proliferation characteristics, biofilm has become an important factor in chronic wound infection [61,62].

The best way to identify the bacteria that cause impediments in healing pathways is to culture pus or a piece of tissue removed from the deepest area of debridement. Another benefit of debridement is the detection of osteomyelitis, which can be detected by probing the bone with a sterile cotton-tipped applicator during a physical examination.

Clinical signs and symptoms are used to predict the presence of bacteria at high loads at the point of care. However, these only reveal the host's reaction to bacterial burdens, not the location (s) of high microbial infections within and around the wound. Even more concerning is the fact that high levels of bacteria in wounds and infections have been shown in various clinical investigations and meta-analyses to be asymptomatic in the majority of clinical cases [57,63,64].

5.5.2.4 Debridement in burn wound

Sepsis, skin grafting requirements, and wound re-epithelization time have all emerged as critical therapeutic outcomes in the treatment of burn wounds. It is assumed that prior debridement will reduce the malignant and bacterial burden from a burn wound [65]. In an adult, re-epithelization time is significantly reduced if treated in a time of 24 hour of injury [66]. Griffin's study is the first to show that performing early non-excisional debridement of traumatic pediatric burns in the operating room under general anesthesia reduces wound re-epithelialization time and the need for skin grafts. When compared to Ketamine PSA in the enzymatic debridement, non-excisional burn wound debridement conducted in the operating room within 24 hour of burn injury led to a wound re-epithelialization period of 8 days. When non-excisional debridement was not accomplished in the dressing room under general anesthetic within 24 hour of damage, the chance of requiring a skin graft was dramatically raised [67].

5.5.2.5 Role of enzymatic debridement in burn wound

Enzymatic debridement is a simple way to begin burn wound treatment. Patients who cannot tolerate frequent surgical debridement can choose enzymatic debridement. However, no clinician has suggested that enzymatic debridement is effective in chemical burns. Furthermore, high voltage electrocution injury, which results in deep muscle damage and increased compartment pressure, does not show any specific benefit from enzymatic debridement; it only has an effect on eschar. In flame burns, enzymatic debridement may have threshold potential, but not in scald burns. The panelists agree that greater surface enzymatic debridement is possible, but it may result in water loss and adaptive resuscitation/volume control on a larger surface. Many statements were written about the efficacy and practice pattern of enzymatic debridement, as well as topics like indications, large surface indications, patients' perspectives, cost-effectiveness, logistical considerations, the timing of application for various causes, preparation and application, post-interventional wound management, skin transplant, result, scar and revision management, and training schemes. The majority of the statement is supported by experts. The use of enzymatic debridement is classified on behalf of timing as immediate/very early (12 hour), early (12–72 hour), or delayed (>72 hour) treatment [68]. Fischer et al. proposed the feasibility and safety of enzymatic debridement in distal upper extremity circumferential deep burns to avoid surgery escharotomy [69]. Berner demonstrated that diabetic patients' peripheral burn injuries can result in eschar even after enzymatic debridement. Those patients required additional surgery and skin grafting. Microangiopathy appears to counteract both enzymatic debridement and the bleeding pattern [70]. A recent study demonstrated and addressed that performing early non-excisional debridement of acute burns in the operating room under general anesthesia reduces wound re-epithelization time and the need for skin grafts. Wound re-epithelialization time will be 7 days faster when non-excisional debridement is done within 24 hour of burn, in contrast to PSA in the enzymatic debridement. When non-excisional debridement was not done in operation theater under general anesthetic within 24 hour of injury, the chance of requiring a skin graft was dramatically increased [67].

5.5.2.6 Arthroscopic debridement

Arthroscopic debridement is a technique that uses specialized instruments to remove damaged cartilage or bone. Arthroscopic debridement is assumed to be the primary therapy for focal cartilage injuries, regardless of injury size or depth. Despite the fact that there is little evidence and research to support arthroscopic debridement [71]. There are several advantages to cartilage debridement. It is less expensive and requires less technical skill than reparative or restorative procedures [72]. Cartilage debridement has a brief recovery period. According to Joshi et al., six weeks after debridement, 91% of high-demand military personnel return to their pre-injury activities [73].

5.6 Biocompatibility and safety evaluation for the debridement process

There are some specific considerations whenever the question "When and how to debride and biocompatibilities of debridement" arises. If debridement is carried over to healthy tissue at any point, this tissue dries and dies. Patients who require revascularization should have their debridement performed after a new blood supply has been delivered into the wound. Debridement will take place days or weeks later. For example, prior to bypass surgery, a minor debridement with pus drainage may be required, with a more extensive debridement considered later, once a new blood supply has been established. For antimicrobial wound dressing debridement, the fluorescence imaging device can confirm the presence of bacteria in the wound due to intrinsic fluorescence characteristics that differ from background tissue.

Larval therapy was observed with an increase in pain in about 38% of patients and hence it can be diminished by using analgesia during therapy [74].

5.7 Postdebridement

After cleaning the wound surface with a different approach, such as debridement, the wound should be processed further to achieve the native structure or function of a specific tissue. Various wound dressing methods are being developed to promote wound healing cascades as well as wound infection defense [8]. A moist occlusive dressing aids the inflammatory phase by creating a low oxygen tension environment (resulting in initiating such factors as hypoxia-inducible factor-1) [75] and in addition, enhances the re-epithelization [76]. Skin substitutes, negative pressure wound therapy, growth factors, and hyperbaric oxygen are some postdebridement aspects.

5.8 Wound dressing

Different kind of wound requires a suitable method of dressing.

5.8.1 Characteristics of an appropriate wound dressing

Several different methods of wound dressing with different materials are present but dressing selection is dependent on its ability to

- provide or sustain a moist environment,
- promote epidermal migration,
- enhance connective tissue synthesis and angiogenesis,
- permits exchanges of gases between wounded tissue and environment,
- defend from bacterial infection,
- the temperature of targeted tissue should be maintained to increase blood flow to the wound bed and enhance epidermal growth,
- be non- reactive, biocompatible, non-allergic, and non-toxic,
- start debridement action which results in leucocyte migration and supports the accumulation of enzymes, and
- be non-adherent to the wound then it will be easy to remove [77].

5.8.2 Concept of TIME

The wound care consensus group decided in 2002 to identify the major barriers to wound healing. Tissues, infection, moisture imbalance, and edge advancement are examples (TIME) [2].

5.8.2.1 Tissue

Devitalized tissue can be found on the wound's external surface. To heal the wound, this devitalized tissue must be removed. Debridement is the process of removing devitalized tissue through mechanical, surgical, enzymatic, and biological means. A wound bed can be prepared by removing non-viable tissue that is necessary for healing. Dead cells on the wound's surface impede healing and keratinocyte migration over the wound bed.

5.8.2.2 Infection

1. Cleansing agents can be used for local infection and healing can be improved by using antimicrobials such as cadexomer iodine.
2. Dilute vinegar is used to reduce bacterial colonization in a chronic wound.
3. The wound that is at high risk of infection can be addressed by silver-impregnated dressing.
4. For intense infection, advanced treatment is required.

5.8.2.3 Moisture balance

1. Keratinocyte migration and healing of wounds can be modulated via the moisture content of the microenvironment.
2. Wound moisture will be provided by choosing a suitable dressing method.
3. Among the various types of moisture retentive dressing, the basic class is films, foams, hydrogels, and hydrocolloids.

5.8.2.4 Edge of wound

1. Observation of wound margins can reveal whether or not wound contraction and epithelialization are proceeding, confirming either the efficacy of the wound treatment or the necessity for re-evaluation.
2. To improve wound healing and thereby impact the "edge" effect, a growing number of therapeutic techniques are being proposed.
3. Electromagnetic therapy (EMT), laser therapy, ultrasound therapy, systemic oxygen therapy, and negative pressure wound therapy (NPWT) are some of the therapies we can use [78].

5.9 New technique other than traditional debridement

5.9.1 Cold plasma treatment

Cold atmospheric pressure plasma treatment (CAPT) has emerged as an alternative option for medical indications in recent years. CAPT is used in a variety of medical services, including disinfection, wound therapy, pain relief, and the treatment of specific skin diseases [79]. CAPT upgrades the healing process in both cases, chronic as well as acute wounds [80–82]. The plasma applied to the wound cause non-inflammatory modification and could inactivate bacteria [83]. CAPT is made up of ROS such as H_2O_2, OH, and ROO, which increase the synthesis of growth hormones and cytokines such as FGF2. Furthermore, many researchers have demonstrated that angiogenesis is a feature of CAPT by the release of nitric oxide, which ultimately results in wound healing. CAPT would easily increase re-epithelialization and neovascularization in mice and increase porcine endothelial cell proliferation. [37]. One of the major benefits of CAPT is that it reduces the patient's pain level during treatment. Plasma has been proposed as a new method of debridement because of its efficacy in removing dead tissue from the wound bed without causing any complications [84]. So we can conclude that CAPT act as alternate to debridement as well as wound treatment.

5.9.2 Electroporation

Irreversible electroporation (IRE) has been identified as a novel alternative to excisional tissue debridement for wound disinfection and tissue regeneration [85]. IRE kills bacteria and damaged cells in necrotic tissue by creating nanopores on cell membranes through rapid changes in resting membrane potential. An in vitro study involving two major skin cell types (dermal fibroblasts and keratinocytes) revealed that similar electrical parameters (field intensity, frequency, electrode configuration, etc.) can result in different post-electroporation proliferation rates, ECM production, and cytokine secretion profiles [86,87]. This technique has been explored significantly in infected burn injuries and excisional wounds [88]. It has been proposed that electroporation not only disinfects the wound microenvironment, but also aids in increased oxygenation, proliferative cytokine secretion, and cell simulation [89,90]. Thus, IRE has been shown to reduce the formation of scar tissues and the generation of hair follicles following wound healing.

5.10 Concluding remarks

The presence of necrotic tissue or infectious microbes causes the wound to heal slowly. It also causes and worsens inflammation and infection. In this paper, we also looked at some alternatives to traditional debridement methods. They do, however, have some limitations.

CAPT poses some difficulties when used at high doses and causes living cell death. Plasma may not be appropriate for use directly on the wound site. Furthermore, CAPT sources are being used, which are inherently more difficult to control and analyze than low-pressure plasma. The interaction of plasma species with biological structure can be quite complex. In clinical practice, electroporation allowed the use of planar electrodes, and skin wounds can extend deep into tissue, resulting in variable-shaped wound treatment that is ineffective. Furthermore, electroporation requires clinical trials on larger animals with wounds similar to human size. In the case of smart dressing, it is extremely difficult to create a multifunctional wound dressing without jeopardizing each smart property. The design of smart dressings for bacterial studies does not define quantitative monitoring of the infectious wound. Debridement, also known as wound cleansing, aids in the removal of dead tissues and biofilm, revealing a healing surface. We talked about different procedures and the role of debridement in both acute and chronic wound dressing. We can mention that debridement is important in wound management and that every physiotherapist should consider it. To get the desired results, we can suggest more advancement in types of debridement methods, such as elaborative research on biological, enzymatic, and autolytic techniques, as well as effective pain management.

Acknowledgment

The authors acknowledge CSIR (Government of India) for Mohit's research fellowship (File No: 09/1005 (12213)/2021-EMR-I) and SERB (Government of India) for the research grant (SRG/2021/002428).

References

[1] K.L. Andrews, K.M. Derby, T.M. Jacobson, B.A. Sievers, L.J. Kiemele, Prevention and management of chronic wounds, Braddom's Phys. Med. Rehabil. (2021) 469−484. Available from: https://doi.org/10.1016/b978-0-323-62539-5.00024-2.

[2] J. Douglass, Wound bed preparation: a systematic approach to chronic wounds, Br. J. Commun. Nurs. 8 (6 Suppl). Available from: https://doi.org/10.12968/bjcn.2003.8.sup2.11554.

[3] K. Raziyeva, Y. Kim, Z. Zharkinbekov, K. Kassymbek, S. Jimi, A. Saparov, Immunology of acute and chronic wound healing, Biomolecules 11 (5) (2021) 1−25. Available from: https://doi.org/10.3390/biom11050700.

[4] V. Falanga, Chronic wounds: pathophysiologic and experimental considerations, J. Invest. Dermatol. 100 (5) (1993) 721−725. Available from: https://doi.org/10.1111/1523-1747.ep12472373.

[5] R. White, K. Cutting, Critical colonisation of chronic wounds: microbial mechanisms, Wounds UK 4 (1) (2008) 70−78.

[6] S. Guo, L.A. DiPietro, Critical review in oral biology & medicine: factors affecting wound healing, J. Dent. Res. 89 (3) (2010) 219−229. Available from: https://doi.org/10.1177/0022034509359125.

[7] M. Hughes, et al., Digital ulcers: should debridement be a standard of care in systemic sclerosis, Lancet Rheumatol. 2 (5) (2020) e302−e307. Available from: https://doi.org/10.1016/S2665-9913(19)30164-X.

[8] G. Han, R. Ceilley, Chronic wound healing: a review of current management and treatments, Adv. Ther. 34 (3) (2017) 599−610. Available from: https://doi.org/10.1007/s12325-017-0478-y.

[9] J. Choo, J. Nixon, E.A. Nelson, E. Mcginnis, Autolytic debridement for pressure ulcers, Cochrane Database Syst. Rev. 2014 (10) (2014). Available from: https://doi.org/10.1002/14651858.CD011331.

[10] C. Gorecki, et al., Impact of pressure ulcers on quality of life in older patients: a systematic review: clinical investigations, J. Am. Geriatr. Soc. 57 (7) (2009) 1175−1183. Available from: https://doi.org/10.1111/j.1532-5415.2009.02307.x.

[11] S. Mancini, R. Cuomo, M. Poggialini, C. D'Aniello, G. Botta, Autolytic debridement and management of bacterial load with an occlusive hydroactive deressing impregnated with polyhexamethylene biguanide, Acta Biomed. 88 (4) (2017) 409−413. Available from: https://doi.org/10.23750/abm.v88i4.5802.

[12] E.M. Guerrero, et al., The clinical impact of arthroscopic vs. open osteocapsular débridement for primary osteoarthritis of the elbow: a systematic review, J. Shoulder Elb. Surg. 29 (4) (2020) 689−698. Available from: https://doi.org/10.1016/j.jse.2019.12.003.

[13] G. Gravante, et al., Multicenter clinical trial on the performance and tolerability of the Hyaluronic acid-collagenase ointment for the treatment of chronic venous ulcers: a preliminary pilot study, Eur. Rev. Med. Pharmacol. Sci. 17 (20) (2013) 2721−2727.

[14] A. Tallis, T.A. Motley, R.P. Wunderlich, J.E. Dickerson, C. Waycaster, H.B. Slade, Clinical and economic assessment of diabetic foot ulcer debridement with collagenase: results of a randomized controlled study, Clin. Ther. 35 (11) (2013) 1805−1820. Available from: https://doi.org/10.1016/j.clinthera.2013.09.013.

[15] J. Ramundo, M. Gray, Enzymatic wound debridement, J. Wound, Ostomy Cont. Nurs. 35 (3) (2008) 273−280. Available from: https://doi.org/10.1097/01.WON.0000319125.21854.78.

[16] T.N. Demidova-Rice, A. Geevarghese, I.M. Herman, Bioactive peptides derived from vascular endothelial cell extracellular matrices promote microvascular morphogenesis and wound healing in vitro, Wound Repair. Regen. 19 (1) (2011) 59−70. Available from: https://doi.org/10.1111/j.1524-475X.2010.00642.x.

[17] R. Ranno, et al., Italian recommendations on enzymatic debridement in burn surgery, Burns 47 (2) (2021) 408−416. Available from: https://doi.org/10.1016/j.burns.2020.07.006.

[18] D.L. Steed, Debridement, Am. J. Surg. 187 (5) (2004) S71−S74. Available from: https://doi.org/10.1016/S0002-9610(03)00307-6.

[19] T. Almohareb, et al., Clinical efficacy of photodynamic therapy as an adjunct to mechanical debridement in the treatment of per-implantitis with abscess, Photodiagnosis Photodyn. Ther. 30 (2020) 101750. Available from: https://doi.org/10.1016/j.pdpdt.2020.101750.

[20] M. Moelleken, F. Jockenhöfer, S. Benson, J. Dissemond, Prospective clinical study on the efficacy of bacterial removal with mechanical debridement in and around chronic leg ulcers assessed with fluorescence imaging, Int. Wound J. 17 (4) (2020) 1011−1018. Available from: https://doi.org/10.1111/iwj.13345.

[21] L. Nazarko, Advances in wound debridement techniques, Br. J. Commun. Nurs. 20 (2015) S6−S8. Available from: https://doi.org/10.12968/bjcn.2015.20.Sup6.S6.

[22] R. Strohal, et al., EWMA document: debridement: an updated overview and clarification of the principle role of debridement, J. Wound Care 22 (2013) S1−S49. Available from: https://doi.org/10.12968/jowc.2013.22.Sup1.S1.

[23] N. Allan, M. Olson, D. Nagel, R. Martin, The impact of versajet™ hydrosurgical debridement on wounds containing bacterial biofilms, Wound Repair. Regen. 18 (2010) A88.

[24] J.G. Powers, C. Higham, K. Broussard, T.J. Phillips, Wound healing and treating wounds chronic wound care and management, J. Am. Acad. Dermatol. 74 (4) (2016) 607−625. Available from: https://doi.org/10.1016/j.jaad.2015.08.070.

[25] G. Cazander, D.I. Pritchard, Y. Nigam, W. Jung, P.H. Nibbering, Multiple actions of Lucilia sericata larvae in hard-to-heal wounds: larval secretions contain molecules that accelerate wound healing, reduce chronic inflammation and inhibit bacterial infection, BioEssays 35 (12) (2013) 1083−1092. Available from: https://doi.org/10.1002/bies.201300071.

[26] R.J. Linger, et al., Towards next generation maggot debridement therapy: transgenic Lucilia sericata larvae that produce and secrete a human growth factor, BMC Biotechnol. 16 (1) (2016) 1−12. Available from: https://doi.org/10.1186/s12896-016-0263-z.

[27] X. Sun, J. Chen, J. Zhang, W. Wang, J. Sun, A. Wang, Maggot debridement therapy promotes diabetic foot wound healing by up-regulating endothelial cell activity, J. Diabetes Complicat. 30 (2) (2016) 318−322. Available from: https://doi.org/10.1016/j.jdiacomp.2015.11.009.

[28] D. Waniczek, A. Kozowicz, M. Muc-Wierzgoń, T. Kokot, E. Świętochowska, E. Nowakowska-Zajdel, Adjunct methods of the standard diabetic foot ulceration therapy, Evid.-based Complement. Altern. Med. 2013 (2013). Available from: https://doi.org/10.1155/2013/243568.

[29] P.N. Li, et al., Molecular events underlying maggot extract promoted rat in vivo and human in vitro skin wound healing, Wound Repair. Regen. 23 (1) (2015) 65−73. Available from: https://doi.org/10.1111/wrr.12243.

[30] D. Blueman, C. Bousfield, The use of larval therapy to reduce the bacterial load in chronic wounds, J. Wound Care 21 (5) (2012) 244−253. Available from: https://doi.org/10.12968/jowc.2012.21.5.244.

[31] H. Čičková, M. Kozánek, P. Takáč, Growth and survival of blowfly Lucilia sericata larvae under simulated wound conditions: implications for maggot debridement therapy, Med. Vet. Entomol. 29 (4) (2015) 416−424. Available from: https://doi.org/10.1111/mve.12135.

[32] E.L. Anghel, M.V. DeFazio, J.C. Barker, J.E. Janis, C.E. Attinger, Current concepts in debridement: science and strategies, Plast. Reconstr. Surg. 138 (3) (2016) 82S−93S. Available from: https://doi.org/10.1097/PRS.0000000000002651.

[33] C.M. Kirschner, K.S. Anseth, Hydrogels in healthcare: from static to dynamic material microenvironments, Acta Mater. 61 (3) (2013) 931−944. Available from: https://doi.org/10.1016/j.actamat.2012.10.037.

[34] J. Moya-López, V. Costela-Ruiz, E. Garciá-Recio, R.A. Sherman, E. De Luna-Bertos, Advantages of maggot debridement therapy for chronic wounds: a bibliographic review, Adv. Ski. Wound Care 33 (10) (2020) 515−524. Available from: https://doi.org/10.1097/01.ASW.0000695776.26946.68.

[35] L. Grayson, G.W. Gibbons, K. Balogh, E. Levin, A.W. Karchmer, Probing to bone in sign of underlying osteomyelitis in, JAMA 273 (9) (1995) 721.

[36] L. Grimaldi, R. Cuomo, C. Brandi, G. Botteri, G. Nisi, C. D'Aniello, Octyl-2-cyanoacrylate adhesive for skin closure: eight years experience, Vivo (Brooklyn) 29 (1) (2015) 145−148 [Online]. Available from: http://www.embase.com/search/results?subaction = viewrecord&from = export&id = L601736513.

[37] S. Kalghatgi, G. Friedman, A. Fridman, A.M. Clyne, Endothelial cell proliferation is enhanced by low dose non-thermal plasma through fibroblast growth factor-2 release, Ann. Biomed. Eng. 38 (3) (2010) 748−757. Available from: https://doi.org/10.1007/s10439-009-9868-x.

[38] H.L. Brown, A. Clayton, P. Stephens, The role of bacterial extracellular vesicles in chronic wound infections: current knowledge and future challenges, Wound Repair. Regen. 29 (6) (2021) 864−880. Available from: https://doi.org/10.1111/wrr.12949.

[39] R. Cuomo, G. Nisi, L. Grimaldi, C. Brandi, A. Sisti, C. D'Aniello, Immunosuppression and abdominal wall defects: use of autologous dermis, Vivo (Brooklyn) 29 (6) (2015) 753−755.

[40] G. Nisi, et al., Effect of repeated subcutaneous injections of carbon dioxide (CO2) on inflammation linked to hypoxia in adipose tissue graft, Eur. Rev. Med. Pharmacol. Sci. 19 (23) (2015) 4501−4506.

[41] J. Dissemond, et al., Modern wound care - practical aspects of non-interventional topical treatment of patients with chronic wounds, JDDG - J. Ger. Soc. Dermatol. 12 (7) (2014) 541−554. Available from: https://doi.org/10.1111/ddg.12351.

[42] Y.F. Liu, P.W. Ni, Y. Huang, T. Xie, Therapeutic strategies for chronic wound infection, Chinese J. Traumatol.-English Ed. 25 (1) (2022) 11−16. Available from: https://doi.org/10.1016/j.cjtee.2021.07.004.

[43] M. Doerler, S. Reich-Schupke, P. Altmeyer, M. Stücker, Impact on wound healing and efficacy of various leg ulcer debridement techniques, JDDG - J. Ger. Soc. Dermatol. 10 (9) (2012) 624−631. Available from: https://doi.org/10.1111/j.1610-0387.2012.07952.x.

[44] Y. Nigam, C. Morgan, Does maggot therapy promote wound healing? the clinical and cellular evidence, J. Eur. Acad. Dermatol. Venereol. 30 (5) (2016) 776−782. Available from: https://doi.org/10.1111/jdv.13534.

[45] J. Jones, Winter's concept of moist wound healing: a review of the evidence and impact on clinical practice, J. Wound Care 14 (6) (2005) 273−276. Available from: https://doi.org/10.12968/jowc.2005.14.6.26794.

[46] L. Amanzi, et al., Digital ulcers in scleroderma: staging, characteristics and sub-setting through observation of 1614 digital lesions, Rheumatology 49 (7) (2010) 1374−1382. Available from: https://doi.org/10.1093/rheumatology/keq097.

[47] G. Piemonte, L. Benelli, F. Braschi, L. Rasero, The local treatment: methodology, debridement and wound bed preparation, Atlas Ulcers Syst. Scler. (2019) 145−159. Available from: https://doi.org/10.1007/978-3-319-98477-3_18.

[48] H.C. Flemming, J. Wingender, U. Szewzyk, P. Steinberg, S.A. Rice, S. Kjelleberg, Biofilms: an emergent form of bacterial life, Nat. Rev. Microbiol. 14 (9) (2016) 563−575. Available from: https://doi.org/10.1038/nrmicro.2016.94.

[49] K. Opletalová, et al., Maggot therapy for wound debridement: a randomized multicenter trial, Arch. Dermatol. 148 (4) (2012) 432−438. Available from: https://doi.org/10.1001/archdermatol.2011.1895.

[50] A. Malekian, et al., Efficacy of maggot therapy on Staphylococcus aureus and Pseudomonas aeruginosa in diabetic foot ulcers: a randomized controlled trial, J. Wound, Ostomy Cont. Nurs. 46 (1) (2019) 25−29. Available from: https://doi.org/10.1097/WON.0000000000000496.

[51] C. Wilasrusmee, et al., Maggot therapy for chronic ulcer: a retrospective cohort and a meta-analysis, Asian J. Surg. 37 (3) (2014) 138−147. Available from: https://doi.org/10.1016/j.asjsur.2013.09.005.

[52] E. Shi, D. Shofler, Maggot debridement therapy: a systematic review, Br. J. Commun. Nurs. 19 (2014) S6−S13. Available from: https://doi.org/10.12968/bjcn.2014.19.Sup12.S6.

[53] D. Leaper, O. Assadian, C.E. Edmiston, Approach to chronic wound infections, Br. J. Dermatol. 173 (2) (2015) 351−358. Available from: https://doi.org/10.1111/bjd.13677.

[54] G. Kallstrom, Are quantitative bacterial wound cultures useful, J. Clin. Microbiol. 52 (8) (2014) 2753–2756. Available from: https://doi.org/10.1128/JCM.00522-14.

[55] L.R. Copeland-Halperin, A.J. Kaminsky, M.M.P.H. Bluefeld, R. Miraliakbari, Sample procurement for cultures of infected wounds: a systematic review swab; microbe; z-swab, Levine, anti-bacterial agent, J. Wound Care Nort. Hameric An. Suppl. 25 (4) (2016) 6–11.

[56] M.Y. Rennie, L. Lindvere-Teene, K. Tapang, R. Linden, Point-of-care fluorescence imaging predicts the presence of pathogenic bacteria in wounds: a clinical study, J. Wound Care 26 (8) (2017) 452–460. Available from: https://doi.org/10.12968/jowc.2017.26.8.452.

[57] M.Y. Rennie, D. Dunham, L. Lindvere-Teene, R. Raizman, R. Hill, R. Linden, Understanding real-time fluorescence signals from bacteria and wound tissues observed with the MolecuLight i:X™, Diagnostics 9 (1) (2019). Available from: https://doi.org/10.3390/diagnostics9010022.

[58] R. Plastic, S. Trainee, P. Surgery, C. Fax, P. No, Title page complete manuscript title Utility of MolecuLight i:X for managing bacterial burden in paediatric burns, [Online]. Available from: https://academic.oup.com/jbcr/advance-article-abstract/doi/10.1093/jbcr/irz167/5572412.

[59] R. Wolcott, J.W. Costerton, D. Raoult, S.J. Cutler, The polymicrobial nature of biofilm infection, Clin. Microbiol. Infect. 19 (2) (2013) 107–112. Available from: https://doi.org/10.1111/j.1469-0691.2012.04001.x.

[60] R. Edwards, K.G. Harding, Bacteria and wound healing, Curr. Opin. Infect. Dis. 17 (2) (2004) 91–96. Available from: https://doi.org/10.1097/00001432-200404000-00004.

[61] S.L. Percival, K.E. Hill, D.W. Williams, S.J. Hooper, D.W. Thomas, J.W. Costerton, A review of the scientific evidence for biofilms in wounds, Wound Repair. Regen. 20 (5) (2012) 647–657. Available from: https://doi.org/10.1111/j.1524-475X.2012.00836.x.

[62] R.D. Wolcott, et al., Biofilm maturity studies indicate sharp debridement opens a timedependent therapeutic window, J. Wound Care 19 (8) (2010) 320–328. Available from: https://doi.org/10.12968/jowc.2010.19.8.77709.

[63] M. Reddy, S.S. Gill, W. Wu, S.R. Kalkar, P.A. Rochon, Does this patient have an infection of a chronic wound, JAMA - J. Am. Med. Assoc. 307 (6) (2012) 605–611. Available from: https://doi.org/10.1001/jama.2012.98.

[64] T.E. Serena, J.R. Hanft, R. Snyder, The lack of reliability of clinical examination in the diagnosis of wound infection: preliminary communication, Int. J. Low. Extrem. Wounds 7 (1) (2008) 32–35. Available from: https://doi.org/10.1177/1534734607313984.

[65] D. Wilder, H.O. Rennekampff, Débridement von verbrennungswunden - Nutzen und möglichkeiten, Handchirurgie Mikrochirurgie Plast. Chir. 39 (5) (2007) 302–307. Available from: https://doi.org/10.1055/s-2007-989227.

[66] F. Shao, W.J. Ren, W.Z. Meng, G.Z. Wang, T.Y. Wang, Burn wound bacteriological profiles, patient outcomes, and tangential excision timing: a prospective, observational study, Ostomy Wound Manag. 64 (9) (2018) 28–36. Available from: https://doi.org/10.25270/owm.2018.9.2836.

[67] B. Griffin, A. Bairagi, L. Jones, Z. Dettrick, M. Holbert, R. Kimble, Early non-excisional debridement of paediatric burns under general anaesthesia reduces time to re-epithelialisation and risk of skin graft, Sci. Rep. 11 (1) (2021) 1–8. Available from: https://doi.org/10.1038/s41598-021-03141-x.

[68] C. Hirche, et al., Eschar removal by bromelain based enzymatic debridement (Nexobrid®) in burns: European consensus guidelines update, Burns 46 (4) (2020) 782–796. Available from: https://doi.org/10.1016/j.burns.2020.03.002.

[69] S. Fischer, et al., Feasibility and safety of enzymatic debridement for the prevention of operative escharotomy in circumferential deep burns of the distal upper extremity, Surg. (U S) 165 (6) (2019) 1100–1105. Available from: https://doi.org/10.1016/j.surg.2018.11.019.

[70] J.E. Berner, D. Keckes, M. Pywell, B. Dheansa, Limitations to the use of bromelain-based enzymatic debridement (NexoBrid®) for treating diabetic foot burns: a case series of disappointing results, Scars, Burn. Heal. 4 (2018). Available from: https://doi.org/10.1177/2059513118816534.

[71] T. Totlis, T. Marín Fermín, G. Kalifis, I. Terzidis, N. Maffulli, E. Papakostas, Arthroscopic debridement for focal articular cartilage lesions of the knee: a systematic review, Surgeon 19 (6) (2021) 356–364. Available from: https://doi.org/10.1016/j.surge.2020.11.011.

[72] P. Blatnik, M. Tušak, Š. Bojnec, A.B. Masten, Economic evaluation of knee arthroscopy treatment in a general hospital, Med. Glas. 14 (1) (2017) 33–40. Available from: https://doi.org/10.17392/887-16.

[73] A. Joshi, N. Kayastha, R. Maharjan, P. Chand, B.R.K.C. Return to preinjury status after routine knee arthroscopy in military population, J. Nepal. Health Res. Counc. 12 (26) (2014) 14–18.

[74] K.Y. Mumcuoglu, E. Davidson, A. Avidan, L. Gilead, Pain related to maggot debridement therapy, J. Wound Care 21 (8) (2012) 400–405. Available from: https://doi.org/10.12968/jowc.2012.21.8.400.

[75] Q. Ke, M. Costa, Hypoxia-inducible factor-1 (HIF-1), 70 (5) (2006) 1469–1480. Available from: https://doi.org/10.1124/mol.106.027029.

[76] V. Jones, J.E. Grey, K.G. Harding, ABC of wound healing: wound dressings, Br. Med. J. 332 (7544) (2006) 777–780.

[77] S. Dhivya, V.V. Padma, E. Santhini, Wound dressings - a review, Biomed 5 (4) (2015) 24–28. Available from: https://doi.org/10.7603/s40681-015-0022-9.

[78] D.J. Leaper, G. Schultz, K. Carville, J. Fletcher, T. Swanson, R. Drake, Extending the TIME concept: what have we learned in the past 10 years? Int. Wound J. 9 (2012) 1–19. Available from: https://doi.org/10.1111/j.1742-481X.2012.01097.x.

[79] T. Bernhardt, M.L. Semmler, M. Schäfer, S. Bekeschus, S. Emmert, L. Boeckmann, Plasma medicine: applications of cold atmospheric pressure plasma in dermatology, Oxid. Med. Cell. Longev. 2019 (2019) 10–13. Available from: https://doi.org/10.1155/2019/3873928.

[80] A. Chuangsuwanich, T. Assadamongkol, D. Boonyawan, The healing effect of low-temperature atmospheric-pressure plasma in pressure ulcer: a randomized controlled trial, Int. J. Low. Extrem. Wounds 15 (4) (2016) 313–319. Available from: https://doi.org/10.1177/1534734616665046.

[81] H.R. Metelmann, et al., Scar formation of laser skin lesions after cold atmospheric pressure plasma (CAP) treatment: a clinical long term observation, Clin. Plasma Med. 1 (1) (2013) 30–35. Available from: https://doi.org/10.1016/j.cpme.2012.12.001.

[82] G. Isbary, et al., Cold atmospheric argon plasma treatment may accelerate wound healing in chronic wounds: results of an open retrospective randomized controlled study in vivo, Clin. Plasma Med. 1 (2) (2013) 25–30. Available from: https://doi.org/10.1016/j.cpme.2013.06.001.

[83] E. Stoffels, Y. Sakiyama, D.B. Graves, Cold atmospheric plasma: charged species and their interactions with cells and tissues, IEEE Trans. Plasma Sci. 36 (4) (2008) 1441–1457. Available from: https://doi.org/10.1109/TPS.2008.2001084.

[84] C. Trial, A. Brancati, O. Marnet, L. Téot, Coblation technology for surgical wound debridement: principle, experimental data, and technical data, Int. J. Low. Extrem. Wounds 11 (4) (2012) 286–292. Available from: https://doi.org/10.1177/1534734612466871.

[85] B. Das, F. Berthiaume, Irreversible electroporation as an alternative to wound debridement surgery, Surg. Technol. Int. 39 (2021) 67–73. Available from: https://doi.org/10.52198/21.STI.39.WH1452.

[86] B. Das, A. Shrirao, A. Golberg, F. Berthiaume, R. Schloss, M.L. Yarmush, Differential cell death and regrowth of dermal fibroblasts and keratinocytes after application of pulsed electric fields, Bioelectricity 2 (2) (2020) 175–185. Available from: https://doi.org/10.1089/bioe.2020.0015.

[87] S. Gouarderes, L. Doumard, P. Vicendo, A.F. Mingotaud, M.P. Rols, L. Gibot, Electroporation does not affect human dermal fibroblast proliferation and migration properties directly but indirectly via the secretome, Bioelectrochemistry 134 (2020) 107531. Available from: https://doi.org/10.1016/j.bioelechem.2020.107531.

[88] A. Golberg, et al., Pulsed electric fields for burn wound disinfection in a murine model, J. Burn. Care Res. 36 (1) (2015) 7−13. Available from: https://doi.org/10.1097/BCR.0000000000000157.

[89] A. Golberg, et al., Preventing scars after injury with partial irreversible electroporation, J. Invest. Dermatol. 136 (11) (2016) 2297−2304. Available from: https://doi.org/10.1016/j.jid.2016.06.620.

[90] X. Li, et al., Rejuvenation of aged rat skin with pulsed electric fields, J. Tissue Eng. Regen. Med. 12 (12) (2018) 2309−2318. Available from: https://doi.org/10.1002/term.2763.

Different methods for nanomaterial-based immobilization of enzymes

6

Satyabrat Gogoi[1], Jejiron M. Baruah[2], Geetanjali Baruah[3] and Jayanta K Sarmah[1]

[1]Department of Chemistry, School of Basic Sciences, The Assam Kaziranga University, Jorhat, Assam, India
[2]Department of Chemistry, North Lakhimpur College (Autonomous), Lakhimpur, Assam, India [3]Department of Biotechnology, School of Health Sciences, The Assam Kaziranga University, Jorhat, Assam, India

6.1 Introduction

Technological advancements improve the efficiency and quality of industrial processes. Industries seek highly efficient products in terms of robustness, longer shelf-life, and low production costs to meet the needs of a broader segment of society. Industrial enzymes are one type of material that is widely used in a variety of such applications. In recent years, the global market share of industrial enzymes has increased many folds [1]. As a result, high-throughput enzymes for industrial applications are required. Enzyme immobilization is a popular strategy for increasing the efficiency of industrial enzymes. However, there have been several reports of enzyme activity being reduced in traditional immobilization methods. Recently, researchers have focused on the strategy of immobilizing enzymes on various nanoparticles (NPs), and as a result, a large number of methods with merit have been developed. The use of NPs for immobilizing enzymes can provide stability in a variety of pH and temperature conditions, as well as reusability and low production costs. This is due primarily to their high specific surface area to volume ratio, exceptional chemical, mechanical, thermal, and cost-effective properties, high surface energy, improved magnetic properties, and highly porous surface architecture. Various novel methods for successfully immobilizing enzymes onto NPs have been developed to date. The most important factor and challenge is the proper selection of the support materials onto which the intended enzymes are to be immobilized. Some factors to consider before selecting support materials include chemical or heat resistance, relatively high strength, low cost, ease of separation, and ease of functional derivatization of the support material. To be useful in applications, the interaction between the enzyme and the NPs must be moderate. The interaction must be neither too weak nor too strong. Most importantly, the enzyme's structure should not be deformed as a result of its interaction with the NPs.

Carbon-based materials such as carbon nanotubes (CNTs), graphene, graphene oxide (GO), and reduced graphene oxide (rGO) are widely used in enzyme immobilization [2]. Metals, metal oxides, and metal hydroxide nanomaterials have also been employed. In enzyme immobilization, gold, titanium dioxide, zinc oxide, and layered double hydroxide nanomaterials are being highlighted [2].

Another exciting strategy for designing immobilized enzymes in the biocatalytic industries is multi-enzyme co-immobilization. Encapsulation, covalent binding, cross-linking, and adsorption are

Antimicrobial Dressings. DOI: https://doi.org/10.1016/B978-0-323-95074-9.00008-7

important immobilization methods for co-immobilization strategies [3]. Co-immobilization of multi-enzyme constructs has successfully addressed the limitations of individual enzyme loaded constructs, resulting in the fabrication of co-immobilized enzyme systems capable of performing specific reactions with high catalytic turnover. Multi-enzyme co-immobilization methods include random co-immobilization, compartmentalization, and positional co-immobilization. As a support for co-immobilization, carbon-based nanocarriers, polymer-based nanocarriers, silica-based nanocarriers, and metal-based nanocarriers have been used.

Despite its many benefits, enzyme immobilization has some drawbacks. Lower affinity to substrates may result from conformational changes in the enzyme molecule. Porous nanomaterials that encapsulate enzyme molecules in the pores may be promising in this regard. Furthermore, the separation (ultrafiltration or centrifugation) of nanomaterial immobilized enzymes is laborious and time-consuming, making nanomaterials unsuitable for enzyme immobilization. Magnetic nanomaterials are thought to be advantageous for enzyme immobilization due to their ease of separation [4].

Nanomaterial immobilized enzymes have been used in a variety of fields, including inhibitor screening, wastewater treatment, target compound detection, and industrial production, due to their favorable properties. The easy recovery of immobilized enzymes with high catalytic activity is critical for their practical applications. Immobilized enzyme reactors, such as packed bed reactors, magnetically stabilized fluidized bed reactors, microchannel reactors, and so on, should thus be a viable method for producing green and sustainable chemicals [5].

Until now, large-scale production of immobilized enzyme has been a challenge. Nonetheless, enzyme immobilization is currently a gifted technology for commercial availability of highly efficient industrial enzymes. In this context, we have provided the most recent methods for preparing immobilized-enzyme NPs.

6.2 Enzyme immobilization

Based on their catalytic properties for substrate conversions in various chemical reactions, enzymes are typically derived from a living organism or cell culture. Under favorable reaction conditions, the enzymes work at mild temperatures, optimal pH, pressure, and substrate specificity. The most advantageous aspect of using enzymes for suitable conversions is that no unwanted intermediates are produced as contamination, only the desired product. As a result, enzymes have a wide range of applications in the cosmetic, leather, paper, textile, food, and drug industries, as well as in laundry and detergent preparation [6−10]. The cost of producing enzymes is extremely high. There are also other factors that contribute to the efficiency of the enzyme as well as the cost effectiveness of downstream processing, which leads to a preference for immobilized enzymes [11−14]. Immobilized enzymes have emerged as highly efficient for profit-making uses in the current era, as they offer many advantages over enzymes in solution, resulting in the isolation of pure end product. To increase resistance to environmental changes such as pH or temperature, an immobilized enzyme is attached to an inert, insoluble, organic or inorganic material such as calcium alginate or silica [15].

Immobilized enzymes were first used in 1916, when Nelson and Griffin discovered that the enzyme invertase absorbed in charcoal has the ability to hydrolyze sucrose [16]. When invertase was absorbed on a solid, such as charcoal or aluminum hydroxide, at the bottom of the reaction vessel, it revealed the same reactivity as when it was distributed uniformly throughout the solution. This breakthrough was later applied to current enzyme immobilization techniques. Grubhofer and Schelth identified the possibility of immobilized enzyme for reuse and stability in 1953 when they reported the covalent immobilization of several enzymes [17].

Immobilized enzymes have proven to be more robust and less susceptible to environmental changes than free enzyme formulations. More importantly, the immobilized enzyme systems' heterogeneity enables seamless retrieval of both enzymes and end products, multiple reuse of the enzymes, rapid termination and continuous operation of reactions, and compatibility with a wide range of bioreactors.

In comparison to available surface areas, early immobilization processes delivered very low enzyme loadings. Various covalent methods of enzyme immobilization were developed from the 1950s to the 1960s. There are currently over 5000 publications and patents focusing on enzyme immobilization techniques. Several hundred enzymes have been immobilized in various forms, including penicillin G acylase, invertase, lipases, proteases, and others, and have been used as catalysts in various industrial processes for several decades [18].

6.3 Properties of immobilized enzymes

Constant advancements in industrial techniques have also increased the demand for suitable low-cost raw materials. In industries where the commercial application of enzymes is common, the demand for cost-effective enzyme formulations that contribute little or no pollution to the environment has increased. Modern biotechnology has aided in the widespread use of these enzymes in approximately 500 products with over 50 applications in industries ranging from detergent to beer production [15]. It is due to the numerous advantages provided by the use of immobilized enzymes, such as reusability, cost savings for effluent treatment, increased operational stability, and so on. Immobilized enzymes can overcome the disadvantages of using free enzymes, such as poor stability under varying conditions, process cost, difficulty in recovering enzymes for reuse, and pH instability of free enzymes. As a result, enzymes attached to non-reacting substrates are in high demand in the current scenario, particularly nanomaterials. Because of their specific surface area and effective enzyme loading capacity, nanomaterials are suitable surface materials for the immobilization process [19,20]. CNTs, nanofibers, NPs, nanocomposites, and magnetic nanoparticles (MNPs) are examples of nanomaterials that are frequently used in the immobilization process [19].

The properties of the immobilized enzymes have diverse factors to count on. These include three wide aspects such as [21]:

1. The method used for immobilization.
2. The physical properties of the support used for the immobilization process.
3. Properties of the free enzyme.

6.3.1 **Methods used for immobilization**

The effectiveness of immobilized enzymes is primarily determined by the method used for immobilization. Furthermore, the type of application determines the method to be used as well as the support material. Adsorption, covalent binding, affinity immobilization, entrapment, and other immobilization techniques are currently in use.

6.3.2 **Physical properties of the support used**

The physical properties of the support material play an important role in the functionality and stability of the immobilized enzyme. However, predicting an ideal support material in advance is extremely difficult. Natural polymers include alginate from brown algae, which is used in conjunction with xanthan, polyacrylamide gel, and calcium alginate beads. Divalent ions such as Ca^{2+} and glutaraldehyde crosslink with alginate, improving enzyme stability [22–24]. Other natural polymers used for immobilization include chitosan and chitin, collagen, Carrageenan, gelatin, starch, pectin, and sepharose in combination with other inert support materials. Depending on the application, synthetic polymers such as amberlite and inorganic materials such as ceramics, silica, celite, zeolites, and glass are used in the immobilization process. Because of their high adsorptive capacity and low fine particulate matter release, activated carbon and charcoal are also preferred materials for the process [25].

A good support material must be insoluble in water, have a high capacity to bind enzymes, be chemically inert, and mechanically stable. The surface area determines the enzyme binding capacity, the ease with which the support can be activated, and the density of enzyme binding sites as a result [21].

6.3.3 **Properties of the free enzyme**

Each enzyme has its own distinct catalytic property, which is used to classify the enzyme and assign it a unique systemic name. It is also critical to select the appropriate enzyme class for the specific job at hand.

Enzymes, nature's solution to regulating chemical reactions in all living organisms, provide a green solution to an industrialized world in the face of growing environmental concerns. Continued growth of the industrial enzyme application is heavily reliant on advancements in techniques for identifying and characterizing novel enzymes from natural sources, followed by optimization of these enzymes for the required performance in selected applications, and high-level expression of the enzyme [15]. We can conclude from the extensive research on enzyme immobilization that it is one of the most promising techniques for highly efficient and economically competent biotechnological processes in the fields of environmental monitoring, biotransformation, diagnostics, pharmaceutical, and food industries. Enzyme-based approaches are gaining traction and, as a result, are replacing traditional chemical-based methods in laboratories and industries, owing to key characteristics such as efficiency, speed of action, and versatility. However, commercialization of immobilized enzymes is still slow due to high costs and storage issues, which could potentially be solved by the use of nanomaterials.

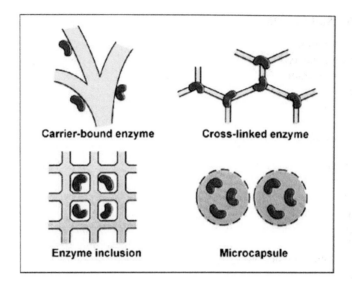

FIGURE 6.1

Different methods of enzyme immobilization [26].

6.4 Methods of immobilization

We can find heterogeneity in the techniques for enzyme immobilization, though none of them can be addressed as the ultimate one. However, it is easily understandable that the support or adherence which makes the enzyme immobilize, fabricate the same up to the category of heterogeneous catalyst as both the support and the enzyme are in different physical state as shown in Fig. 6.1.

The followings are a few major methodologies adopted towards the implementation of immobilization in the enzymes [27]:

6.4.1 Binding support with solid material

The restricted mobility of enzymes is primarily managed through physical and chemical bonding with a solid surface. This is further subdivided into three sections: (1) adsorption, hydrogen bonding, or hydrophobic interactions, (2) ionic bonding, and (3) covalent bonding. All three categories are used and adopted based on their utility and necessity.

6.4.1.1 By physical adsorption (by hydrophobic interactions or hydrogen bonding)

This type of enzyme attachment is caused by physical bonding, which can be caused by either electrostatic attractive force between the host (the solid surface) and the guest molecule (the enzyme) or hydrophobic interaction (passive interaction). Because of the weak force of interaction between the substrate and the enzyme, the uniqueness has the advantage of being easily detached. As a result, the enzyme's immobilization can be terminated at any time [26]. Among discrete examples, the production of biodiesel via ester hydrolase is one of most suitable examples in which the

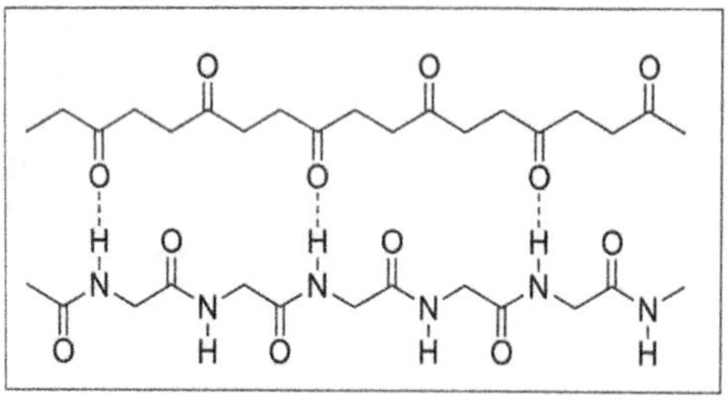

FIGURE 6.2

Hydrogen bond generation between —NH groups and carbonyl groups [26].

enzyme *Candida antarctica* lipase B (CALB) is hosted by acrylic resin: Novozym 435 [28]. Importantly, because there is a possibility of disengagement of the enzyme in the media in this type of interaction, though the possibility of the enzyme becoming dissolved in organic media is almost unworkable, denaturation of the enzyme due to polar-polar interaction cannot be ruled out. Nonetheless, this enzyme is extremely effective in the process. Furthermore, Novozym 435 is efficient in the conversion of fatty acids to peracids when hydrogen peroxide is used as the substrate [29]. In modern applications, cost effectiveness is another criterion for the fulfillment of any industrial production, so researchers propose mica as a low-cost adsorbent matrix [30]. Being a porous silicate grid, mica can easily be utilized in bulk production and can also be modified by alkoxysilane and used for the adsorption of *Candida rugosa lipase* (CRL) [31]. Lipase from *Candida* is another important enzyme used in the commercial production of biodiesel in China. It is immobilized over a textile membrane by adsorption and has been found to be simple and inexpensive [32].

A similar type of physical bonding can be found in the immobilization of enzymes due to charged surfaces via hydrogen bonding. For example, polyketone polymer, a product of carbon monoxide and ethane, can be used for the immobilization of different enzymes, such as amine oxidase and peroxidase, without the inclusion of any bi-functional agent and based solely on numerous hydrogen bonds between the —NH group of the polypeptide chain and the polymer's carbonyl group (Fig. 6.2) [33].

The bond between the host and the guest in this type of non-covalent or non-ionic bonding is not as strong and can be broken or reversed more easily than in traditional bonding methods. As a result, the enzyme may contaminate the media and become inactive due to disintegration from the solid surface [26].

6.4.1.2 Ionic binding

Because of the previously mentioned weak interaction, which causes unsteady immobilization of enzymes, ionic bonded immobilization is preferred over adsorption because ionic bond is considered to be stronger than both hydrophobic interaction and hydrogen bond [34,35]. On that account ion exchange resins are prominent as solid supports for enzyme immobilization. Thus, Lipozyme RM IM [*Rhizomucor miehei* lipase (1, 3-specific lipase)], which is immobilized onto a macroporous anion exchange resin, is applied for wax esters [36], synthesis of fatty acid amides [37], sugar esters [38] and biodiesel production [39]. Another robust ion exchange support for enzyme immobilization is MANA-agarose, which has been successfully practiced for the immobilization of aspartase (aspartic acid ammonia-lyase) [40]. Apart from glucose isomerase, immobilized over ion exchange resins, which is manufactured yearly over 10 million tons (world's highest) to convert concentrated glucose solution to high fructose corn syrup [27], an interesting silica-based support for immobilization is set up which holds both the hydrophobic as well as charged groups [41]. In the latter case, co-bonded octyl and sulfonic acid silica serves as a bed for CRL, which eventually yielded sulfonic acid groups after oxidation with hydrogen peroxide [42]. Theoretically, this system should allow any enzyme to amuse itself inside the pores by extending its hydrophobic or charged portions towards the corresponding strips of the support.

6.4.1.3 Covalent binding

For a stable attachment, covalent bonding is preferred, which is normally accomplished through the linkage of functional groups on the surface of the protein group. Porous materials have an advantage over non-porous materials in terms of binding affinity because they add covalent characteristics to the immobilization phenomenon of enzymes. Furthermore, the specificity in terms of size, shape, volume, and large surface area secures the preference of such porous material far ahead of temporary physical adsorption of enzymes, where enzyme dysfunction is quite a valid story due to their easy detachment. In this context, a notable example is the modification of mesoporous activated carbon with excess glutaraldehyde and ethylene diamine, which results in the covalent binding of *Pseudomonas gessardii* lipase [42]. This was found to be excellent for the ease of doing hydrolysis of olive oil and lipid waste up to 50 cycles, with a subsequent reserve activity of 65%. Furthermore, Manganese peroxidase has demonstrated improved temperature stability with its comparable size to the pore dimensions of the mesoporous silica FSM-16, which was not observed with high concentrations of H_2O_2 as substrate or under free enzymatic conditions [43].

These three adsorption examples clearly justify the role of different immobilization conditions of enzymes and thus their applicability in various circumstances based on their functional efficacy.

6.4.2 Entrapment

The fracturable delicate structure of enzymes and proteins necessitates encapsulation, as they may be invaded by peptidase or proteinase if not. Thus, entrapment comes into play because enzyme denaturation is impossible until the supporting surface acts as a matrix. Covalent bonding or porous materials may be used for capping in this environment. However, in order to make a sensible and realistic distinction between a laboratory method and a bulk industrial application of enzyme immobilization, low cost with high efficiency is a must [44]. It has been observed that this procedure is frequently only applicable to high molecular weight materials because their size fits the pore size of the matrix (e.g., calcium alginate) [45]. But, in spite of having such appropriateness, calcium alginate can be used only for enzymes having high molecular weight (> 100 kDa), due to substantial size of the material. Moreover, it is established that 1-ethyl-(dimethylaminopropyl) carbodiimide hydrochloride (EDC) and sulfo-NHS together get activated with carboxyl groups of Alginate and can react with amino group. In today's cutting-edge technological world, in lieu of Alginate, similar lens shaped matrix composed of polyvinyl alcohol (LentiKats) is used for large enzymes (> 50 kDa) [9]. Impressively, this matrix can also be utilized for aspartase for the conversion of fumaric acid to L-aspartic acid [40] and, because of possible reversibility of the reaction, fumaric acid can again be achieved in a biofriendly manner.

Another well-known material in this class is sol-gel, which is essentially a silica material [26]. Although the standard sol-gel material is not very strong and may cause problems due to restricted substrate diffusion towards the enzyme, the high porosity and thermal and mechanical stability provided by this material make it exceptionally efficient [46]. Furthermore, lipid vehicles can be included in the efficient activity of enzyme entrapment. Furthermore, sol-gel materials can be classified as applied biocatalysis processes. The primary experiment produced xerogels and aerogels with a high surface area and tremendous porosity, as well as small pores [47]. Hydrophobic sol-gels have a very unique and important utility in that they can activate the enzyme by linking with the hydrophobic part of it, which is phenomenal because entrapment of enzymes restricts their diffused movement, and thus the enzyme approaches inactivity under those conditions. In addition, co-immobilization of different enzymes in the same matrix is noteworthy, establishing a reduced rate of substrate production as well as the removal of harmful by-products.

MNPs are another important in situ entrapping agent for enzymes. The use of magnetic particles, particularly MNPs, is critical for the selective binding of proteins to specific metal atoms in the core. When ferrous chloride was combined with urea, urease, and α-amylase in an oxygen atmosphere, excellent results were obtained [48]. In this case, Urease hydrolyzed Urea, increasing the pH of the solution while simultaneously oxidizing Fe^{2+} and producing $Fe(OH)_2$ particles. It is worth noting that this process parallels the encapsulation of Urease and -amylase. The amylase was then used to hydrolyze starch, and it was discovered that it acts as an enzyme that is free in nature and does not denature after five continuous reactions.

Entrapment of enzymes by membrane reactors is another recently discovered pathway. It has been reported that *Bacillus circulans*, specifically β-galactosidase, demonstrated their unique applicability in the production of galacto-oligosaccharides (GOS) from lactose; however, the membrane, which was capable of blocking the enzyme but acted as a permeable one for galactose and glucose produced during the process, was not. This execution enhanced the yield of GOS by 33% [49]. Furthermore, ultrafiltration membrane has been used as a membrane reactor in the fructification of cellobiose, where the membrane itself separated the substrate (sugar beep pulp) and enzyme [50]. Ultrafiltration membrane have also been utilized in the decoloration of azo dyes with manganese peroxide [51]. Furthermore, the utility of the membrane reactor is significant in the production of biodiesel, where it assisted in keeping methanol levels low. Thus, enzyme entrapment can occur in a variety of ways, though the purpose is found to be excellent in each and every instance.

6.4.3 Affinity binding and metal-link immobilization

This section can be devoted to biological interactions between proteins or small biomolecules and proteins because these types of binding offer high selectivity and specificity in terms of size, shape, and location type. As a result, this is also referred to as bio-affinity or biomolecule interactions. The chelation of histidine residue with several transition metal elements is one of the most prominent examples of such affinity [e.g., Co(II), Cu(II) or Ni(II)] [52]. This metal-supported histidine residue tag is widely used for protein immobilization. Despite such sincere attachment of the desired protein, this is essentially to be purified prior to performance because several proteins have similar structures and can easily bind with the metal portion. It is worth noting that antibodies can also perform the task of protein immobilization due to their high selectivity in binding with enzymes.

All of these techniques for immobilizing enzymes are based on peripheral molecular interactions, including both physical and chemical bonding. However, it is easy to conclude from the preceding discussion that, while each interaction has advantages and disadvantages, chemical bindings are much more specific and stronger than the adsorption process. Because physical binding is a reversible process, whereas chemical binding is not, the former synergy is more capable of preventing enzyme denaturation or contamination of the reaction medium.

6.5 Nanoparticles as immobilization matrix

For enzyme immobilization, various nanostructures have been investigated. Different immobilization techniques can be used depending on the morphology, topology, and chemical and physical properties of the nanomaterials. We present a concise description of various enzyme immobilized nanostructures in this paper.

6.5.1 Carbon-based nanomaterials

Recent research shows that carbon nanomaterials are suitable for enzyme immobilization. Carbon nanostructures include fullerene, GO, CNTs, nanodiamond, graphitic quantum dots, carbon dots (CDs), and other zero, one, and two-dimensional nanostructures. It is difficult to achieve effective immobilization of enzymes on carbon nanomaterials due to structural diversity. Nonetheless, recent advances are promising, paving the way for various applications of enzyme immobilized carbon-based nanomaterials [2].

Covalent and non-covalent conjugation are both possible with CNT. Sidewall oxidation generates $-COOH$ groups, which are then activated by carbodiimide for covalent immobilization of enzymes on CNTs. These activated $-COOH$ groups can be cross-linked with enzyme $-NH_2$ groups. Using EDC/NHS coupling, Liu et al. demonstrated effective covalent conjugation of lysosome with $-COOH$ activated CNT [53] (Fig. 6.3A). Lysosome is an enzyme that promotes wound healing. Non-conventional immobilization with CNT, on the other hand, is thought to be more promising because it preserves the structure of the enzyme intake, ensuring high biological activity. Because CNT is a two-dimensional nanostructure, it has a large available surface area and can provide sites for enzyme adsorption.

Because of their favorable properties such as high surface area, high thermal and chemical stability, and notable biodegradability, graphene-based nanomaterials can also be used for enzyme immobilization. Graphene is made up of an array of combinable aromatic pi structures with monoatomic structures, with all C atoms being sp^2 hybridized. The fundamental issue with enzyme immobilization in graphitic structures is a lack of proper functionality that can bind with enzymes.

This barrier can be overcome by modifying graphitic structures. It contains nanoscale objects and macromolecules on the surface of the oxidative form, GO, which can act as a site for both

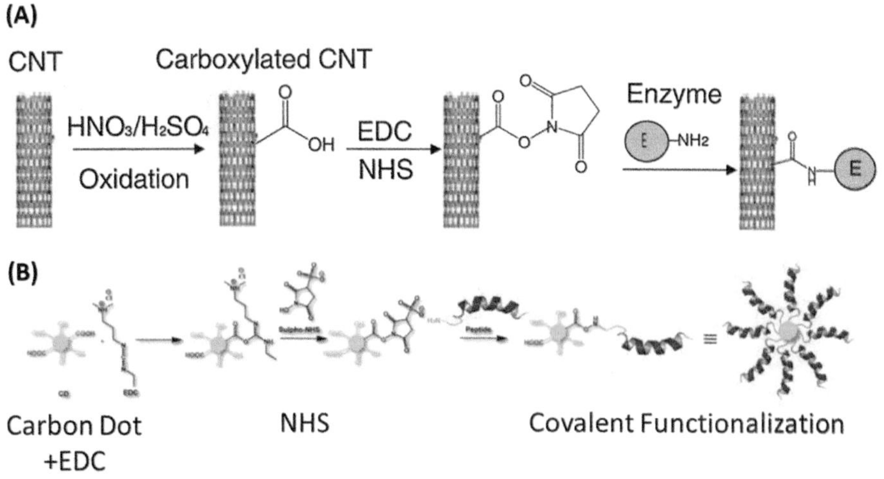

FIGURE 6.3

(A) Covalent immobilization of enzyme on carboxylated carbon nanotube [54]; (B) Covalent immobilization of enzyme on carbon dots [55].

covalent and non-covalent functionalization of enzymes. Covalent functionalization of graphitic nanomaterials is primarily based on three types of reactions: (1) ring opening reaction of epoxy at the basal plan, (2) cycloaddition or diazonium reactions of RGO, and (3) nucleophilic reaction of the −COOH group at the edge of GO sheets [56]. The covalent immobilization of enzyme onto GO sheet is depicted in Fig. 6.4. The enzyme was first immobilized on GO by activating −COOH groups with thionyl chloride and then functionalizing octadecylamine. A large number of reports on enzyme immobilized graphene derivatives are available. Carbodiimides/NHS cross-linkers are the most commonly used cross-linkers in the majority of these reports. Glutaraldehyde is also an effective cross-linker [2]. Non-covalent approaches to enzyme immobilization on graphitic surfaces, on the other hand, rely on secondary interactions such as pi−pi interactions, H-bonding, van der Walls interactions, and cation-pi interactions [57,58]. A novel hybrid approach that relies on both covalent and non-covalent functionalization is also known. Glutamic acid dehydrogenase and glucose oxidase were effectively immobilized to graphitic structures using a hybrid approach involving pi-pi stacking and NHS cross-linker.

CDs, as opposed to CNT and graphene nanomaterials, are more easily immobilized by enzymes. CDs, the newest member of the carbon nano-family, are made up of a sp^2 hybridized graphitic core and oxygen functional groups at the surface. The presence of surface groups, which can be used for enzyme conjugation without any prior modification, is the most obvious advantage of CDs. Another advantage of CDs for functionalization of various biomolecules is their high aqueous solubility. Covalent conjugation is most commonly used for enzyme immobilization with CDs. EDC/NHS coupling is a simple method for enzyme immobilization. Gogoi et al. successfully conjugated osteogenic peptides with CDs via EDC/NHS coupling [55]. This is depicted in

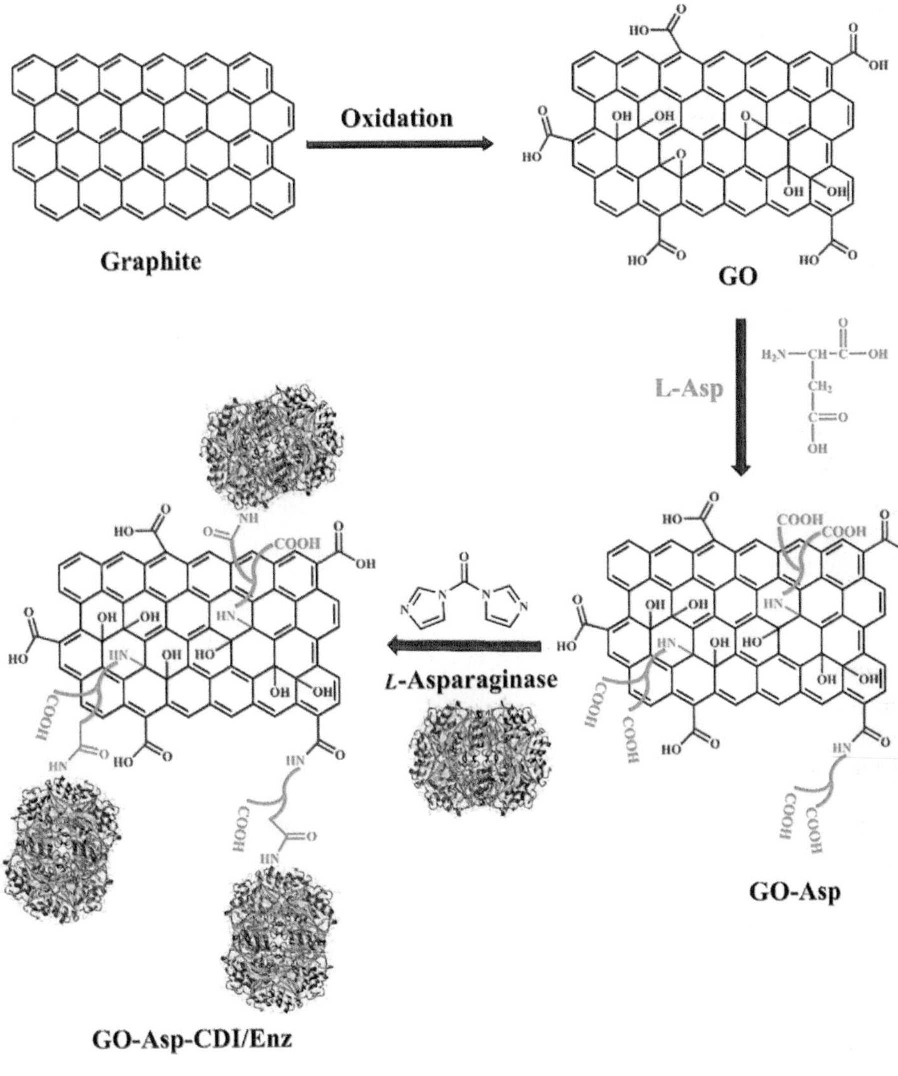

FIGURE 6.4

Covalent immobilization of enzyme on graphene oxide [59].

Fig. 6.4B. This bio-nanohybrid was embedded in a waterborne polyurethane matrix that served as a scaffold for bone tissue regeneration. The effectiveness of peptide immobilized CDs in bone tissue engineering was demonstrated in both in vivo and in vitro studies. Gogoi and Khan reported covalent conjugation of Troponin T antibody, a cardiovascular biomarker [60]. Such immobilized CD was used as a probe for the sensitive detection of Troponin T.

6.5.2 Metal-based nanomaterials

Metals are also suitable for enzyme immobilization. Metallic nanoparticles can be used for enzyme conjugation in the form of metal, metal oxide, or metal hydroxide. Au-NPs are the most studied metallic nanoparticles for enzyme immobilization. Au, as a nanomaterial, has properties such as biocompatibility, thermal and chemical stability, and a large surface area. The most significant advantage is that enzyme immobilization on Au-NPs can be approached in a variety of ways. For enzyme immobilization, Au- NPs can be easily functionalized/modified. It is possible to use naked Au-NPs. The most common modification of Au-NP is carboxylic modification, which is accomplished by using a modifier containing both −COOH and thiol groups. Thiol groups can interact with Au-NPs, freeing up −COOH groups for enzyme immobilization. For enzyme immobilization with carboxylic modified Au-NPs, methods such as carbodiimide/NHS coupling are easily applicable. For non-covalent immobilization of Au-NPs, different physical adsorptions have been reported. Physical adsorption on Au-NPs is primarily electrostatic, hydrophobic, or hydrophilic. Oxides of Fe, Ti, Cu, and Zn have been reported. In the following section, the oxides of Fe and Cu are discussed separately as MNPs. TiO_2 NPs can be immobilized with enzymes through covalent conjugation, cross-linking, or physical adsorption [61−63]. TiO_2 NPs' surfaces can be chemically modified. This modification enables various chemical functionalities, including covalent and cross-linking conjugations. Similarly, nanomaterials can be charged, which aids in the electrostatic adsorption of enzyme on the TiO_2 nano-surface. Immobilization of ZnO NPs can be accomplished through electrostatic adsorption. ZnO NPs can be positively charged in immobilization solution by maintaining a high isoelectric point (around 9.5) [2]. As a result, an enzyme with a low isoelectric point that is negatively charged can be easily immobilized via electrostatic adsorption. Metal hydroxides are also used to immobilize enzymes. Layered double hydroxides (LDHs) have recently gained a lot of attention. Electrostatic attraction, ion exchange, van der Waals interaction, or hydrogen bonding can all be used to functionalize enzymes on the LDH surface [64]. Immobilization of Lysozyme has been reported on LDH with profound antimicrobial activity for which exhibits excellent wound healing characteristics [65].

6.5.3 Magnetic nanoparticles

NPs' rise as an emerging candidate in the domain of enzyme immobilization [66,67]. The incorporation of magnetic properties within these NPs has increased the efficiency of materials at the nanoscale even further. In addition to having a high catalytic activity due to a large surface-to-volume ratio, NPs that can use a magnetic field can vastly increase binding capacity while also increasing specificity towards catalytic activity. As a result, MNPs can immobilize enzymes by encapsulating them with magnetic properties, and the advancement of magnetic field susceptibility has provided an added feature of enzyme recovery, eliminating contamination in the final adduct. Interestingly, these NP categories have already demonstrated promising results in biomaterials. It is worth noting that, in addition to the remarkable stability and possible modulation of catalytic specificity achieved by enzymes while their mobility is restricted by MNPs, this elite group of tiny particles can provide cost effectiveness as well as greater diffusibility to the enzymes by lowering transfer resistance [68].

It is obvious that transition metal-assisted/-doped/-coated NPs, with special reference to silica-based composites, can be used to fabricate NPs with magnetic properties because the majority of the elements in the series can demonstrate magnetic behavior. In this regard, it is first necessary to supplement the capsule's biocompatibility towards the enzyme; thus, the publication of iron-silica composites as biocompatible enzyme immobilizers was undeniably a significant contribution in this domain [69]. Later, pronounced catalytic activity under the umbrella of biocompatible matrix of nickel suffused silica composites has reported the immobilization of enzyme along with the advantage of [70] guarding it from get deactivated in both lower and higher pH. It was also noticed that such unprecedented activity of these immobilized enzymes could be achieved due to alteration of configuration of enzymes [71,72] and besides that the thermostability of such immobilized enzymes were reported to be extraordinarily better than those of free enzymes [73]. In similar research [74], a composite made up of silica (40 nm) coated MNPs (10 nm) (SiMN), charged with Cu^{2+} by iminodiacetic acid (multidentate ligand) (Cu^{2+} charged-SiMN), was used for the immobilization of *Bacillus stearothermophilus* L1 lipase, in which the comparison was done with microporous silica gel (115 μm) (SG and Cu^{2+} charged-SG). Interestingly, though Cu^{2+} was used in both the cases as the charged particle with a little variation in input ratio (1:1.14 and 1:1.99 successively), the specific activity of the composite holding MNPs showed unparalleled results (Cu^{2+} charged-SiMN> SiMN> Cu^{2+} charged-SG> SG). It is also worth mentioning that Cu^{2+} charged-SiMN immobilized lipase could retain their activity up to 70% till 5th repetitions, while others lost their efficiency significantly in the similar experimental ambience.

CRL demonstrated improved immobility with Fe_2O_3 NPs and smooth separation of the substrate or enzyme by simply placing the system in a magnetic environment with a magnet [75,76].

The establishment of MNP immobilization has become so high that several enzymes have found their utility through immobilization in various biological applications, *viz*, Trypsin and chymotrypsin on Fe and Au [77,78] etc. In fact, Glucose oxidase, peroxidase and streptokinase were also immobilized over MNPs [79−81].

Binding of cholesterol oxidase via carbodiimide activation with Fe_3O_4 NPs reveals that MNPs provide both superior activity and stability as, with around 100% binding efficiency, these NPs provide the platform for tremendous efficacy to these immobilized enzymes than that of free ones along with temperature, substrate concentration, and pH [82]. This incredible advancement has significantly improved enzyme efficiency in the entrapped form by MNPs in the domain of clinical and biological applications. Furthermore, the improved thermal stability of immobilized *Pseudomonas cepacia* lipase over MNPs with carbodiimide activation demonstrated biodiesel production capability, with binding efficiency and activity recovery of 90% and 70%, respectively [83]. Furthermore, laccase in the immobilized form has demonstrated excellent storage, operational, and thermal stability in the field of enzyme catalyzed biosensor for fiber optics [84].

The enhanced effectivity of an enzyme when embraced by MNPs can be attributed to thermal and kinetic stability, as well as increased activity due to quality bonding between the host and the guest. As a result, much scientific work in the field of biotechnological applications has been directed toward enzyme immobilization by MNPs to achieve better developments in terms of both qualitative and quantitative results.

6.5.4 Other nanomaterials

Other nanomaterials known for enzyme immobilization include carbon-based, metallic, and MNPs. We can refer to silica-based nanomaterials in this context. Silica-based nanomaterials can be used for both covalent and non-covalent enzyme conjugation. Silica NPs can be chemically modified by using a silane blocking agent. Through such modifications, different chemical functionalities such as amine, carboxylic, and thiol groups can be easily introduced on the silica surface. These functional groups serve as the active site for enzyme covalent immobilization [2]. Physical adsorption can also be used to immobilize silica-based nanomaterials with enzymes. Similarly, cellulose nanofibers (CNF) can be used to immobilize enzymes [85]. CNF is regarded as an appealing option because it can adapt to both hydrophobic and hydrophilic conditions as required by the application. Enzyme covalent immobilization on CNF is a possibility. CNF's surface is primarily composed of −OH groups. However, for enzyme immobilization, these −OH groups must first be activated. Cyanogen bromide is the most commonly used coupling agent [85]. In the activated state, −OH groups can couple to the enzyme via the N-terminus, allowing for effective covalent conjugation. Other coupling agents, such as cyanuric chloride and carbodiimide, can be used in addition to cyanogen bromide.

6.6 Conclusions

The catalytic activity of enzymes smoothens lifesaving biochemical reactions, and through this projection, the importance of enzymatic processes inside living organisms can be easily understood. Modern state-of-the-art research reveals that the elegancy of enzymes can be further advanced by integrating enzymes with a platform (e.g., various breeds of NPs, viz., MNPs, CDs, etc.) so that the mobilization of the enzymes can be restricted. In terms of immobilization, the techniques used to restrict the motion of enzymes must be unique, as the process of entrapment of the enzyme must be accurate and specific in nature, and biocompatibility must be addressed. As a result, many of the techniques discussed in this review are scrutinized. There is conclusive evidence that all of the processes that have been approved for the incorporation of enzymes are fully capable of meeting the requirements of various applications, whether it is a physical adsorption technique (e.g., hydrogen bonding) or ionic or covalent bonding. This chapter also discusses a new category of binding support that is an enhancement to this program, namely, NPs. Because of their high surface-to-volume ratio, these tiny materials can provide a massive amount of surface for enzymes to adhere to. One of the most unusual applications in this context is the use of CDs as enzyme immobilizers. CD, as the primary building constituent element of CDs, creates the scaffolding for the enzyme, making it naturally biocompatible. Surprisingly, we discovered that the goal of immobilizing enzyme with NPs can be strengthened further by tuning the magnetic nature within the NPs, and promising results have already been reported when the NPs are armored with magnetic susceptibility. Thus, it is easily recognized that enzyme immobilization techniques are based on various physico-chemical techniques, though the inclusion of NPs, with appropriateness, could be a game changer in the field of enzyme encapsulation with the inclusion of improved binding ability as well as biocompatibility throughout the process.

Acknowledgment

Jejiron M. Baruah is grateful to FIRST scheme, North Lakhimpur College (Autonomous) for financial assistance (Ref No.: NLC/RC/MRP/2020/01).

References

[1] L. Alvarado-Ramírez, M. Rostro-Alanis, J. Rodríguez-Rodríguez, C. Castillo-Zacarías, J.E. Sosa-Hernández, D. Barceló, et al., Exploring current tendencies in techniques and materials for immobilization of laccases—a review, Int. J. Biol. Macromol. 181 (2021) 683−696.

[2] D.M. Liu, C. Dong, Recent advances in nano-carrier immobilized enzymes and their applications, Process. Biochem. 92 (2020) 464−475.

[3] M. Bilal, N. Hussain, J.H.P. Américo-Pinheiro, Y.Q. Almulaiky, H.M. Iqbal, Multi-enzyme co-immobilized nano-assemblies: bringing enzymes together for expanding bio-catalysis scope to meet biotechnological challenges, Int. J. Biol. Macromol. 186 (2021) 735−749.

[4] S. Sharma, P.S. Gupta, S.K. Arya, A. Kaur, 2022. Enzyme immobilization: implementation of nanoparticles and an insight into polystyrene as the contemporary immobilization matrix. Process Biochem.

[5] R.A. Sheldon, S. van Pelt, Enzyme immobilisation in biocatalysis: why, what and how, Chem. Soc. Rev. 42 (15) (2013) 6223−6235.

[6] W. Aehle (Ed.), Enzymes in Industry: Production and Applications, John Wiley & Sons, 2007.

[7] B.M. Berna, F. Batista, Enzyme immobilization literature survey methods in biotechnology: immobilization of enzymes and cells, 2006.

[8] S.A. Costa, H.S. Azevedo, R.L. Reis, Enzyme immobilization in biodegradable polymers for biomedical applications, 2005.

[9] J.M. Guisan, Immobilization of enzymes as the 21st century begins, Immobil. Enzymes Cell (2006) 1−13.

[10] R.A. Sheldon, Enzyme immobilization: the quest for optimum performance, Adv. Synth. Catalysis 349 (8-9) (2007) 1289−1307.

[11] S. Nisha, S. Arun Karthick, N. Gobi, A review on methods, application prop. Immobilized Enzyme, Chem. Sci. Rev. Lett. 1 (3) (2012) 148−155.

[12] L. Cao, Carrier-Bound Immobilized Enzymes: Principles, Applications and Design, Wiley-VCH Verlag GmbH & Co. KGaA, Weinheim, Germany, 2005, pp. ISBN 3−527-31232-3.

[13] K. Hernandez, R. Fernandez-Lafuente, Control of protein immobilization: coupling immobilization and site-directed mutagenesis to improve biocatalyst or biosensor performance, Enzyme Microb. Technol. 48 (2) (2011) 107−122.

[14] B. Krajewska, Application of chitin-and chitosan-based materials for enzyme immobilizations: a review, Enzyme Microb. Technol. 35 (2−3) (2004) 126−139.

[15] J.R. Cherry, A.L. Fidantsef, Directed evolution of industrial enzymes: an update, Curr. Opin. Biotechnol. 14 (4) (2003) 438−443.

[16] J.M. Nelson, E.G. Griffin, Adsorption of invertase, J. Am. Chem. Soc. 38 (5) (1916) 1109−1115.

[17] N. Grubhofer, L. Schleith, Modifizierte ionenaustauscher als spezifische adsorbentien, Naturwissenschaften 40 (19) (1953) 508. −508.

[18] T.H. Richardson, X. Tan, G. Frey, W. Callen, M. Cabell, D. Lam, et al., A novel, high performance enzyme for starch liquefaction: discovery and optimization of a low pH, thermostable α-amylase, J. Biol. Chem. 277 (29) (2002) 26501−26507.

[19] M. Sarno, M. Iuliano, Immobilization of horseradish peroxidase on Fe_3O_4/au_go nanoparticles to remove 4-chlorophenols from waste water, Chem. Eng. Trans. 73 (2019) 217−222.

[20] L.R. Wilken, Z.L. Nikolov, Recovery and purification of plant-made recombinant proteins, Biotechnol. Adv. 30 (2) (2012) 419−433.

[21] P.J. Worsfold, Iupac classification and chemical characteristics of immobilized enzymes, Pure Appl (1995).

[22] Y.M. Elçin, Encapsulation of urease enzyme in xanthan-alginate spheres, Biomaterials 16 (15) (1995) 1157−1161.

[23] A.B. Vermelho, C.T. Supuran, J.M. Guisan, Microbial enzyme: applications in industry and in bioremediation, Enzyme Res. (2012) 2012.

[24] S. Datta, L.R. Christena, Y.R.S. Rajaram, Enzyme immobilization: an overview on techniques and support materials, 3 Biotech. 3 (2013) 1.

[25] G.D. Kibarer, G. Akovali, Optimization studies on the features of an activated charcoal-supported urease system, Biomaterials 17 (15) (1996) 1473−1479.

[26] A.A. Homaei, R. Sariri, F. Vianello, R. Stevanato, Enzyme immobilization: an update, J. Chem. Biol. 6 (4) (2013) 185−205.

[27] M.C. Franssen, P. Steunenberg, E.L. Scott, H. Zuilhof, J.P. Sanders, Immobilised enzymes in biorenewables production, Chem. Soc. Rev. 42 (15) (2013) 6491−6533.

[28] E.M. Anderson, K.M. Larsson, O. Kirk, One biocatalyst−many applications: the use of Candida antarctica B-lipase in organic synthesis, Biocatal. Biotransfor. 16 (3) (1998) 181−204.

[29] S. Warwel, F. Brüse, C. Demes, M. Kunz, M. Rüsch gen, Klaas. J. Mol. Catal. B: Enzym. 1 (1995) 29−35.

[30] S. Miao, Z. Liu, H. Ma, B. Han, J. Du, Z. Sun, et al., Synthesis and characterization of mesoporous aluminosilicate molecular sieve from K-feldspar, Microporous Mesoporous Mater. 83 (1−3) (2005) 277−282.

[31] U.H. Zaidan, M.B.A. Rahman, M. Basri, S.S. Othman, R.N.Z.R.A. Rahman, A.B. Salleh, Silylation of mica for lipase immobilization as biocatalysts in esterification, Appl. Clay Sci. 47 (3−4) (2010) 276−282.

[32] J. Lu, K. Nie, F. Xie, F. Wang, T. Tan, Enzymatic synthesis of fatty acid methyl esters from lard with immobilized Candida sp. 99−125, Process. Biochem. 42 (9) (2007) 1367−1370.

[33] E. Agostinelli, F. Belli, G. Tempera, A. Mura, G. Floris, L. Toniolo, et al., Polyketone polymer: a new support for direct enzyme immobilization, J. Biotechnol. 127 (4) (2007) 670−678.

[34] M. Yakup Arica, H. Soydogan, G. Bayramoglu, Reversible immobilization of Candida rugosa lipase on fibrous polymer grafted and sulfonated p (HEMA/EGDMA) beads, Bioprocess. Biosyst. Eng. 33 (2) (2010) 227−236.

[35] S. Gao, W. Wang, Y. Wang, G. Luo, Y. Dai, Influence of alcohol treatments on the activity of lipases immobilized on methyl-modified silica aerogels, Bioresour. Technol. 101 (19) (2010) 7231−7238.

[36] A. Salis, V. Solinas, M. Monduzzi, Wax esters synthesis from heavy fraction of sheep milk fat and cetyl alcohol by immobilised lipases, J. Mol. Catal. B: Enzymatic 21 (4−6) (2003) 167−174.

[37] E.A.J. Al-Mulla, W.M.Z.W. Yunus, N.A.B. Ibrahim, M.Z.A. Rahman, Enzymatic synthesis of fatty amides from Palm Olein, J. Oleo Sci. 59 (2) (2010) 59−64.

[38] R.A. Khan, D. Perin, E. Murano, M. Bergamin, WO2012/053017, 26 April 2012.

[39] N.R. Sonare, V.K. Rathod, Transesterification of used sunflower oil using immobilized enzyme, J. Mol. Catal. B: Enzymatic 66 (1−2) (2010) 142−147.

[40] M. Cárdenas-Fernández, C. López, G. Alvaro, J. López-Santín, Immobilized L-aspartate ammonia-lyase from Bacillus sp. YM55-1 as biocatalyst for highly concentrated L-aspartate synthesis, Bioprocess. Biosyst. Eng. 35 (8) (2012) 1437−1444.

[41] M.M. Zheng, Y. Lu, L. Dong, P.M. Guo, Q.C. Deng, W.L. Li, et al., Immobilization of Candida rugosa lipase on hydrophobic/strong cation-exchange functional silica particles for biocatalytic synthesis of phytosterol esters, Bioresour. Technol. 115 (2012) 141−146.

[42] X.T. Peng, Z.G. Shi, Y.Q. Feng, Rapid and high-throughput determination of melamine in milk products and eggs by full automatic on-line polymer monolith microextraction coupled to high-performance liquid chromatography, Food Anal. Methods 4 (3) (2011) 381−388.

[43] T. Sasaki, T. Kajino, B. Li, H. Sugiyama, H. Takahashi, New pulp biobleaching system involving manganese peroxidase immobilized in a silica support with controlled pore sizes, Appl. Environ. Microbiol. 67 (5) (2001) 2208−2212.

[44] T.M. Lammens, D. De Biase, M.C. Franssen, E.L. Scott, J.P. Sanders, The application of glutamic acid α-decarboxylase for the valorization of glutamic acid, Green. Chem. 11 (10) (2009) 1562−1567.

[45] B.L. Tee, G. Kaletunç, Immobilization of a thermostable α-amylase by covalent binding to an alginate matrix increases high temperature usability, Biotechnol. Prog. 25 (2) (2009) 436−445.

[46] T.K. Hakala, T. Liitiä, A. Suurnäkki, Enzyme-aided alkaline extraction of oligosaccharides and polymeric xylan from hardwood kraft pulp, Carbohydr. Polym. 93 (1) (2013) 102−108.

[47] M.T. Reetz, Entrapment of biocatalysts in hydrophobic sol-gel materials for use in organic chemistry, Adv. Mater. 9 (12) (1997) 943−954.

[48] M.G. Bellino, A.E. Regazzoni, Immobilization of enzymes into self-assembled iron (III) hydrous oxide nanoscaffolds: a bio-inspired one-pot approach to hybrid catalysts, Appl. Catal. A: Gen. 408 (1−2) (2011) 73−77.

[49] R. Das, D. Sen, A. Sarkar, S. Bhattacharyya, C. Bhattacharjee, A comparative study on the production of galacto-oligosaccharide from whey permeate in recycle membrane reactor and in enzymatic batch reactor, Ind. Eng. Chem. Res. 50 (2) (2011) 806−816.

[50] A. Kohl, T. Bruck, J. Gerlach, M. Zavrel, L. Rocher, M. Kraus, WO2012/001102, 2012.

[51] C. Lopez, M.T. Moreira, G. Feijoo, J.M. Lema, Dye decolorization by manganese peroxidase in an enzymatic membrane bioreactor, Biotechnol. Prog. 20 (1) (2004) 74−81.

[52] C.S. Rao, T. Sathish, P. Ravichandra, R.S. Prakasham, Characterization of thermo-and detergent stable serine protease from isolated Bacillus circulans and evaluation of eco-friendly applications, Process. Biochem. 44 (3) (2009) 262−268.

[53] Y. Liu, J.V. Edwards, N. Prevost, Y. Huang, J.Y. Chen, Physico-and bio-activities of nanoscale regenerated cellulose nonwoven immobilized with lysozyme, Mater. Sci. Eng. C. 91 (2018) 389−394.

[54] W. Feng, P. Ji, Enzymes immobilized on carbon nanotubes, Biotechnol. Adv. 29 (6) (2011) 889−895.

[55] S. Gogoi, S. Maji, D. Mishra, K.S.P. Devi, T.K. Maiti, N. Karak, Nano-bio engineered carbon dot-peptide functionalized water dispersible hyperbranched polyurethane for bone tissue regeneration, Macromol. Biosci. 17 (3) (2017) 1600271.

[56] M. Adeel, M. Bilal, T. Rasheed, A. Sharma, H.M. Iqbal, Graphene and graphene oxide: functionalization and nano-bio-catalytic system for enzyme immobilization and biotechnological perspective, Int. J. Biol. Macromol. 120 (2018) 1430−1440.

[57] X. Yang, X. Zhang, Z. Liu, Y. Ma, Y. Huang, Y. Chen, High-efficiency loading and controlled release of doxorubicin hydrochloride on graphene oxide, J. Phys. Chem. C. 112 (45) (2008) 17554−17558.

[58] C.H. Lu, H.H. Yang, C.L. Zhu, X. Chen, G.N. Chen, A graphene platform for sensing biomolecules, Angew. Chem. 121 (26) (2009) 4879−4881.

[59] M. Monajati, S. Borandeh, A. Hesami, D. Mansouri, A.M. Tamaddon, Immobilization of L-asparaginase on aspartic acid functionalized graphene oxide nanosheet: enzyme kinetics and stability studies, Chem. Eng. J. 354 (2018) 1153−1163.

[60] S. Gogoi, R. Khan, Fluorescence immunosensor for cardiac troponin T based on Förster resonance energy transfer (FRET) between carbon dot and MoS_2 nano-couple, Phys. Chem. Chem. Phys. 20 (24) (2018) 16501−16509.

[61] S. Shoabargh, A. Karimi, G. Dehghan, A. Khataee, A hybrid photocatalytic and enzymatic process using glucose oxidase immobilized on TiO_2/polyurethane for removal of a dye, J. Ind. Eng. Chem. 20 (5) (2014) 3150−3156.

[62] J. Li, X. Li, Q. Zhao, Z. Jiang, M. Tadé, S. Wang, et al., Polydopamine-assisted decoration of TiO_2 nanotube arrays with enzyme to construct a novel photoelectrochemical sensing platform, Sens. Actuators B: Chem. 255 (2018) 133−139.

[63] W. Zhuang, Y. Zhang, J. Zhu, R. An, B. Li, L. Mu, et al., Influences of geometrical topography and surface chemistry on the stable immobilization of adenosine deaminase on mesoporous TiO_2, Chem. Eng. Sci. 139 (2016) 142−151.

[64] Y. Wang, C. Liu, Y. Zhang, B. Zhang, J. Liu, Facile fabrication of flowerlike natural nanotube/layered double hydroxide composites as effective carrier for lysozyme immobilization, ACS Sustain. Chem. Eng. 3 (6) (2015) 1183−1189.

[65] Z. Wang, H. Yu, K. Ma, Y. Chen, X. Zhang, T. Wang, et al., Flower-like surface of three-metal-component layered double hydroxide composites for improved antibacterial activity of lysozyme, Bioconjugate Chem. 29 (6) (2018) 2090−2099.

[66] J.M. Baruah, J. Narayan, Aqueous-mediated synthesis of group IIB-VIA semiconductor quantum dots: challenges and developments, Solar Cells (2020).

[67] J.M. Baruah, S. Kalita, J. Narayan, Green chemistry synthesis of biocompatible ZnS quantum dots (QDs): their application as potential thin films and antibacterial agent, Int. Nano Lett. 9 (2) (2019) 149−159.

[68] S.A. Ansari, Q. Husain, Potential applications of enzymes immobilized on/in nano materials: a review, Biotechnol. Adv. 30 (3) (2012) 512−523.

[69] T.R. Besanger, Y. Chen, A.K. Deisingh, R. Hodgson, W. Jin, S. Mayer, et al., Screening of inhibitors using enzymes entrapped in sol-gel-derived materials, Anal. Chem. 75 (10) (2003) 2382−2391.

[70] R.S. Prakasham, G.S. Devi, C.S. Rao, V.S.S. Sivakumar, T. Sathish, P.N. Sarma, Nickel-impregnated silica nanoparticle synthesis and their evaluation for biocatalyst immobilization, Appl. Biochem. Biotechnol. 160 (7) (2010) 1888−1895.

[71] S.V. Ramakrishna, R.S. Prakasham, Microbial fermentations with immobilized cells, Curr. Sci. (1999) 87−100.

[72] B. Srinivasulu, R.S. Prakasham, A. Jetty, S. Srinivas, P. Ellaiah, S.V. Ramakrishna, Neomycin production with free and immobilized cells of Streptomyces marinensis in an airlift reactor, Process. Biochem. 38 (4) (2002) 593−598.

[73] R.S. Prakasham, G.S. Devi, K.R. Laxmi, C.S. Rao, Novel synthesis of ferric impregnated silica nanoparticles and their evaluation as a matrix for enzyme immobilization, J. Phys. Chem. C. 111 (10) (2007) 3842−3847.

[74] H. Kim, H.S. Kwon, J. Ahn, C.H. Lee, I.S. Ahn, Evaluation of a silica-coated magnetic nanoparticle for the immobilization of a His-tagged lipase, Biocatal. Biotransfor. 27 (4) (2009) 246−253.

[75] A. Dyal, K. Loos, M. Noto, S.W. Chang, C. Spagnoli, K.V. Shafi, et al., Activity of Candida rugosa lipase immobilized on γ-Fe_2O_3 magnetic nanoparticles, J. Am. Chem. Soc. 125 (7) (2003) 1684−1685.

[76] H.M. Gardimalla, D. Mandal, P.D. Stevens, M. Yen, Y. Gao, Superparamagnetic nanoparticle-supported enzymatic resolution of racemic carboxylates, Chem. Commun. 35 (2005) 4432−4434.

[77] M.J. Hansen, N. Natale, P. Kornacki, A.J. Paszczynski, Conjugating magnetic nanoparticles for application in health and environmental research, in: 48th annual Idaho Academy of Science Meeting and Symposium, Moscow, Idaho, 2006, June, Vol. 4.

[78] B.J. Jordan, R. Hong, B. Gider, J. Hill, T. Emrick, V.M. Rotello, Stabilization of α-chymotrypsin at air−water interface through surface binding to gold nanoparticle scaffolds, Soft Matter 2 (7) (2006) 558−560.

[79] L.M. Rossi, A.D. Quach, Z. Rosenzweig, Glucose oxidase–magnetite nanoparticle bioconjugate for glucose sensing, Anal. Bioanal. Chem. 380 (4) (2004) 606–613.

[80] E.G.R. Fernandes, A.A.A.D. Queiroz, G.A. Abraham, J.S. Román, Antithrombogenic properties of bioconjugate streptokinase-polyglycerol dendrimers, J. Mater. Sci.: Mater. Med. 17 (2) (2006) 105–111.

[81] M. Koneracka, P. Kopčanský, M. Timko, C.N. Ramchand, A. De Sequeira, M. Trevan, Direct binding procedure of proteins and enzymes to fine magnetic particles, J. Mol. Catal. B: Enzymatic 18 (1–3) (2002) 13–18.

[82] G.K. Kouassi, J. Irudayaraj, G. McCarty, Examination of cholesterol oxidase attachment to magnetic nanoparticles, J. Nanobiotechnol. 3 (1) (2005) 1–9.

[83] M. Kai-Ho, Y. Chi-Yang, I. Kuan, L. Shiow-Ling, Some properties of immobilized Pseudomonas cepacia lipase onto magnetic nanoparticles, J. Biosci. Bioeng. 108 (2009) S112–S113.

[84] J. Huang, C. Liu, H. Xiao, J. Wang, D. Jiang, G.U. Erdan, Zinc tetraaminophthalocyanine-Fe_3O_4 nanoparticle composite for laccase immobilization, Int. J. Nanomed. 2 (4) (2007) 775.

[85] S. Sulaiman, M.N. Mokhtar, M.N. Naim, A.S. Baharuddin, A. Sulaiman, A review: potential usage of cellulose nanofibers (CNF) for enzyme immobilization via covalent interactions, Appl. Biochem. Biotechnol. 175 (4) (2015) 1817–1842.

Kinetic, stability, and activity of the nanoparticles-immobilized enzymes

Marcela Slovakova

Department of Biological and Biochemical Sciences, Faculty of Chemical Technology, University of Pardubice, Pardubice, Czech Republic

List of abbreviations

K_M	Michaelis constant
k_{cat}	enzyme turnover number
NP	nanoparticles
PAA	poly(acrylic acid)
PGA	polyglycolic acid
PLA	polylactic acid
PLGA	polylactic-co-glycolic acid
PVA	polyvinyl alcohol
V_{max}	maximum reaction rate

7.1 Introduction

The use of enzymes immobilized on nanoparticles (NPs) in antimicrobial dressings and wound care products differs from other bioanalytical operations involving immobilized enzymes in that enzymes are not reused for multiple reaction cycles. It is their sustained activity and long-term stability in vivo under specific conditions. The properties of the immobilized enzyme have a significant impact on the type of carrier, nanomaterial, and immobilization methodology used. It was possible to evaluate how the binding of enzymes to carriers affected their activity when the enzyme kinetics of immobilized enzymes with antimicrobial effects were studied. Antimicrobial hydrolytic enzymes degrade the main structural components of bacterial and fungal cell walls, whereas antimicrobial oxidoreductases work by forming reactive molecules on the spot. Another healing effect of immobilized hydrolytic enzymes is that they were developed as protease supports for necrotic tissue degradation, also in burns [1−3] and for various tissue regeneration [4]. It promotes the formation and migration of keratinocytes and fibroblasts and aids in the restoration of wound base conditions by releasing stimulant peptide fragments. Payne et al. [5] showed that collagenase and papain urea, used for debridement, even in cases of high bacterial loads, were beneficial and safe. Earlier, Drago et al. [1] reported on enzymatic scarolysis in which dead residues were easily removed while living tissues were preserved. Proteolytic enzyme ointments are already being used to break

Antimicrobial Dressings. DOI: https://doi.org/10.1016/B978-0-323-95074-9.00001-4

down inanimate tissue [6,7]. The immobilization of enzymes on a solid support allows the enzyme's action to be localized while also increasing its stability after isolation from the natural environment and during various wound processes. When compared to native enzymes, immobilized enzymes eliminate the need for continuous enzyme replacement [8]. During the epithelialization phase of healing, local application of enzymes bound to various solid carriers promotes cell migration [6,9,10]. Today, a well-established area of polymeric nanofibers with fixed proteolytic enzymes has shown the removal of necrotic deposits in the wound that did not require surgical wound cleaning [11−13]. In 2020, Siritapetawee et al. [14] described an efficient method for fixing NP-carrying enzymes on cotton gauze. Furthermore, silicone protease emulsions and nanoemulsions have been described [15,16], numerous polymeric carriers [17−19] and hydrogels [20].

7.2 Active nanoparticles-immobilized enzymes

Enzymes used in therapies have shown great application potential in medicine in the 21st century [7,21]. Wound healing and antimicrobial effects were combined with extended enzyme stability when enzymes were immobilized on nanomaterials. These were microbial and other enzyme sources [22]. The enzymes used for antimicrobial and healing therapies fall into three categories: (1) hydrolytic and proteolytic enzymes that can degrade the main structural components of bacteria and/or fungi cell walls, (2) proteolytic enzymes that have healing effects by degrading necrotic tissue, and (3) antimicrobial oxidoreductases that act by forming reactive molecules on site. The three groups of enzymes, listed in Table 7.1, were successfully immobilized in NP, and further characterized along with their antimicrobial area.

7.3 Immobilized antimicrobial hydrolytic enzymes

With the rise of antibiotic resistance, a new class of antibacterials known as enzybiotics has been studied. These enzymes, such as lysins, bacteriocins, and lysozyme, were capable of degrading the bacterial cell wall. According to their specificity, in 2021 Vachher et al. [22] mentioned division of lysins into five main categories: N-acetylmuramoyl-L-alanine amidases, N-acetylmuramidases/lysozymes, endo-β-N acetylglucosaminidases, endopeptidases and lytic transglycosylases. The antimicrobial activity of lysozyme, which is already immobilized in a variety of NP, is based on the degradation of bacterial membrane structures. The immobilized lysozyme catalyzed the hydrolysis of 1,4-beta-linkages in peptidoglycans between N-acetylmuramic acid and N-acetyl-D-glucosamine residues [44,45]. In 2019, Han et al. [55] published a work based on pyruvate dehydrogenase depletion of pyruvate. This method had the potential to be therapeutic in the control of *Pseudomonas aeruginosa* biofilm infections. The enzymatic activity of the encapsulated pyruvate dehydrogenase in NP poly(lactic-co-glycolic) acid successfully depleted the available pyruvate and thus dissolved the biofilm *P. aeruginosa*, as confirmed by the authors (Fig. 7.1).

Table 7.1 Various active nanoparticle-immobilized enzymes developed for antimicrobial and healing dressings.

Enzyme	E.C. number[a]	References
Alkaline metalloprotease (*Bacillus subtilis*)	E.C.3.4.21	[23]
Alkaline serine protease (*Bacillus* sp.)	E.C.3.4.21.14	[14]
Alkaline protease (*Aspergillus oryzae*)	E.C.3.4.21.63	[24]
Serine protease (*Euphorbia cf. lactea latex*)	E.C.3.4.21	[25]
Alpha-amylase (*Aspergillus oryzae*)	E.C. 3.2.1.1	[26]
L-Asparaginase (Bionase)	E.C.3.5.1.1	[27,28]
Bromelain	EC. 3.4.22.33	[16,18]
Caseinase (*Bacillus megaterium*)	E.C.3.4.21.50	[29]
Cellobiose Dehydrogenase	E.C.1.1.99.18	[30]
Collagenase	E.C.3.4.24.3	[19,31]
Ficin	E.C.3.4.22.3	[32]
Glucose oxidase	E.C.1.1.3.4	[28,33−37]
Horseradish peroxidase	E.C.1.11.1.7	[37]
Alpha-Chymotrypsin	E.C.3.4.21.1	[38,39]
Inulinase (*Aspergillus niger*)	E.C.3.2.1.7	[28]
Laccase	E.C.1.10.3.2	[40,41]
Lysosomal enzymes, Lysozyme	E.C.3.2.1.17	[42−51]
Papain	E.C.3.4.22.2	[52−54]
Pyruvate dehydrogenase	E.C.1.2.1.51	[55]
Subtilisin and subtilisin like protease LG12	E.C.3.4.21.62	[15,56]
Trypsin	E.C. 3.4.21.4	[57,58]

[a]*From Brenda [59]*

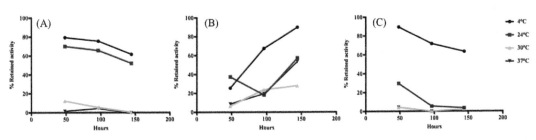

FIGURE 7.1

Storage stability of pyruvate dehydrogenase nanoparticle (NP) at various temperatures, (A) native, (B) immobilized in NP in a 50:50 ratio, and (C) in a 75:25 ratio.

Reproduced with permission from C. Han, J. Goodwine, N. Romero, K.S. Steck, K. Sauer, A. Doiron, Enzyme-encapsulating polymeric nanoparticles: a potential adjunctive therapy in Pseudomonas aeruginosa biofilm-associated infection treatment, Colloids Surf. B Biointerfaces 184 (2019) 110512; published by Elsevier, 2019.

7.4 Immobilized proteases nanoparticles

Naturally occurring proteolytic enzymes such as Ficin [32], bromelain [13,16], trypsin, lysozyme, and lysosomal enzymes [42,60], streptokinase-streptodornase, papain and papain-urea [1,54,61], or microbial enzyme such as clostridial collagenase [3,22] and proteinases from the greenbottle larva *Lucilia sericata* [62] were recently studied for wound healing. Individual enzymes' effects in antibacterial dressings differ depending on their specificity. Protease treatment classified wounds based on their type. A purulent wound with a high load of microorganisms, such as bacteria, necessitates the use of modern antimicrobial dressings containing effective agents and enzyme activity. Under physiological conditions, the use of dialdehyde cellulose resulted in significant hydrolysis and biological degradation in the human body. The byproducts of dialdehyde cellulose hydrolysis had beneficial antioxidant properties [63]. Several other cases of NP have been reported, which could have a significant impact on wound healing. Topically administered bromelain, a mixture of pineapple proteolytic enzymes (papain and cysteine protease), with its rapid effect significantly reduced morbidity and mortality in severely burned patients. It enabled earlier skin transplants and reduced the risk of sepsis, thereby shortening the convalescence period. In comparison, bromelain demonstrated a faster rate of complete debridement than collagenase. When compared to pure bromelain, nanoemulsions containing encapsulated bromelain in the phase of vitamin E acetate oil reduced wound contraction more effectively [16]. The in vitro release profiles showed that the maximum bromelain released of NP occurred in the first 4 hour. The decomposition effect of bromelain NP in chitosan hydrogel was the most beneficial in terms of necrotic tissue reduction and reepithelialization [18]. A 150-nm nanotriangular NP with preserved papain proteolytic activity reduced phagocytic cell activation while causing no toxicity [64]. The papain of *Carica papaya* latex immobilized in chitosan matrixes of medium molecular weight (200 kDa) and high molecular weight (350 kDa) exhibited antibiofilm activity and increased antimicrobial efficiency against bacteria embedded in biofilms. The preservation of 90% of the initial activity predicted an effective biofilm destruction wound treatment, with an increase in antimicrobial treatment (Fig. 7.2) [53]. Collagenase therapy improved the treatment of pressure ulcers and diabetic feet while decreasing the need for frequent surgical rupture. It aided in the restoration of wound conditions, stimulating the proliferation and migration of keratinocytes and fibroblasts while also releasing stimulatory peptide fragments [6,65]. Early in burns, treatment with clostridial collagenase in hydroxyethylcellulose as a vehicle delivery reduced necrosis and apoptosis. This increased cell infiltration while protecting the tissue from conversion [66]. At therapeutic doses, collagenase ointment could only destroy collagen but not keratin, fat, or fibrin, so it had no effect on intact skin. It also hastens wound closure [67]. Collagenase nanocapsules were used to treat excessive fibrotic collagen accumulation. The nanocapsules' shape allowed for the slow release of encapsulated collagenase for up to 10 days, and its efficacy in the murine model confirmed a greater reduction in fibrosis [19]. Our group investigated the enzyme activity, stability, and kinetics of collagenase immobilized in chitosan NP with antimicrobial and collagenolytic effects, and we investigated the enzyme activity, stability, and kinetics of collagenase immobilized in chitosan NP with antimicrobial and collagenolytic effects [31].

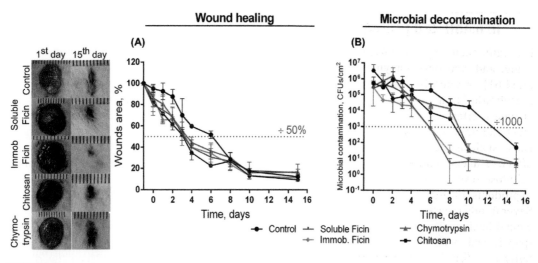

FIGURE 7.2

Immobilized Ficin activity was demonstrated for the healing of skin wounds infected with *Staphylococcus aureus* (A) and their microbial decontamination (B).

Reproduced with permission from (D.R. Baidamshina, V.A. Koroleva, E.Y. Trizna, S.M. Pankova, M.N. Agafonova, M.N. Chirkova, et al., Anti-biofilm and wound-healing activity of chitosan-immobilized Ficin, Int. J. Biol. Macromol. 164 (2020) 4205–4217); published by Elsevier, 2020.

7.5 Immobilized antimicrobial oxidoreductases

Immobilized oxidoreductases, which include dehydrogenases, oxidases, oxygenases, and hydroxylases, are increasingly being used in antimicrobial dressings. Laccases are oxidases that oxidize a variety of phenolic substrates in nature. Laccase immobilized in bacterial nanocellulose has also been studied for antibacterial properties and enzyme kinetic studies [40]. The enzyme-induced dispersion of mature *P. aeruginosa* biofilms was prolonged by NP-immobilized pyruvate dehydrogenase, which fought denaturing conditions. Pyruvate dehydrogenase was encapsulated in various polylactic-co-glycolic acid (PLGA) NP formulas, resulting in improved stability [55]. Another particles-immobilized enzyme with antibacterial effect was described by [30]. Cellobiose dehydrogenase immobilized in chitosan NP via various methods produced H_2O_2 in situ, completely inhibiting bacteria growth for a defined time. Several authors concluded that the disinfection effect of glucose oxidase NP was antimicrobial and promoted wound healing. Zhao et al. [35] presented glucose oxidase encapsulated in NP-integrated glucose-responsive coordination polymer nanozyme systems. The production of H_2O_2 and gluconic acid by glucose oxidase, which was expected to consume glucose and oxygen, activated the peroxidase activity of the coordination polymer NP in the catalytic cascade production of a highly active hydroxyl radical (•OH). The antioxidant and antimicrobial properties of glucose oxidase immobilized in zeolites coated with chitosan and crosslinked using natural phenol caffeic acid were demonstrated. Tegl et al. [34], recommended the observed effect of the developed system for topical wound treatment. Chitosan NP was used as the carrier for several glucose oxidase immobilization procedures; covalent, sorption, and precipitation sorption called enzyme coating. NP provided enzyme generated H_2O_2 in situ and inhibited microbe proliferation in bacterial cultures and biofilms [33]. Another antibacterial tool containing glucose oxidase was described by Qin et al. [36]. The glucose oxidase was loaded into hollow-sphere mesoporous metal oxide nanoreactors with biocompatible ceria (Fig. 7.3). Nanozyme formed by peroxidase activity entered a cascade reaction and converted glucose to toxic hydroxyl radicals.

7.6 Choice of immobilization method and nanoparticles

Biocompatibility, desired absorption rate, sufficient permeability and high specific surface area, strength and suitable particle form, chemical reactivity and stability that allow enzyme binding, adsorption, and release, and hydrophilic character are all essential properties of carriers suitable for enzyme immobilization in antimicrobial and healing coatings [52,61,68]. The hydrophilic nature of the solid support is desirable due to the need to reduce not only enzyme inactivation but also enzyme stability due to denaturation. Proteases, for example, were incorporated into hydrophobic

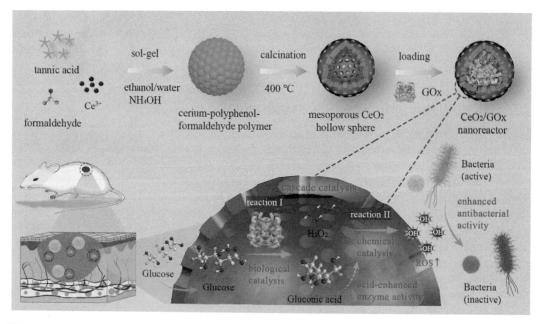

FIGURE 7.3

Formation of the active glucose oxidase nanoreactor in a hollow mesoporous CeO_2 sphere for cascade catalytic antibacterial therapy.

Reprinted with permission from (J. Qin, Y. Feng, D. Cheng, B. Liu, Z. Wang, Y. Zhao, et al., Construction of a mesoporous ceria hollow sphere/enzyme nanoreactor for enhanced cascade catalytic antibacterial therapy, ACS Appl. Mater. Interfaces 13 (2021) 40302–40314). Copyright 2021 American Chemical Society.

silicone matrices in the presence of hydrophilic polyvinyl alcohol (PVA) [15]. In comparison to large particles, the small dimensions of NP provide a larger surface area. As a result, the binding capacity should increase as particle size decreases. However, Ahmed et al. [29] found that with increasing particle size, the yield of adsorbed and covalently bound caseinase also increased, presumably for steric reasons. Different functional groups, such as hydroxyl, amine, and carboxylic, are found in commonly used NP, either naturally or as a result of modification. These factors influence the surface charge of NP (charged or neutral, hydrophilic or hydrophobic) as well as how enzymes bind via covalent or physical bonds.

The immobilized enzyme's primary requirement is long-term activity preservation. The active steric availability of the enzyme active site is required for this. Enzyme immobilization methods have a significant impact on enzyme activity due to structural molecule changes, particularly secondary structure [44,58]. The increased stability based on immobilization, on the other hand, is frequently attributed to the internal characteristics of individual immobilization processes. Individual factors and the cumulative effects of several factors typically determine immobilized enzyme stability. Temperature and pH changes can be significantly stabilized by immobilization [23,29,39,54]. Increased antimicrobial effects can lead to increased activity and long-term stability of the NP-immobilized enzyme [45]. Furthermore, the long-term stability of immobilized proteolytic enzymes prevents autoproteolysis compared to native enzymes [32,39,69].

Tens of techniques for immobilizing enzymes were developed; for those with antimicrobial and healing effects, approximately five were used (Fig. 7.4), which can be divided into two groups. Reversible immobilization techniques include adsorption, and irreversible immobilization

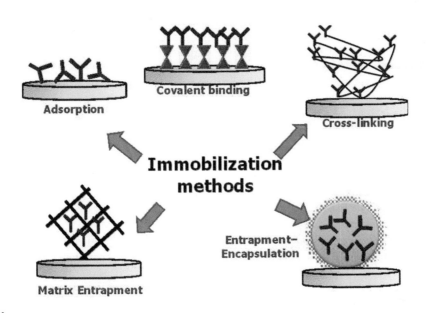

FIGURE 7.4

Schematic representation of the modes of antimicrobial enzymes immobilization [70].

techniques include covalent attachment and intermolecular crosslinking, polymeric gels entrapment, and encapsulation.

The most common immobilization of NP enzymes in the field was physical adsorption [14,23,25,26,28,32,33,40,42−46,48,49,53,54]. This straightforward method is based on simple interactions between the enzyme molecule and the NP functional groups via hydrophobic, electrostatic, van der Waals, specific/affinity, and hydrogen bonding. In some ways, its disadvantage may be the rapid desorption and release of enzymes. However, in theory, the gradual and controlled release of the enzyme, as well as the antimicrobial and healing applications of its effect, are appropriate [15]. At the same time, the advantage of sorption is that the enzyme's structure and thus activity are almost unchanged. For example, comparing physical adsorption of the caseinase enzyme to glutaraldehyde intermolecular crosslinking resulted in higher immobilization yields [29].

The strong bonds between the enzyme and the NP are represented by the covalent method and intramolecular crosslinking, most commonly glutaraldehyde [31,38,39,50,52,54,56,57]. As a result of protease stabilization against proteolysis, multipoint and multisubunit covalent binding may be more stable [52]. The attachment of enzymes to functional groups of the support via linkers can provide rigidity to the enzyme's immobilized structure, improving enzyme stability [29,34,71]. Treatment with glutaraldehyde had little effect on enzyme activity (recovery was 70%) but significantly improved enzyme stability [24]. In another study, glutaraldehyde crosslinking had less of an effect on enzyme activity and stability than simple covalent immobilization [58]. Covalent immobilization with different adsorption techniques produced less effective results in enzyme activity in some cases [30,33].

Physical trapping was a gentle way to avoid damaging effects on the structure of immobilized enzymes. Encapsulation techniques have previously been shown to involve, in addition to enzyme molecule stabilization, the slow release of active ingredients from plant extracts and transport to the target site [15]. Polymers such as chitosan [27,51], sodium alginate [18], starch [47], PLGA [55], derivatives of acrylamide and methacrylate [19] and PVA cryogel [56] were frequently used for enzyme encapsulation. Inorganic materials included coordination polymer NP formed by ferrous ions and guanosine monophosphate [35], hollow-sphere nanoreactors of mesoporous metal oxide containing biocompatible ceria [36], and silver-functionalized silk fabric adsorbed lysozyme encapsulated in silica or titania [46]. Another type of lysozyme encapsulated in liposomes has also been shown to have antimicrobial properties [47]. Another method for encapsulating enzymes was to use a nanoemulsion or a dried silicon protease emulsion [15]. In an animal model, the effect of bromelain stabilized by incorporation in a nanoemulsion composed of vitamin E acetate as an oil phase and a hydrophilic gel matrix was evaluated in wound healing of burned skin [16]. The information on the methods of binding enzymes with antimicrobial effects on NP is extended in Table 7.2 by the materials used. NP materials were metallic, inorganic, siliceous, and polymeric, or mixtures thereof, referred to as hybrid.

Silver (Ag) NP has been used as an antimicrobial agent in biomedical fields since the twentieth century, as reviewed by Gunasekaran et al. [72]. Dickerson et al. [46] enriched their natural antimicrobial effect with lysozyme adsorbed on a complex of silver NP and silk fabric fibers. Other authors verified the unaffected catalytic site of lysozyme adsorbed in silver NP on the antimicrobial effect against *Escherichia coli* [44]. Antimicrobial action of hybrid silver NPs containing polyethylene glycol (PEG), PVA, and chitosan polymers caring oxidoreductase laccase in synergistic performance against antimicrobial resistance [41].

Table 7.2 Nanoparticle type of material for antimicrobial enzymes immobilization.

Type of material	Nanoparticle material	References
Metal NP	Silver, gold	[14,25,28,44,45,52]
Inorganic-based NP	Titanium oxide, alumina, hydroxyapatite, copper nanocrystals, zeolite, glass ceramics, ZnO_2, cerium oxide silica	[23,26,29,34,36−38,42,43,46,57]
Polymer NP	Natural polymers (alginates, chitosan, cellulose) biodegradable polyesters (PLA, poly(acrylic) acid, PLGA), lipids, PVA	[18,19,24,27,30,31,35,40,47,48,50,51,53−56]
Hybrid NP	Polymer-inorganic NP polymer-magnetic NP silica-magnetic NP metal-polymer NP	[34,39−42,49,57,58]

Gold (Au) NP, which are nontoxic and biocompatible in both in vivo and in vitro environments, have a wide range of applications, including antimicrobials [73]. Because of their high surface energy, Au NPs are extremely reactive and an excellent choice for immobilizing enzymes with antibacterial properties [52]. Ionic-liquid-modified Au NPs were used to immobilize the lysozyme while preserving its molecular native structure and catalytic activity. Against a suspended solution of *Micrococcus lysodeikticusce* cells, there was an increase in antimicrobial activity and a significant measurement of the thermal stability of lysozyme adsorbed in Au NP [45].

Antimicrobial enzymes were immobilized using a variety of inorganic solids. Because of their ease of synthesis, large surface area, and controlled pore diameter, silica-based supports are excellent matrices for immobilizing enzymes [57]. The high biocompatibility of silica NPs is a significant advantage [46]. Its low in vivo biodegradability could be improved by doping with strontium ion [74]. Titanium oxide has photocatalytic properties, and photocatalysis with titanium oxide particles has confirmed the elimination of bacteria, fungi, and algae [75]. Lysosomal enzymes adsorbed in titanium oxide particles in an optimal ratio demonstrated increased enzyme stability for up to one month [43].

Inorganic minerals with self-antimicrobial activity for the preparation of enzyme-inorganic mineral hybrid composites have proven to be a viable alternative to bacterial contamination of NP. For example, the synthesis of copper hydrosulfate nanocrystals and the encapsulation of glucose oxidase in the nanocrystals consistently demonstrated long-term active antimicrobial NP [37].

Depending on the carrier preparation method used, polymeric nanostructures can be synthesized from biocompatible and biodegradable polymers. Natural polymer wound dressing is appealing to the pharmaceutical industry because it is cost effective, nontoxic to the human body, and environmentally friendly [76]. Alginate, a polysaccharide derived from brown algae, was investigated as a wound dressing material due to its favorable biocompatibility and biodegradability [13]. Because natural polymers are derived from natural sources, they may contain impurities (heavy metals, endotoxins, and proteins) [40]. The composition of synthetic polymers is precisely defined. Synthetic polymers commonly used are polylactic acid, polyglycolic acid, PLGA, and PVA [55−57]. Individual polymers can be combined to form a variety of copolymers with varying properties that determine their biocompatibility and biodegradability [55]. Polymeric materials' microstructures are typically amorphous; however, they can accommodate high enzyme absorption, preventing deformation of the enzyme catalytic site [47]. Chitin nanowhiskers were nanocarriers of natural biological macromolecules that were small and light, chemically stable, biodegradable, non-cytotoxic, and had renewable antimicrobial activity due to the presence of lysozyme [48].

7.7 Kinetics of nanoparticles-immobilized enzymes

In the field of enzyme kinetics, the rate of enzyme-catalyzed reactions was investigated, which is influenced by the concentration of reactants in the main factory, temperature, pH, mass transfer effects, the presence of reaction inhibitors, and so on [77,78]. The reaction kinetics of native and immobilized enzymes may differ due to conformational changes in enzymes after immobilization, according to various authors [58,79,80]. Catalysis occurred at the active sites of the enzyme and did not involve only a small or linear portion of the enzyme molecule; thus, it was critical not to interfere with the formation of the enzyme's active conformation for the substrate during immobilization [81]. Immobilized enzyme kinetic properties included, but were not limited to, steric limitations caused by the barrier of materials, their chemical properties, and the microenvironment of the bound enzyme molecules [82].

The Michaelis−Menten theories, which are referred to as the substrate enzyme saturation and saturation curve, describe how enzyme reaction rates depend on substrate concentrations. The rate of the reaction increased as the substrate concentration increased until the system was saturated, at which point it stopped increasing. This dependence can be displayed graphically in Fig. 7.5A [82].

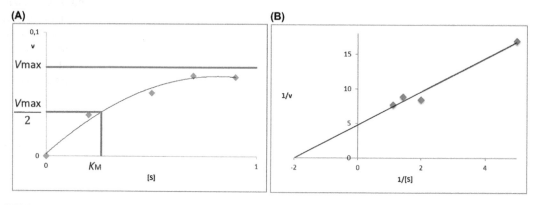

FIGURE 7.5

Michaelis−Menten representation of enzyme activity data. (A) The initial velocity rate V of a given enzyme reaction is calculated from the initial linear part of the measurement. Substrate concentration [S] is the limit substrate concentration for each rate experiment V. If the substrate concentration is low, the rate of change curve corresponds approximately to the first order kinetics (linear increase). However, this almost linear increase cannot continue indefinitely due to the limited number of binding sites on the enzyme molecules. At a sufficiently high substrate concentration, the maximum rate (V_{max}) is reached after saturation of the binding sites of all available enzyme molecules. Michaelis constant (K_M) is defined as the concentration of substrate at which the rate of the enzymatically catalyzed reaction is equal to half the maximum rate (i.e., $V = \frac{V_{max}}{2}$). (B) Linearization by using reciprocal values v and [S] according to Lineweaver and Burk.

This kinetic model suited the experimental conditions for many enzymes and assumes very simple kinetics [77,83]. Equations were usually linearized to calculate kinetic parameters. The modification proposed in 1934 by Lineweaver and Burke [84] used reciprocal values and was used by the vast majority of authors, as shown as an example in Fig. 7.5B. The numerical value of the K_M corresponded to the substrate concentration at which the rate of the enzymatic reaction rate V decreased to half the value of the V_{max}. The V_{max} of the enzyme reactions reflected the speed with which the biocatalyst could catalyze biotransformation. V_{max} and K_M are constants.

Recently, in 2021 Aledo [85] experimentally compared other methods for calculating the K_M and V_{max} constants for beta-galactosidase, using Eadie—Hofstee and Hanes—Woolf transformations. The axes in the Eadie graph were reversed from the Lineweaver—Burk graph; the calculation principle remained the same. The Hanes—Woolf graph depicted the dependence of substrate concentration on the substrate concentration to reaction rate ratio. According to [24], when calculating the kinetic constants for alkaline protease (*Aspergillus oryzae*) immobilized on chitosan NP from the Hanes—Woolf graph, they avoided the misleading impression of experimental error and uneven distribution of points in the Lineweaver—Burk graph, while not including angular error distortion from the Eadie—Hofstee graph.

The kinetic parameters K_M, V_{max}, and k_{cat} were important parameters of the enzymes immobilized in the NP to assess whether the immobilization changed the enzyme-substrate affinity and thus the steric availability of the active site of the enzyme. Table 7.3 presents a quantitative comparison of the kinetic parameters K_M, k_{cat}, and V_{max} of NP-immobilized antimicrobial enzymes with native enzymes in solution.

Comparison of K_M and V_{max} values indicated changes in specific activity and steady-state kinetic parameters for the immobilized enzyme. It was previously reported that much lower K_M values for trypsin immobilized in NP have been reported to indicate higher affinities for the substrate [79]. Other authors also reported similar observations with reduced K_M values for enzymes immobilized with NP with antimicrobial effects [29,37]. Several authors have explained this as an increase in the local concentration of the substrate in the vicinity of the immobilized enzyme molecules [52,86], and therefore identified K_M as an apparent constant.

However, some authors reported unchanged [24,52,53,56] or even higher K_M values compared to native enzymes [32,40,58]. Liu et al. [52] found that the K_M values for gold NP-conjugated papain and native papain were in the same order. Vice versa, the turnover numbers k_{cat} were significantly higher for papain conjugated with gold NP. As the authors stated, for native enzymes in solution, K_M was generally independent of enzyme concentration. The enzyme's conformation and the formation of the enzyme-substrate complex were most likely affected after immobilization. The active site should be more accessible at low substrate concentrations than at high ones. Another finding was that V_{max} increased in free-solution reactions with the same trend as enzyme concentration. Therefore, a better parameter for defining and also comparing the reaction rate of a number of biocatalysts was the k_{cat} value. k_{cat} expresses the number of substrate molecules that are converted to product per enzyme molecule per second [52]. In another case, the authors reported no change in K_M after immobilization of papain in medium- and high-molecular-weight chitosan NP. There were no changes in enzyme affinity for the substrate, but the significant decrease in V_{max} and k_{cat} values indicated limited flexibility of enzyme molecules due to immobilization. [53]. Sun et al. [58] reported lower K_M values in a kinetic experiment with trypsin immobilized in carboxymethyl chitosan-functionalized magnetic NP with EDC or glutaraldehyde activating agents. Their kinetic

Table 7.3 Kinetic constants of immobilized enzymes with antimicrobial effects compared to native.

Enzyme	Immobilization method, carrier	K_M (mol/l)	V_{max} (M min^{-1})	k_{cat} (s^{-1})	References
		native / immobilized			
Alkaline protease	Covalent with GA, chitosan NP	0.011×10^{-3}/ 0.013×10^{-3}	4.56×10^{-3} g^{-1}/ 1.38×10^{-3} g^{-1}	2.6/0.81	[24]
Alkaline metalloprotease	Adsorption, hydroxyapatite NP	3.125×10^{-3}/ 2×10^{-3}	40/50		[23]
Ficin	Adsorption, chitosan NP	25×10^{-6}/ 38×10^{-6}	1.314×10^{-3} mg^{-1}/ 0.779×10^{-3} mg^{-1}	1.01/0.605	[32]
Caseinase	Adsorption, mica glass-ceramics NP	0.36 mg/mL/ 0.19 mg/mL	476.19 U/mg/ 1176.47 U/mg		[29]
Glucose oxidase	Encapsulation, copper nanocrystals	19.01×10^{-3}/ 14.39×10^{-3}		210.32/9.42	[37]
Laccase	Adsorption, bacterial nanocellulose	0.421×10^{-3}/ 0.776×10^{-3}	8.676×10^{-3}/ 5.292×10^{-3}		[40]
Lysozyme	Adsorption, gold NP	0.34×10^{-6}/ $0.15 - 0.89 \times 10^{-6}$	1.423×10^{-3}/ $1.104 - 1.796 \times 10^{-3}$		[45]
Papain	Covalent with EDC, S-NHS, gold NP	2.71×10^{-3}/ 1.61×10^{-3}	12.18×10^{-6}/ 17.46×10^{-6}	0.475×10^{-4}/0.2	[52]
Papain	Adsorption, chitosan NP	22.3×10^{-6}/ 24.7 and 23.1×10^{-6}	1.253×10^{-3} mg^{-1}/ 0.432×10^{-3} and 0.349×10^{-3} mg^{-1}	0.963/0.113 and 0.075	[53]
Subtilisin	Encapsulation, poly(acrylic) acid	0.25×10^{-3}/ 0.18×10^{-3}		69.3/51	[56]
Trypsin	Covalent with EDC or GA, CM chitosan magnetic NP	0.99×10^{-3}/1.59 and 1.35×10^{-3}	1.537×10^{-3}/ 1.221×10^{-3} and 1.3789×10^{-3}		[58]

studies showed that while the biocatalytic activity of immobilized trypsin was preserved, the K_M value was lower and the affinity of trypsin for the BAEE substrate was reduced by immobilization. The explanation was based on FTIR spectra of immobilized trypsin, which showed changes in secondary structures in both types of trypsin NP carriers.

Measurements and calculation of kinetic constants of native and immobilized enzymes were usually performed at optimal pH and temperature [24,32,58]. Temperature and pH changes in immobilized enzyme kinetic reactions were reflected in activity and stability. The purified metalloprotease of *Bacillus subtilis* supplemented with hydroxyapatite NP increased the native enzyme's thermostability over a wide alkaline pH range at 90°C [23]. It should be noted that increasing temperature increased the rate of enzyme reactions only in a well-defined temperature range [87].

It can be concluded that NP immobilized enzymes, along with other benefits, provided controlled antimicrobial effects. Immobilization on NP, in addition to accelerating reaction rates, could be a viable strategy for preserving valuable enzymes for antimicrobial actions.

Acknowledgment

This research was supported by OP RDE project "Strengthening interdisciplinary cooperation in research of nanomaterials and their effects on living organisms," reg. no. CZ.02.1.01/0.0/0.0/17_048/0007421 and by the Ministry of Education Youth and Sports (the program InterExcellence-InterAction, grant no. LTAUSA-19003).

References

[1] H. Drago, G.H. Marín, F. Sturla, G. Roque, K. Mártire, V. Díaz Aquino, et al., The next generation of burns treatment: intelligent films and matrix, controlled enzymatic debridement, and adult stem cells, Transplant. Proc 42 (2010) 345–349. Available from: https://doi.org/10.1016/j.transproceed.2009.11.031.

[2] S. Tandon, A. Sharma, S. Singh, S. Sharma, S.J. Sarma, Therapeutic enzymes: discoveries, production and applications, J. Drug. Deliv. Sci. Technol 63 (2021) 102455. Available from: https://doi.org/10.1016/j.jddst.2021.102455.

[3] H. Alipour, A. Raz, S. Zakeri, N. Dinparast Djadid, Therapeutic applications of collagenase (metalloproteases): a review, Asian Pac. J. Trop. Biomed 6 (2016) 975–981. Available from: https://doi.org/10.1016/j.apjtb.2016.07.017.

[4] T.U. Wani, A.H. Rather, R.S. Khan, M.A. Beigh, M. Park, B. Pant, et al., Strategies to use nanofiber scaffolds as enzyme-based biocatalysts in tissue engineering applications, Catalysts 11 (2021) 536. Available from: https://doi.org/10.3390/catal11050536.

[5] W.G. Payne, R.E. Salas, F. Ko, D.K. Naidu, G. Donate, T.E. Wright, et al., Enzymatic debriding agents are safe in wounds with high bacterial bioburdens and stimulate healing, Eplasty 8 (2008) e17.

[6] L. Shi, D. Carson, Collagenase santyl ointment, J. Wound, Ostomy Cont. Nurs 36 (2009) S12–S16. Available from: https://doi.org/10.1097/WON.0b013e3181bfdd1a.

[7] A.R. Sheets, T.N. Demidova-Rice, L. Shi, V. Ronfard, K.V. Grover, I.M. Herman, Identification and characterization of novel matrix-derived bioactive peptides: a role for collagenase from Santyl® ointment in post-debridement wound healing? PLoS One 11 (2016) e0159598. Available from: https://doi.org/10.1371/journal.pone.0159598.

[8] S. Dekina, I. Romanovska, O. Sevastyanov, Y. Shesterenko, A. Ryjak, L. Varbanets, et al., Development and characterization of chitosan/polyvinyl alcohol polymer material with elastolytic and collagenolytic activities, Enzyme Microb. Technol 132 (2020) 109399. Available from: https://doi.org/10.1016/j.enzmictec.2019.109399.

[9] F. Qu, J.-M.G.M.G. Lin, J.L. Esterhai, M.B. Fisher, R.L. Mauck, Biomaterial-mediated delivery of degradative enzymes to improve meniscus integration and repair, Acta Biomater 9 (2013) 6393–6402. Available from: https://doi.org/10.1016/j.actbio.2013.01.016.

[10] S.K. McCallon, D. Weir, J.C. Lantis, Optimizing wound bed preparation with collagenase enzymatic debridement, J. Am. Coll. Clin. Wound Spec 6 (2014) 14–23. Available from: https://doi.org/10.1016/j.jccw.2015.08.003.

[11] C. Serrano, L. García-Fernández, J.P. Fernández-Blázquez, M. Barbeck, S. Ghanaati, R. Unger, et al., Nanostructured medical sutures with antibacterial properties, Biomaterials 52 (2015) 291−300. Available from: https://doi.org/10.1016/j.biomaterials.2015.02.039.

[12] A. Tallis, T.A. Motley, R.P. Wunderlich, J.E. Dickerson, C. Waycaster, H.B. Slade, Clinical and economic assessment of diabetic foot ulcer debridement with collagenase: results of a randomized controlled study, Clin. Ther. 35 (2013) 1805−1820. Available from: https://doi.org/10.1016/j.clinthera.2013.09.013.

[13] S. Bayat, N. Amiri, E. Pishavar, F. Kalalinia, J. Movaffagh, M. Hashemi, Bromelain-loaded chitosan nanofibers prepared by electrospinning method for burn wound healing in animal models, Life Sci 229 (2019) 57−66. Available from: https://doi.org/10.1016/j.lfs.2019.05.028.

[14] J. Siritapetawee, W. Limphirat, P. Pakawanit, C. Phoovasawat, Application of Bacillus sp. protease in the fabrication of silver/silver chloride nanoparticles in solution and cotton gauze bandages, Biotechnol. Appl. Biochem (2020) 1−10. Available from: https://doi.org/10.1002/bab.2075.

[15] R. Bott, J. Crissman, C. Kollar, M. Saldajeno, G. Ganshaw, X. Thomas, et al., A silicone-based controlled-release device for accelerated proteolytic debridement of wounds, Wound Repair. Regen 15 (2007) 227−235. Available from: https://doi.org/10.1111/j.1524-475X.2007.00209.x.

[16] H. Rachmawati, E. Sulastri, M. Immaculata Iwo, D. Safitri, A. Rahma, Bromelain encapsulated in self assembly nanoemulsion exhibits better debridement effect in animal model of burned skin, J. Nano Res 40 (2016) 158−166. Available from: https://doi.org/10.4028/http://www.scientific.net/JNanoR.40.158.

[17] E.E. Dosadina, E.E. Savelyeva, A.A. Belov, The effect of immobilization, drying and storage on the activity of proteinases immobilized on modified cellulose and chitosan, Process. Biochem 64 (2018) 213−220. Available from: https://doi.org/10.1016/j.procbio.2017.10.002.

[18] S. Bayat, A.R. Zabihi, S.A. Farzad, J. Movaffagh, E. Hashemi, S. Arabzadeh, et al., Evaluation of debridement effects of bromelain-loaded sodium alginate nanoparticles incorporated into chitosan hydrogel in animal models, Iran. J. Basic. Med. Sci. 24 (2021) 1404−1412. Available from: https://doi.org/10.22038/ijbms.2021.58798.13060.

[19] M.R. Villegas, A. Baeza, A. Usategui, P.L. Ortiz-Romero, J.L. Pablos, M. Vallet-Regí, Collagenase nanocapsules: an approach to fibrosis treatment, Acta Biomater 74 (2018) 430−438. Available from: https://doi.org/10.1016/j.actbio.2018.05.007.

[20] P.T.S. Kumar, N.M. Raj, G. Praveen, K.P. Chennazhi, S.V. Nair, R. Jayakumar, In vitro and in vivo evaluation of microporous chitosan hydrogel/nanofibrin composite bandage for skin tissue regeneration, Tissue Eng. Part. A 19 (2013) 380−392. Available from: https://doi.org/10.1089/ten.tea.2012.0376.

[21] V.E. Bosio, G.A. Islan, Y.N. Martínez, N. Durán, G.R. Castro, Nanodevices for the immobilization of therapeutic enzymes, Crit. Rev. Biotechnol 36 (2015) 1−18. Available from: https://doi.org/10.3109/07388551.2014.990414.

[22] M. Vachher, A. Sen, R. Kapila, A. Nigam, Microbial therapeutic enzymes: a promising area of biopharmaceuticals. Curr, Res. Biotechnol 3 (2021) 195−208. Available from: https://doi.org/10.1016/j.crbiot.2021.05.006.

[23] A. Mukhopadhyay, K. Chakrabarti, Enhancement of thermal and pH stability of an alkaline metalloprotease by nano-hydroxyapatite and its potential applications, RSC Adv 5 (2015) 89346−89362. Available from: https://doi.org/10.1039/C5RA16179G.

[24] K. Sangeetha, Emilia Abraham, T., Investigation on the development of sturdy bioactive hydrogel beads, J. Appl. Polym. Sci 107 (2008) 2899−2908. Available from: https://doi.org/10.1002/app.27445.

[25] J. Siritapetawee, W. Limphirat, N. Nantapong, D. Songthamwat, Fabrication of silver chloride nanoparticles using a plant serine protease in combination with photoactivation and investigation of their biological activities, Biotechnol. Appl. Biochem 65 (2018) 572−579. Available from: https://doi.org/10.1002/bab.1638.

[26] O. Długosz, J. Matysik, W. Matyjasik, M. Banach, Catalytic and antimicrobial properties of α-amylase immobilised on the surface of metal oxide nanoparticles, J. Clust. Sci 32 (2021) 1609−1622. Available from: https://doi.org/10.1007/s10876-020-01921-5.

[27] A. Vimal, A. Kumar, Antimicrobial potency evaluation of free and immobilized l-asparaginase using chitosan nanoparticles, J. Drug. Deliv. Sci. Technol 61 (2021) 102231. Available from: https://doi.org/10.1016/j.jddst.2020.102231.

[28] R. Choukade, A. Jaiswal, N. Kango, Characterization of biogenically synthesized silver nanoparticles for therapeutic applications and enzyme nanocomplex generation, 3 Biotech. 10 (2020) 462. Available from: https://doi.org/10.1007/s13205-020-02450-8.

[29] S.A. Ahmed, S.A.A. Saleh, S.A.M. Abdel-Hameed, A.M. Fayad, Catalytic, kinetic and thermodynamic properties of free and immobilized caseinase on mica glass-ceramics, Heliyon 5 (2019) e01674. Available from: https://doi.org/10.1016/j.heliyon.2019.e01674.

[30] G. Tegl, B. Thallinger, B. Beer, C. Sygmund, R. Ludwig, A. Rollett, et al., Antimicrobial cellobiose dehydrogenase-chitosan particles, ACS Appl. Mater. Interfaces 8 (2016) 967−973. Available from: https://doi.org/10.1021/acsami.5b10801.

[31] M. Slováková, V. Kratochvilová, J. Palarčík, R. Metelka, P. Dvořáková, J. Srbová, et al., Chitosan nanofibers and nanoparticles for immobilization of microbial collagenase, Vlakna a Text 23 (2016) 193−198.

[32] D.R. Baidamshina, V.A. Koroleva, E.Y. Trizna, S.M. Pankova, M.N. Agafonova, M.N. Chirkova, et al., Anti-biofilm and wound-healing activity of chitosan-immobilized Ficin. Int. J, Biol. Macromol 164 (2020) 4205−4217. Available from: https://doi.org/10.1016/j.ijbiomac.2020.09.030.

[33] K.-M. Yeon, J. You, M.D. Adhikari, S.-G. Hong, I. Lee, H.S. Kim, et al., Enzyme-immobilized chitosan nanoparticles as environmentally friendly and highly effective antimicrobial agents, Biomacromolecules 20 (2019) 2477−2485. Available from: https://doi.org/10.1021/acs.biomac.9b00152.

[34] G. Tegl, V. Stagl, A. Mensah, D. Huber, W. Somitsch, S. Grosse-Kracht, et al., The chemo enzymatic functionalization of chitosan zeolite particles provides antioxidant and antimicrobial properties, Eng. Life Sci 18 (2018) 334−340. Available from: https://doi.org/10.1002/elsc.201700120.

[35] J. Zhao, X. Bao, T. Meng, S. Wang, S. Lu, G. Liu, et al., Fe(II)-driven self-assembly of enzyme-like coordination polymer nanoparticles for cascade catalysis and wound disinfection applications, Chem. Eng. J 420 (2021) 129674. Available from: https://doi.org/10.1016/j.cej.2021.129674.

[36] J. Qin, Y. Feng, D. Cheng, B. Liu, Z. Wang, Y. Zhao, et al., Construction of a mesoporous ceria hollow sphere/enzyme nanoreactor for enhanced cascade catalytic antibacterial therapy, ACS Appl. Mater. Interfaces 13 (2021) 40302−40314. Available from: https://doi.org/10.1021/acsami.1c10821.

[37] Z. Li, Y. Ding, S. Li, Y. Jiang, Z. Liu, J. Ge, Highly active, stable and self-antimicrobial enzyme catalysts prepared by biomimetic mineralization of copper hydroxysulfate, Nanoscale 8 (2016) 17440−17445. Available from: https://doi.org/10.1039/C6NR06115J.

[38] L. Derr, S. Steckbeck, R. Dringen, L. Colombi Ciacchi, L. Treccani, K. Rezwan, Assessment of the proteolytic activity of α-chymotrypsin immobilized on colloidal particles by matrix-assisted laser desorption ionization time-of-flight mass spectrometry, Anal. Lett 48 (2015) 424−441. Available from: https://doi.org/10.1080/00032719.2014.951449.

[39] P. Prikryl, J. Lenfeld, D. Horak, M. Ticha, Z. Kucerova, Magnetic bead cellulose as a suitable support for immobilization of α-chymotrypsin, Appl. Biochem. Biotechnol 168 (2012) 295−305. Available from: https://doi.org/10.1007/s12010-012-9772-y.

[40] L.M.P. Sampaio, J. Padrão, J. Faria, J.P. Silva, C.J. Silva, F. Dourado, et al., Laccase immobilization on bacterial nanocellulose membranes: antimicrobial, kinetic and stability properties, Carbohydr. Polym 145 (2016) 1−12. Available from: https://doi.org/10.1016/j.carbpol.2016.03.009.

[41] M.N.M. Cunha, H.P. Felgueiras, I. Gouveia, A. Zille, Synergistically enhanced stability of laccase immobilized on synthesized silver nanoparticles with water-soluble polymers, Colloids Surf. B Biointerfaces 154 (2017) 210−220. Available from: https://doi.org/10.1016/j.colsurfb.2017.03.023.

[42] S.H. Bang, S.S. Sekhon, S.-J. Cho, S.J. Kim, T.-H. Le, P. Kim, et al., Antimicrobial properties of lysosomal enzymes immobilized on NH_2 functionalized silica-encapsulated magnetite nanoparticles, J. Nanosci. Nanotechnol 16 (2016) 1090−1094. Available from: https://doi.org/10.1166/jnn.2016.10660.

[43] S.H. Bang, A. Jang, J. Yoon, P. Kim, J.S. Kim, Y.-H. Kim, et al., Evaluation of whole lysosomal enzymes directly immobilized on titanium (IV) oxide used in the development of antimicrobial agents, Enzyme Microb. Technol 49 (2011) 260−265. Available from: https://doi.org/10.1016/j.enzmictec.2011.06.004.

[44] V. Ernest, S. Gajalakshmi, A. Mukherjee, N. Chandrasekaran, Enhanced activity of lysozyme-AgNP conjugate with synergic antibacterial effect without damaging the catalytic site of lysozyme. Artif. Cells Nanomed, Biotechnol 42 (2014) 336−343. Available from: https://doi.org/10.3109/21691401.2013.818010.

[45] S. Kumar, A. Sindhu, P. Venkatesu, Ionic liquid-modified gold nanoparticles for enhancing antimicrobial activity and thermal stability of enzymes, ACS Appl. Nano Mater 4 (2021) 3185−3196. Available from: https://doi.org/10.1021/acsanm.1c00401.

[46] M.B. Dickerson, C.L. Knight, M.K. Gupta, H.R. Luckarift, L.F. Drummy, M.L. Jespersen, et al., Hybrid fibers containing protein-templated nanomaterials and biologically active components as antibacterial materials, Mater. Sci. Eng. C. 31 (2011) 1748−1758. Available from: https://doi.org/10.1016/j.msec.2011.08.006.

[47] P. Matouskova, I. Marova, J. Bokrova, P. Benesova, Effect of encapsulation on antimicrobial activity of herbal extracts with lysozyme, Food Technol. Biotechnol 54 (2016) 304−316. Available from: https://doi.org/10.17113/ftb.54.03.16.4413.

[48] S. Jiang, Y. Qin, J. Yang, M. Li, L. Xiong, Q. Sun, Enhanced antibacterial activity of lysozyme immobilized on chitin nanowhiskers, Food Chem 221 (2017) 1507−1513. Available from: https://doi.org/10.1016/j.foodchem.2016.10.143.

[49] K. Erol, D. Tatar, A. Veyisoğlu, A. Tokatlı, Antimicrobial magnetic poly(GMA) microparticles: synthesis, characterization and lysozyme immobilization, J. Polym. Eng 41 (2021) 144−154. Available from: https://doi.org/10.1515/polyeng-2020-0191.

[50] A.A. Cerón, L. Nascife, S. Norte, S.A. Costa, J.H. Oliveira do Nascimento, F.D.P. Morisso, et al., Synthesis of chitosan-lysozyme microspheres, physicochemical characterization, enzymatic and antimicrobial activity, Int. J. Biol. Macromol 185 (2021) 572−581. Available from: https://doi.org/10.1016/j.ijbiomac.2021.06.178.

[51] Y. Wang, S. Li, M. Jin, Q. Han, S. Liu, X. Chen, et al., Enhancing the thermo-stability and anti-bacterium activity of lysozyme by immobilization on chitosan nanoparticles, Int. J. Mol. Sci 21 (2020). Available from: https://doi.org/10.3390/ijms21051635.

[52] S. Liu, M. Höldrich, A. Sievers-Engler, J. Horak, M. Lämmerhofer, Papain-functionalized gold nanoparticles as heterogeneous biocatalyst for bioanalysis and biopharmaceuticals analysis, Anal. Chim. Acta 963 (2017) 33−43. Available from: https://doi.org/10.1016/j.aca.2017.02.009.

[53] D.R. Baidamshina, V.A. Koroleva, S.S. Olshannikova, E.Y. Trizna, M.I. Bogachev, V.G. Artyukhov, et al., Biochemical properties and anti-biofilm activity of chitosan-immobilized papain, Mar. Drugs 19 (2021) 197. Available from: https://doi.org/10.3390/md19040197.

[54] N.F. Vasconcelos, A.P. Cunha, N.M.P.S. Ricardo, R.S. Freire, L. Vieira, A.P. de, et al., Papain immobilization on heterofunctional membrane bacterial cellulose as a potential strategy for the debridement of skin wounds, Int. J. Biol. Macromol. 165 (2020) 3065−3077. Available from: https://doi.org/10.1016/j.ijbiomac.2020.10.200.

[55] C. Han, J. Goodwine, N. Romero, K.S. Steck, K. Sauer, A. Doiron, Enzyme-encapsulating polymeric nanoparticles: a potential adjunctive therapy in Pseudomonas aeruginosa biofilm-associated infection treatment, Colloids Surf. B Biointerfaces 184 (2019) 110512. Available from: https://doi.org/10.1016/j.colsurfb.2019.110512.

[56] A.V. Bacheva, O.V. Baibak, A.V. Belyaeva, E.S. Oksenoit, T.I. Velichko, E.N. Lysogorskaya, et al., Activity and stability of native and modified subtilisins in various media, Biochem 68 (2003) 1261−1266. Available from: https://doi.org/10.1023/B:BIRY.0000009142.88974.85.

[57] F. Bucatariu, F. Simon, C. Bellmann, G. Fundueanu, E.S. Dragan, Stability under flow conditions of trypsin immobilized onto poly(vinyl amine) functionalized silica microparticles. Colloids Surfaces A Physicochem, Eng. Asp 399 (2012) 71−77. Available from: https://doi.org/10.1016/j.colsurfa.2012.02.030.

[58] J. Sun, L. Yang, M. Jiang, Y. Shi, B. Xu, H. Ma, Stability and activity of immobilized trypsin on carboxymethyl chitosan-functionalized magnetic nanoparticles cross-linked with carbodiimide and glutaraldehyde, J. Chromatogr. B 1054 (2017) 57−63. Available from: https://doi.org/10.1016/j.jchromb.2017.04.016.

[59] https://www.brenda-enzymes.org/index.php

[60] P.I. Tolstykh, T.E. Ignatyuk, V.K. Gostishchev, Morphological study of the effects of textile-immobilized enzymes on an experimental purulent wound, Bull. Exp. Biol. Med 118 (1994) 1027−1029. Available from: https://doi.org/10.1007/BF02445804.

[61] L. Shi, R. Ermis, B. Kiedaisch, D. Carson, The effect of various wound dressings on the activity of debriding enzymes, Adv. Skin. Wound Care 23 (2010) 456−462. Available from: https://doi.org/10.1097/01.ASW.0000383224.64524.ae.

[62] L. Chambers, S. Woodrow, A.P. Brown, P.D. Harris, D. Phillips, M. Hall, et al., Degradation of extracellular matrix components by defined proteinases from the greenbottle larva Lucilia sericata used for the clinical debridement of non-healing wounds, Br. J. Dermatol 148 (2003) 14−23. Available from: https://doi.org/10.1046/j.1365-2133.2003.04935.x.

[63] A.A. Vaniushenkova, S.N. Ivanova, S.V. Kalenov, N.S. Markvichev, A.A. Belov, The hydrolytic destruction of modified cellulosic materials in conditions simulating a purulent wound, Biointerface Res. Appl. Chem 10 (2020) 7265−7277. Available from: https://doi.org/10.33263/BRIAC106.72657277.

[64] A.M.B.F. Soares, L.M.O. Gonçalves, R.D.S. Ferreira, J.M. de Souza, R. Fangueiro, M.M.M. Alves, et al., Immobilization of papain enzyme on a hybrid support containing zinc oxide nanoparticles and chitosan for clinical applications, Carbohydr. Polym 243 (2020) 116498. Available from: https://doi.org/10.1016/j.carbpol.2020.116498.

[65] J. Dreyfus, G. Delhougne, R. James, J. Gayle, C. Waycaster, Clostridial collagenase ointment and medicinal honey utilization for pressure ulcers in US, hospitals. J. Med. Econ 21 (2018) 390−397. Available from: https://doi.org/10.1080/13696998.2017.1423489.

[66] R.E. Frederick, R. Bearden, A. Jovanovic, N. Jacobson, R. Sood, S. Dhall, Clostridium collagenase impact on zone of stasis stabilization and transition to healthy tissue in burns, Int. J. Mol. Sci 22 (2021) 8643. Available from: https://doi.org/10.3390/ijms22168643.

[67] J.D. Miller, E. Carter, D.C. Hatch, M. Zhubrak, N.A. Giovinco, D.G. Armstrong, Use of collagenase ointment in conjunction with negative pressure wound therapy in the care of diabetic wounds: a case series of six patients, Diabet. Foot Ankle (2015) 6. Available from: https://doi.org/10.3402/dfa.v6.24999.

[68] L. Shi, S. Ramsay, R. Ermis, D. Carson, pH in the bacteria-contaminated wound and its impact on clostridium histolyticum collagenase activity, J. Wound Ostomy Cont. Nurs 38 (2011) 514−521. Available from: https://doi.org/10.1097/WON.0b013e31822ad034.

[69] S.C. Pinto, A.R. Rodrigues, J.A. Saraiva, J.A. Lopes-da-Silva, Catalytic activity of trypsin entrapped in electrospun poly(ε-caprolactone) nanofibers, Enzyme Microb. Technol. 79–80 (2015) 8–18. Available from: https://doi.org/10.1016/j.enzmictec.2015.07.002.

[70] Rodriguez, B.A.G., Trindade, E.K.G., Cabral, D.G.A., Soares, E.C.L., Menezes, C.E.L., Ferreira, D.C. M., et al., Nanomaterials for advancing the health immunosensor, in: Biosensors - Micro and Nanoscale Applications, 2015. InTech. https://doi.org/10.5772/61149.

[71] Raushan Singh, M. Tiwari, Ranjitha Singh, J.-K. Lee, From protein engineering to immobilization: promising strategies for the upgrade of industrial enzymes, Int. J. Mol. Sci 14 (2013) 1232–1277. Available from: https://doi.org/10.3390/ijms14011232.

[72] T. Gunasekaran, T. Nigusse, M.D. Dhanaraju, Silver nanoparticles as real topical bullets for wound healing, J. Am. Coll. Clin. Wound Spec 3 (2011) 82–96. Available from: https://doi.org/10.1016/j.jcws.2012.05.001.

[73] E.P. Cipolatti, M.J.A. Silva, M. Klein, V. Feddern, M.M.C. Feltes, J.V. Oliveira, et al., Current status and trends in enzymatic nanoimmobilization, J. Mol. Catal. B Enzym 99 (2014) 56–67. Available from: https://doi.org/10.1016/j.molcatb.2013.10.019.

[74] X. Guo, H. Shi, W. Zhong, H. Xiao, X. Liu, T. Yu, et al., Tuning biodegradability and biocompatibility of mesoporous silica nanoparticles by doping strontium, Ceram. Int 46 (2020) 11762–11769. Available from: https://doi.org/10.1016/j.ceramint.2020.01.210.

[75] I.B. Ditta, A. Steele, C. Liptrot, J. Tobin, H. Tyler, H.M. Yates, et al., Photocatalytic antimicrobial activity of thin surface films of TiO_2, CuO and TiO_2/CuO dual layers on Escherichia coli and bacteriophage, Appl. Microbiol. Biotechnol 79 (2008) 127–133. Available from: https://doi.org/10.1007/s00253-008-1411-8.

[76] M.E. Okur, I.D. Karantas, Z. Şenyiğit, N. Üstündağ Okur, P.I. Siafaka, Recent trends on wound management: new therapeutic choices based on polymeric carriers, Asian J. Pharm. Sci 15 (2020) 661–684. Available from: https://doi.org/10.1016/j.ajps.2019.11.008.

[77] E.M. Papamichael, H. Stamatis, P.-Y. Stergiou, A. Foukis, O.A. Gkini, Enzyme kinetics and modeling of enzymatic systems, in: R.S. Singh, R.R. Singhania, A. Pandey, C. Larroche (Eds.), Advances in Enzyme Technology. Biomass, Biofuels, Biochemicals, Elsevier, 2019, pp. 71–104. Available from: https://doi.org/10.1016/B978-0-444-64114-4.00003-0.

[78] C.H. Evans, Interactions of tervalent lanthanide ions with bacterial collagenase (clostridiopeptidase A), Biochem. J 195 (1981) 677–684. Available from: https://doi.org/10.1042/bj1950677.

[79] Z. Bílková, M. Slováková, N. Minc, C. Fütterer, R. Cecal, D. Horák, et al., Functionalized magnetic micro- and nanoparticles: optimization and application to μ-chip tryptic digestion, Electrophoresis 27 (2006) 1811–1824. Available from: https://doi.org/10.1002/elps.200500587.

[80] P. Valencia, F. Ibañez, Estimation of the effectiveness factor for immobilized enzyme catalysts through a simple conversion assay, Catalysts 9 (2019) 930. Available from: https://doi.org/10.3390/catal9110930.

[81] U. Eckhard, E. Schönauer, P. Ducka, P. Briza, D. Nüss, H. Brandstetter, Biochemical characterization of the catalytic domains of three different clostridial collagenases, Biol. Chem. 390 (2009) 11–18. Available from: https://doi.org/10.1515/BC.2009.004.

[82] M.J. Cooney, Kinetic measurements for enzyme immobilization, in: S.D. Minteer (Ed.), Enzyme Stabilization and Immobilization. Methods and Protocols, Humana Press, New York, NY, 2017, pp. 215–232. Available from: https://doi.org/10.1007/978-1-4939-6499-4_17.

[83] V. Podzemná, M. Slováková, L. Kourková, L. Svoboda, Utilization of the IC-calorimeter for study of enzymatic reaction, J. Therm. Anal. Calorim 101 (2010) 715–719. Available from: https://doi.org/10.1007/s10973-010-0951-1.

[84] H. Lineweaver, D. Burk, The determination of enzyme dissociation constants, J. Am. Chem. Soc 56 (1934) 658–666. Available from: https://doi.org/10.1021/ja01318a036.

[85] J.C. Aledo, Enzyme kinetic parameters estimation: a tricky task? Biochem. Mol, Biol. Educ 49 (2021) 633−638. Available from: https://doi.org/10.1002/bmb.21522.

[86] M. Slovakova, N. Minc, Z. Bilkova, C. Smadja, W. Faigle, C. Fütterer, et al., Use of self assembled magnetic beads for on-chip protein digestion, Lab. Chip 5 (2005) 935. Available from: https://doi.org/10.1039/b504861c.

[87] Murray, R.K., Granner, D.K., Mayes, P.A., Rodwell, V.W., Harper's Biochemistry, 23rd edition. H + H 4th Czech, Prague, 2002.

Antibacterial hydrogel dressings and their applications in wound treatment

Xinyu Song[1], Jorge Padrão[2], Marta Fernandes[2], Ana Isabel Ribeiro[2], Liliana Melro[2], Cátia Alves[2], Liangmin Yu[1] and Andrea Zille[2]

[1]*Key Laboratory of Marine Chemistry Theory and Technology, Ministry of Education, Ocean University of China, Qingdao, Shandong, P.R. China* [2]*Centre for Textile Science and Technology (2C2T), University of Minho, Guimarães, Portugal*

8.1 Introduction

An open wound is caused by traumatic skin damage, such as cuts, lacerations, abrasions, avulsions, penetrations, surgery, bite, burns, or diseases (ulcers from diabetes, hypertension, rheumatoid arthritis, and so on) [1]. Open wound management remains a serious global challenge, resulting in severe morbidity, mortality, and a significant economic burden [2]. Wound healing is frequently characterized by a dynamic series of overlapping phases, including inflammation, proliferation (including coagulation, granulation tissue formation, and re-epithelialization), and extracellular matrix (ECM) formation [3]. Open wounds are classified as acute or chronic based on their healing time and progress [4]. Acute wounds caused by common trauma or surgical procedures are expected to progress through an orderly series of wound healing phases, which typically last 8 to 12 weeks depending on their size and depth [5]. If the inflammation phase lasts for an extended period of time, an acute wound may be classified as a chronic wound, which often results in a delayed and incomplete healing process [6,7]. Intrinsic and extrinsic factors such as a lack of nutrition, pathogenic infections, co-morbidities, medications, or inappropriate wound dressing selection or management can all delay wound healing [8,9]. When the defensive skin integrity of an open wound is compromised, pathogenic microorganisms can invade and cause local infections or even sepsis [10,11]. Accurate wound assessment and appropriate wound management are required for wound treatment to achieve effective healing [12,13].

Advanced biomaterials have been developed to accelerate healing and reduce infection risk. Cloths, foams, films, hydrofibers, hydrocolloids, and hydrogels are examples of biomaterial wound dressings [14−16]. Hydrogels are three-dimensional network structure gels that contain more than 70% water and are crosslinked hydrophilic polymers by physical interactions or covalent bonds. Semi-stiff sheets and amorphous forms of hydrogels are common [17,18]. Several hydrogels have been extensively and successfully used as wound dressings due to their high water content, lower wound adherence, promoted autolysis debridement, epithelial migration, granulation growth, biocompatibility, biodegradability, and easy loading and release of bioactive agents [19,20]. For improved wound management, advances in materials science and biotechnology have resulted in the development of a wide range of antimicrobial hydrogels that include a sensor, imaging,

Antimicrobial Dressings. DOI: https://doi.org/10.1016/B978-0-323-95074-9.00010-5

debridement, microbial infection control, and wound healing functions (including immunoregulation, angiogenesis, and ECM remodeling) [21−24]. Antimicrobial hydrogels are classified into two types based on their modes of action: contact-killing and release-killing. Pathogenic microorganisms can be eradicated only through contact or close contact with the hydrogel in the case of contact-killing [21]. In the release-killing method, bioactive agents are leached into the wound site and act against pathogens near the hydrogel contact area as well as those in the wound's infected surrounding area (Fig. 8.1). It is worth noting that the contact-killing category includes both naturally active and drug-loaded antimicrobial hydrogels [21,25]. The inherently active antimicrobial hydrogels are generally fabricated by functional polymers, such as polypeptides, polysaccharides (e.g., chitosan, alginate, cellulose, gellan gum, dextran, hyaluronic acid, starch), and proteins (e.g., silk protein, fibrins, collagen) [26,27]. While drug-loaded antimicrobial hydrogels are versatile due to the various active agents loaded in the hydrogels, such as antimicrobial agents (e.g., antibiotics, silver nanoparticles, gold nanoparticles, antimicrobial peptides, essential oils), growth factors (e.g., vascular endothelial growth factor, epidermal growth factor, fibroblast growth factor, transforming growth factor, platelet-derived growth factor), and antioxidant and anti- (e.g., vitamin, trace elements, therapeutic gases, heparin, curcumin, capsaicin, cytokines) [28−32,33]. Furthermore, as the next generation of wound dressings, multifunctional antimicrobial hydrogels with stimuli-responsive drug release features (pH, enzyme, temperature, light, magnetic, etc.) have received a lot of attention [34−37]. Furthermore, antimicrobial hydrogels with sensor and imaging functions are being developed. These functions may aid in the detection of real-time moisture, pH, temperature, or other infection biomarkers for accurate wound etiology and current status assessments [38,39].

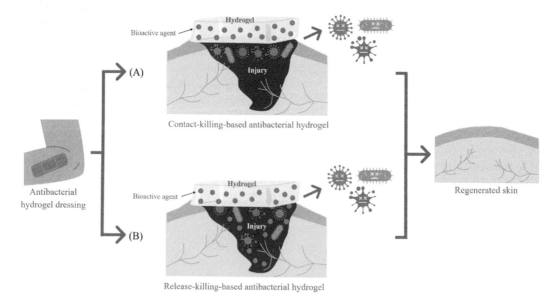

FIGURE 8.1

Distinct antibacterial hydrogel modus operandi: (A) Contact-killing-based antibacterial hydrogel; (B) Release-killing based antibacterial hydrogel.

This chapter will provide a comprehensive overview of advanced antimicrobial hydrogels in research and commercially available hydrogel dressings, including their design, fabrication method, and in vitro or in vivo wound management efficacy. The antimicrobial hydrogels were classified based on their mode of action and drug versatility mechanisms. Their operational details and critical discussion, which include their benefits and drawbacks, may provide insights into the future design of antimicrobial hydrogels for better wound management in clinical applications.

8.2 Contact-killing-based hydrogels

Hydrogels can be used as matrix materials, emulsions, or injectables. The most commonly used are matrixes, which are prepared and tailored ex situ from the wound site and are easily manipulated. Emulsion templates make it easier to create porous polymeric networks. In contrast, injectable hydrogels are advantageous carriers because they are biocompatible, non-invasive, and adaptable to a wide range of wound types [40]. The active antibacterial agent may be designed to be firmly attached to the matrix regardless of the type of hydrogel used as a wound dressing. As a result, no antibacterial agent is expected to be released. This strategy has two distinct advantages and two distinct disadvantages. Without release, the duration of antibacterial activity may be significantly longer. Furthermore, if the antibacterial agent has exceptional bactericidal properties but unacceptable cytotoxicity (particularly systemic toxicity), this strategy may significantly mitigate this critical issue. However, because the active agent is strongly entrapped in the hydrogel matrix, it may be ineffective in its vicinity. Unless it is capable of performing an indirect antibacterial activity, such as the production of reactive oxygen species that are released in the wound site and kill the invading bacteria [41]. Furthermore, if the active agent requires "unrestricted" interaction with the bacteria, the tight connection with the hydrogel matrix may significantly reduce its effectiveness. Padro et al. reported on this issue, in which the antibacterial agent, bovine lactoferrin, was covalently bound to bacterial nanocellulose via periodate oxidation. This process not only reduced the mobility of bovine lactoferrin, but also rendered the protein's N-terminus, which contains the most antimicrobial domain, bio-inaccessible, reducing its antibacterial activity [42]. Baus et al. reported the use of copper and calcium ions as crosslinkers of cellulose nanofibrils and as an active antibacterial agent. It displayed a bacteriostatic and antifouling effect against *Staphylococcus epidermidis* (*S. epidermidis*) and *Pseudomonas aeruginosa* (*P. aeruginosa*), respectively [43]. Want et al. developed a polyethylene glycol dimethacrylate, N,N-methylene-bis-acrylamide, methyl methacrylate, 1-vinyl-3-butylimidazolium, and acrylamide hydrogel using an ionic liquid as a crosslinker. It proved efficacy against *Staphylococcus aureus* (*S. aureus*), *Escherichia coli* (*E. coli*), and *Candida albicans* (*C. albicans*). Due to its prompt tackling of the wound infection, this hydrogel accelerated the wound healing process [44]. Finally, it should be noted that some hydrogels have inherent antimicrobial properties when the main matrix components are biocidal. Chitosan is the most well-known naturally antibacterial compound [45].

8.3 Release-killing-based hydrogels

Hydrogels that are intended to eliminate microorganisms throughout the wound site environment must release antimicrobial agents. Because of the administration of a high dose of antimicrobial agents in its location and vicinity, this strategy is expected to manage infections effectively and safely. Antimicrobial agents, on the other hand, are frequently released with or without the presence of infection, resulting in a rapid depletion of their drug supply. Because of the presence of sublethal doses, this facilitates the development of resistant pathogen strains. Furthermore, due to being weakly absorbed or bonded to the hydrogels, diffusional-release hydrogels are prone to burst releases, which are defined as an abrupt release of a high concentration of antimicrobial agents. Burst release may pose a serious threat to host cell viability and may result in critical organ and tissue damage [46]. As a result, the development of hydrogels with controlled drug release systems or even stimuli response release systems has enormous potential. Drug release kinetics can be described using mathematical models generated from experimental release data. It is possible to design safer materials or adjust their use in specific clinical situations by understanding the various factors that govern the release rate and behavior. Several mathematical models can be used to describe the kinetics of drug release. The main mathematical models of drug release are the zero-order, first order, Higuchi, Hixson-Crowell, Ritger-Peppas, Korsmeyers-Peppas, Brazel-Peppas, Baker-Lonsdale, Hopfenberg, Weibull, and Peppas-Sahlin [47]. The fitness of the release values to each of the referred models must be modeled (or even others, such as modified Gompertz) [48,49]. In a nutshell, the zero-order model applies to materials in which the drug is released at a constant rate over time. When the percentage of release depends on the concentration of active agent, the first-order model is used. The Higuchi model is used to describe three-dimensional structures where release occurs via diffusion. The Hixson-Crowell model is used in drugs whose diameter and surface area change during release. Finally, the Korsmeyer-Peppas and derivative models are the most commonly used to describe drug release systems, particularly in hydrogels [50].

The good of the fitness determines which model provides a more accurate representation of the release behavior when compared to the models Ritger-Peppas, Korsmeyers-Peppas, Brazel-Peppas, and Peppas-Sahlin. The diffusivity exponent is used to describe the release in these models (n). The model data describes the release mechanism, which includes quasi-Fickian, Fickian diffusion, anomalous transport (non-Fickian), and transport case II (non-Fickian). When the degree of swelling governs agent release, it falls into the category of Fickian diffusion. Due to its significantly smaller size in comparison to the hydrogel matrix, quasi-Fickian diffusion describes the rapid release of an agent. The Anomalous transport model is used when diffusion and swelling are the main events influencing the release rate. Transport II includes cases where diffusion is the dominant mechanism (over swelling) and may be related to hydrogel erosion. Table 8.1 depicts the release mechanisms related to the diffusivity exponent (n) and the active agent structure shape.

Table 8.1 Diffusivity exponent (n) value according to the release mechanism of the models [50–52].

Diffusivity exponent (n)			
Film	**Cylinder**	**Spheres**	**Release mechanism**
$n < 0.5$	$n < 0.45$	$n < 0.43$	Quasi-Fickian
$n = 0.5$	$n = 0.45$	$n = 0.43$	Fickian diffusion
$0.5 < n < 1.0$	$0.45 < n < 0.89$	$0.43 < n < 0.85$	Anomalous transport (non-Fickian)
$n > 1.0$	$n > 0.89$	$n > 0.85$	Transport case II

The common strategies for the development of diffusional release of active antimicrobial agents from hydrogels are described in this section, which is divided by active ingredient type: nanoparticles, antibiotics, and other antimicrobial agents (e.g., curcumin, quercetin, dopamine, matrine, and cordycepin). When the mathematical model of drug release is available, it will be described.

8.3.1 Silver nanoparticles

Several studies have reported the use of functional hydrogels containing metal nanoparticles as biocompatible active antimicrobial agents. Metal nanoparticles, in contrast to conventional antimicrobial agents, have multiple mechanisms of action, often impeding microbial resistance. The antimicrobial mechanisms of the metal nanoparticles used are still unknown. Nonetheless, a variety of metal nanoparticle modes of operation have been proposed, including: (1) direct electrostatic interactions with microorganism cell membrane and/or cell wall components; (2) homeostatic stress generation due to an abnormal ion imbalance, which impairs respiration, interrupts energy transduction, and leads to cell death; (3) interactions with sulfur-containing molecules such as proteins; and (4) the production of reactive oxygen species and the release of metal nanoparticles are also said to be capable of preventing biofilm formation and impeding cell wall synthesis [53,54].

Due to their excellent antimicrobial properties against a wide range of bacteria, fungi, and viruses, including some resistant strains, silver nanoparticles have received special attention as antimicrobial agents [55]. Furthermore, silver nanoparticles cytotoxicity is under intensive analysis. Silver nanoparticles concentration of 0.5 μg/mL has been described as highly effective against bacteria. Moreover, 0.25 μg/mL of silver nanoparticles display absent cytotoxicity to various mammalian cells [56]. As a result, crosslinked polymers containing silver nanoparticles in hydrogels, biomembranes, and scaffolds have emerged as a new class of multifunctional biomaterials for skin tissue engineering [57]. Silver nanoparticles-based wound dressings have been designed and are already commercially available as Aquacel Ag, Acticoat Flex, Tegader Ag, Silverce, among others [58].

The hydrogels containing silver nanoparticles were created using a variety of techniques. Silver nanoparticles can be prepared in a single step by chemical or biological synthesis and then embedded in hydrogel matrixes, or they can be synthesized in situ during hydrogel formation. In the reduction step, highly reactive chemicals such as sodium borohydride are commonly used; however, plant extracts or even hydrogel components (without any additives) were also used. As a result, green solvents, eco-friendly reducing agents, and nontoxic hydrogel materials are the primary issues that should be addressed in synthesis protocols [59].

The two-step method, which includes separate production of the hydrogel and silver nanoparticles, followed by their combination, is a simple approach for producing hydrogels containing silver nanoparticles. Several studies report on various strategies for delaying or controlling the release of silver nanoparticles or silver ions.

8.3.1.1 Two-step method hydrogel synthesis

Jiang et al. used cavitation to combine commercial silver nanoparticles with oxidized konjac glucomannan and carboxymethyl chitosan. The hydrogels produced were able to reduce the cytotoxicity of silver nanoparticles and mitigate the burst release of silver ions. Inductively coupled plasma mass spectrometry was used to detect the concentration of diffused silver ions from the hydrogel in a weakly acidic solution, demonstrating gradual and slower release of silver ions in comparison to pure silver nanoparticles. The achieved hydrogel did not impair fibroblast viability in vitro and outperformed the hydrogel without silver nanoparticles in antibacterial effectiveness against *S. aureus* and *E. coli* [60]. Zhao et al. created a polydopamine, polyaniline, polyvinyl alcohol, and silver nanoparticle-containing injectable hydrogel. This hydrogel was created by reducing sodium borohydride and stabilizing it with polyvinylpyrrolidone and trisodium citrate. The hydrogel inhibited *E. coli* and *S. aureus* growth significantly and demonstrated good biocompatibility [59]. Kumar et al. reported the synthesis of silver nanoparticles using *Eucalyptus citriodora* leaf extracts. The silver nanoparticles were added to a polyvinyl alcohol and chitosan solution. Spraying glutaraldehyde and boric acid as crosslinkers resulted in the hydrogel. The release profiles of silver nanoparticles were monitored using atomic absorption spectroscopy, and a clear delay in the maximum cumulative release of silver ions was observed in the various hydrogel formulations tested. Because of the hydrophilic and biocompatible nature of the polymers used in its formulation, the hydrogel demonstrated hemocompatibility. Finally, the hydrogel had a weak and moderate antibacterial activity against *S. aureus* and *E. coli*, respectively [61]. Gou et al. developed an injectable gelatin-carboxylated cellulose hydrogel with aminated-silver nanoparticles. Polyvinyl pyrrolidone and a silver salt precursor were used to create aminated-silver nanoparticles at 120°C, followed by a silver nanoparticles reaction with 3-aminopropyltriethoxysilane. Electrostatic forces caused the negatively charged carboxylated cellulose to interact with the aminated silver nanoparticles. Furthermore, van der Waals interactions and hydrogen bonds were found to promote the formation of a ternary hydrogel with a homogeneous polymeric network. The aforementioned chemical interactions regulated the release of silver nanoparticles with prolonged activities, accelerated wound healing, and demonstrated antibacterial activity against *E. coli* [62]. Gupta et al. created silver nanoparticles by reducing and stabilizing them with a curcumin-hydroxypropyl- β-cyclodextrin complex. Curcumin microencapsulation in cyclodextrins was used to overcome curcumin's hydrophobicity. The paddry method was used to load silver nanoparticles into bacterial nanocellulose hydrogels. The hydrogel's adequate antimicrobial performance against *S. aureus*, *P. aeruginosa*, and *Candida auris*, as well as the absence of cytotoxicity, suggested an adequate release of silver nanoparticles [63].

8.3.1.2 *One-step method hydrogel synthesis*
8.3.1.2.1 Chemical-based methodology
The silver nanoparticles are synthesized within the hydrogel matrix in a single step. During hydrogel applications, this strategy aims to avoid unwanted agglomeration and burst leakage of silver nanoparticles. Tan et al. created boron-catechol and polyaspartamide hydrogels with silver nanoparticles synthesized in situ. Silver nanoparticles were formed after immersing the hydrogel in a silver solution due to the catechol moieties, which can coordinate with metal ions and induce the growth of metal nanoparticles without the use of an additional reducing agent. The hydrolysis of the carbonate ester linkages along the polyaspartamide chains was primarily responsible for the slow release profile observed over a long period of time (40 days to release only 7.5%). It was justified by the hydrogels' high hydrophobicity, which resulted in poor water uptake and slow diffusion through the network. According to this profile, this material could be used for long-term applications. However, cytotoxicity was observed at high concentrations (5 mg/mL). Antimicrobial tests showed a bacteriostatic and bactericidal effect against *S. aureus* and *E. coli*, respectively [64]. Chalitangkoon et al. created a hydroxyethylacryl-chitosan and sodium alginate hydrogel with in

situ silver nanoparticles that were chemically synthesized using sodium borohydride. The presence of silver nanoparticles increased swelling and improved mechanical properties, according to the findings. The in vitro drug release profiles of para-acetylaminophenol, a soluble model drug, were investigated. The drug release was prolonged as the crosslinking density and silver concentration increased. The entire drug was released within 32 hour. The film fit the first-order model well, indicating that the concentration of para-acetylaminophenol affects drug release. The hydrogels had antibacterial activity against *E. coli* and *S. aureus* while causing no cytotoxicity [65]. Chen et al. also created silver nanoparticles using sodium alginate as a reducing and stabilizing agent. The freeze-thaw method and the calcium ion crosslinking method were used to create carboxymethyl chitosan and polyvinyl alcohol hydrogel. The silver nanoparticles were distributed uniformly throughout the hydrogel and demonstrated a synergistic antibacterial effect with chitosan against *E. coli* and *S. aureus*. One possible synergistic mode of operation is that the protonated ammonium on the chitosan chain promotes electrostatic interaction and tropism with the negatively charged cell membrane of bacteria, disrupting membrane stability and interfering with metabolism. The current silver nanoparticles, on the other hand, cause lethal reactive oxygen-catalyzed reactions. Furthermore, the hydrogel exhibited excellent biocompatibility within a 72 hour period [66].

8.3.1.2.2 Polymer-based methodology

Thi et al. created an injectable hydrogel by combining a catechol-rich gelatin solution with a silver salt precursor. Silver nanoparticles were synthesized in situ during hydrogel formation using catechol groups and no other reductants. The hydrogel was created by coupling 1ethyl3(3dimethylaminopropyl)carbodiimide and Nhydroxysuccinimide. The silver was sustainably released over 2 weeks from the hydrogels and prolonged the degradation time of pure hydrogel scaffolds from 18 to 70 hour. The superior crosslinking density obtained in this hydrogel delayed silver nanoparticles diffusion. In fact, solely 8.7% (~ 4.5 µg) of the total silver incorporated inside the hydrogel was released. This dose, sustained by slow silver release, remained within effective bactericidal concentrations and was free of cytotoxicity. As a result, the composite demonstrated excellent antibacterial activity against Gram-negative and Gram-positive bacteria while causing no toxicity to mammalian cells [56]. Capanema et al. used citric acid as a crosslinker to create hybrid membranes of carboxymethyl cellulose using the solvent casting method. The carboxymethyl cellulose functional groups were important in the synthesis of silver nanoparticles and the crosslinking of the polymer network. The results showed that superabsorbent hydrogels were produced with swelling and degradation behaviors dependent on crosslinker concentration, degree of carboxymethylation of carboxymethyl cellulose, and silver nanoparticle content in the formulation. The hybrid nanocomposite was found to be cytocompatible and antibacterial against *S. aureus*, *E. coli*, and *P. aeruginosa* [57]. Wang et al. created hydrogels by combining chitosan, oxidized konjac glucomannan, and silver nanoparticles. Schiff-base linkages were used to connect the aldehyde groups of oxidized konjac glucomannan and the amine groups in the backbones of protonated chitosan and protonated tranexamic acid to create self-adapting hydrogel wound dressings. Histological analysis revealed that this self-adapting hydrogel outperforms the commercial hydrogel dressing, Aquacel Ag, in the in vivo wound-healing process. The cytotoxicity results indicated that the cell viability in the case of the experimental group remained greater than 98% for at least 48 hour. The low cytotoxicity was attributed to the biocompatibility of the natural polymers in the hydrogel formulation. Aquacel Ag and the prepared hydrogel demonstrated an intriguing antibacterial effect against *E. coli* and *S. aureus*. Nonetheless, the chitosan-konjac-silver nanoparticles hydrogel demonstrated superior efficacy [67].

Masood et al. created a chitosan and polyethylene glycol hydrogel containing silver nanoparticles. The silver nanoparticles were created by adding polyethylene glycol during the hydrogel formation process. The addition of polyethylene glycol also aided in the stabilization of silver nanoparticles, which is essential for their prolonged biological activity. In addition, glutaraldehyde was used as a crosslinker. The rate of release of silver nanoparticles was measured using ultraviolet-visible spectrophotometry. The hydrogel released silver nanoparticles slowly and consistently for at least seven days, demonstrating the slow biodegradation of developed hydrogels (there was no burst release). This hydrogel was found to be effective against *P. aeruginosa*, *E. coli*, *S. aureus*, *B. subtilis*, and *B. pumilus* [68]. Zepon et al. described a simple and environmentally friendly method for producing κ-Carrageenan hydrogels via a sol (heating) - gel (cooling) process that allowed for the simultaneous synthesis of silver nanoparticles during the heating time. Based on ultraviolet analysis, the synthesis of silver nanoparticles was found to be faster at higher temperatures, most likely due to the expansion of the κ-Carrageenan, which facilitates the interaction of silver ions with the available functional groups on the polysaccharide. The addition of silver nanoparticles altered the hydrogel's swelling properties, as well as its viscosity and gelling temperature. In the release study, the silver nanoparticles were continuously released for up to 48 hour in a concentration sufficient to prevent bacterial growth as confirmed by antimicrobial tests. The amount of silver released from the hydrogel was unaffected by the pH of the pseudo-extracellular fluid. The hydrogel's antimicrobial activity against *S. aureus* and *P. aeruginosa* was tested, confirming the inhibitory effects of silver nanoparticles [58].

8.3.1.2.3 Irradiation-based methodology

Several works describe the use of gamma irradiation to reduce silver ions present in a polymer matrix to simultaneously generate silver nanoparticles and hydrogels [69−72]. For instance, Lima et al. produced a hydrogel containing polyvinylpyrrolidone, polyethylene glycol, agar, and carboxymethyl cellulose with silver nanoparticles using Cobalt 60 gamma irradiation under an inert atmosphere (nitrogen). Gamma irradiation was concomitantly performed for polymer crosslinking, reduction of silver ions, and sterilization of the obtained hydrogel. The release profiles for silver nanoparticles from the hydrogel exhibited an initial burst release up to 3 hour. A relative stability of release thereafter up to 40 hour followed. No indications of in vitro toxicity were observed [69].

Ultraviolet light was used to reduce silver ions and form the hydrogel [73−75]. Baukum et al. created an alginate-gelatin hydrogel with silver nanoparticles that were crosslinked with glutaraldehyde and calcium chloride using a solvent casting method. Atomic absorbance spectroscopy was used to determine the amount of silver ions released by the hydrogel. Although there was an initial burst release, the diffusion rate eventually slowed and stabilized. The hydrogel was non-toxic and inhibited the growth of *S. aureus*, *P. aeruginosa*, and *E. coli* [73].

Ounkaew et al. created hydrogel membranes containing in situ synthesized silver nanoparticles from carboxymethyl starch and polyvinyl alcohol. Crosslinking into a hydrogel was accomplished through chelation with polyvalent cations, specifically citric acid, which established ester linkages. The hydrogel's release of silver ions was studied using atomic absorbance spectroscopy. Rapid silver ion release was observed during the first 6 hour followed by a slower release. The initial release stage may prevent bacterial adhesion to the wound and, as a result, biofilm formation, whereas the slower release stage provides wound site asepsis during prolonged treatment. The profile release mechanism of silver nanoparticles fit the Kormeyer–Peppas model, indicating a Fickian diffusion mechanism. Furthermore, the hydrogel was found to be non-toxic to human fibroblast cells, indicating good biocompatibility and antibacterial activity against *E. coli* and *S. aureus*. The antibacterial activity could be attributed to citric acid's acidity as well as the release of silver ions from silver nanoparticles. *S. aureus* inhibition zones were larger than *E. coli* inhibition zones [58,68,76].

8.3.2 Additional nanoparticles

Most of the studies in the literature were performed using silver nanoparticles, however, also some different nanoparticles can be found, namely copper [77], gold [78], zinc oxide [79], magnesium [80], cerium [69], silver-titanium [81], and zinc oxide-silver [82] nanoparticles. Ponco et al. used copper nanoparticles to create carboxymethyl cellulose and polyvinyl alcohol. First, copper ions were electrostatically bonded with the hydroxyl and carboxyl groups in polymer matrixes. The nanoparticles were then created chemically using hydrazine. The composite was more effective against *E. coli* than against *S. aureus* [77]. Zinc oxide nanoparticles were incorporated in hydrogels of bacterial cellulose/chitosan [79], xanthan/polyvinyl alcohol [83], and carboxymethyl cellulose/κ-Carrageenan/graphene oxide/konjac glucomannan [82]. Antimicrobial properties of zinc oxide nanoparticles against *S. aureus*, *E. coli*, and *Candida albicans* were discovered. Ahmed et al. created a hydrogel with polyvinyl alcohol and starch that was crosslinked with glutaraldehyde. The authors also included filler graphitic carbon nitride and antibacterial silver-titanium dioxide nanoparticles. The kinetics of drug release were studied using various mathematical models, with the Higuchi model providing the best fit. The non-Fickian diffusion mechanism prevailed, resulting in a sustained and slow release of nanoparticles. The hydrogel allowed for complete healing in seven days and provided good antibacterial properties against *S. aureus* and *E. coli* [81]. The research using various silver nanoparticles demonstrated the potential for the development of novel antimicrobial hydrogels with comparable or even superior properties. The environmental and safety concerns associated with the use of silver nanoparticles necessitate the development of novel materials.

8.3.3 Antibiotics

Antibiotics are essential for the control of microbial infection. Antibiotics are typically administered via injection or oral administration, with significant dosages required due to change/degradation or leakage between tissues and organs until it reaches the target area [84]. They can also be applied directly to wounds as creams or gels, but they are not effective for long periods of time [85]. Antibiotics and their release from hydrogel materials are the most studied active pharmaceutical agents for wound dressing applications, alongside metallic nanoparticles. Drug-loaded wound dressings can deliver drug molecules through the wound site for a longer period of time without the need for frequent dressing replacement [86].

Hydrogel dressings can be made simply by mixing and drying them, or they can be crosslinked to increase their stability and porosity. Porous structures provide an adequate environment for nutrient and oxygen transfer, cell growth and proliferation, evaporation of wound exudates, preventing fluid accumulation and, as a result, lowering the risk of skin maceration and infections, and drug molecule release [85,87]. Huang et al. created an antimicrobial hydrogel film by crosslinking hyaluronic acid with carboxylated chitosan and gentamicin via intermolecular covalent bonds mediated by carbodiimide chemistry. Hyaluronic acid is a key molecule in connective tissue structuration and collagen formation and maintenance, whereas carboxylated chitosan inhibits bacterial growth. Both polymers are water soluble, biocompatible, biodegradable, and easily modified. They were crosslinked to improve mechanical properties and control degradation rate, resulting in the hydrogel film's long-term antibacterial effect. The developed hydrogel demonstrated good water absorption capacity, water vapor permeation rate, mechanical properties, enzymatic hydrolysis resistance, and biocompatibility. Release tests showed that almost 40% of gentamicin was released in 24 hour and after that was released at a slower rate for 9 days. This behavior indicates that gentamicin is being released simultaneously via degradation and diffusion. Furthermore, the antibacterial activity of the hydrogel against *P. aeruginosa* and *S. aureus* was found to be moderate, whereas the commercial dressing Aquacel Ag demonstrated weak activity. In vivo wound healing tests in mice using a full-thickness skin defect revealed that both the hydrogel and Aquacel Ag resulted in relatively intact skin after 16 days [88]. Picone et al. used glutaraldehyde to create a chemically crosslinked xyloglucan-polyvinyl alcohol hydrogel film. Glycerol was added as a plasticizer and to provide adequate water retention properties prior to drying. The hydrogel produced had a porous morphology (pore size of 20 μm), a high gel fraction (93%), 350% swelling in phosphate buffer saline, biocompatibility (viability of 90% for epithelial cells), hemocompatibility, immunogenicity, and partial adhesiveness. To investigate the hydrogel's ability to absorb and release molecules, it was first swollen in water to an equilibrium state before being spectroscopically evaluated with a fluorescent hydrophilic dye. The hydrogel absorbed 80% of the dye and released 98% of the load in 24 hour, following a Fickian diffusion model. Furthermore, antibacterial tests showed that this hydrogel provides mechanical protection against bacterial infiltration, while the hydrogel loaded with ampicillin showed an inhibitory effect against *E. coli* [89]. Polat et al. developed nanocomposite hydrogels made of agar, κ-Carrageenan, and montmorillonite that were crosslinked using free radicals. These hydrogels were used as carriers for analgesic lidocaine hydrochloride and antibiotic chloramphenicol. Montmorillonite contributed to increase the ultimate compressive stress (47.70 kPa) and reducing the amount of lidocaine hydrochloride and chloramphenicol released (90% within 3.5 hour). The hydrogels had a high swelling index in various physiological solutions, were biocompatible (osteosarcoma cells viability was 104%), and had antibacterial activity against *E. coli* and *S. aureus* (strong inhibition zones) [84].

Freeze-drying is used in several studies to create stable crosslinked porous hydrogels. Erdagi et al. created gelatin/diosgenin-carboxylated nanocellulose antibiotic-based hydrogels with genipin as a crosslinking agent. When compared to other synthetic crosslinking agents, this strategy achieved lower cytotoxicity and anti-inflammatory properties. The hydrogels exhibited a porous morphology with interconnected porosity (pore sizes of 10 to 30 μm), good swelling capacity, high gel yield (~90%), and biocompatibility (fibroblast cells viability of ~80%). The loading efficiency of neomycin in the hydrogels was found to be 95.5% and drug release tests showed a release of approximately one-fifth within 15 minutes and was complete after 24 hour. The hydrogels presented weak and moderate antibacterial activity against *E. coli* and *S. aureus*, respectively [90]. Sadeghi et al. used freeze-drying to create an antibacterial wound dressing from carboxymethyl cellulose (carboxymethyl cellulose-human hair keratin and clindamycin loaded halloysite nanotubes, with citric acid as a crosslinker and glycerol as a plasticizer). The obtained hydrogel presented a distinctly interconnected porous structure (mean pore sizes of 98 mm), high compressive modulus

(179 KPa), high water uptake (1.5 g/g), and adequate water vapor permeation rate (1921 g/m^2 day^{-1}). Moreover, the hydrogel showed blood compatibility (protein adsorption of $\sim$37 mg/g, clotting time of 13 seconds) and cell viability ($>$90% of mouse fibroblasts viability). The in vitro release study indicated that the clindamycin release was controlled via a Fickian diffusion mechanism, with a release of nearly 13% in 4 hour, reaching 50% in 7 days. Additionally, the fabricated hydrogel dressing displayed a weak activity against *S. aureus* (79%) [91]. Kaur et al. created poly (vinyl alcohol)-sodium alginate hydrogels that were crosslinked using a chemical/ionic method to create water-insoluble membranes for use in the treatment of burn wound infections. Bacteriophages were loaded to combat antibiotic-resistant pathogens and co-incorporated with minocycline to boost phage treatment efficacy. The hydrogel presented self-adherence, high swelling index ($\sim$850%), high gel fraction ($\sim$52%), high protein absorption (0.1 mg/cm^2), good hemocompatibility, antibacterial activity against methicillin resistant *S. aureus*, *Klebsiella pneumoniae*, and *P. aeruginosa*, and biocompatibility (skin epithelial cells viability $>$94.25%). Elution assay showed nearly all the antibiotic and phage particles were released within the first 15 minutes. In vivo tests using an induced murine burn wound model confirmed the potential of the combinational treatment with phage and antibiotic showing complete regeneration on the 14th day of treatment [92]. Tao et al. created a stable drug carrier hydrogel by repeatedly freezing and thawing a sericin and polyvinyl alcohol hydrogel. The hydrogel had a porous morphology (pore size of 38.19 µm, porosity $>$85%), a high swelling index (1200%), good mechanical properties, and biocompatibility (viability of mouse fibroblasts and human epithelial cells of 130% and 140%, respectively). The hydrogel has also the ability to load and release drugs (gentamicin release of approximately 80% within 10 hour) and antibacterial activity against *E. coli*, *S. aureus*, and *P. aeruginosa* [87]. Ahmed et al. prepared calcium alginate hydrogels through several freeze-drying cycles to deliver ciprofloxacin directly to the wound site of infected diabetic foot ulcers. Dressings with 0.005%−0.025% of ciprofloxacin presented high porosity ($\sim$98%−95%), moisture content ($\sim$17%), equilibrium water content ($\sim$94%−92%), swelling ($\sim$1520%−1087%), water vapor transmission rate ($\sim$3446−3499 g/ m^2 day^{-1}), and biocompatibility (human keratinocytes cell viability $>$85%). The hydrogel with 0.025% ciprofloxacin showed the fastest release rate (nearly 68% within 5 minutes), followed by sustained Fickian drug release. It also had extremely potent antibacterial activity against *E. coli*, *S. aureus*, and *P. aeruginosa*. Overall, the new hydrogel dressing outperformed the commercial dressing Algisite Ag [85].

Using multilayers to integrate the benefits of each constituent is a different approach for the development of hydrogels. Tamahkara et al. created a multilayer wound dressing using layer-by-layer self-assembly via electrostatic interactions between four polymeric layers. The upper layers, carboxylated polyvinyl alcohol and gelatin, were in charge of moisture control and providing a physical barrier to microorganisms. The hyaluronic acid was used as an antibiotic-loaded layer, and the lower layer, which was also gelatin, was used to control antibiotic release and remove excess exudate from the wound site. The swelling ratios of the multilayer hydrogels were approximately 518% and 339% for pH 5.5 and 7.4, respectively. After 15 days, the hydrolytic degradation test revealed complete degradation, indicating a long-term degradation profile with good stability. Regarding drug release, ampicillin exhibited a burst release of 34.5% within 6 hour followed by a 65% zero-order model release within 7 days. Furthermore, multilayer hydrogels demonstrated no toxicity (mouse fibroblast cells) and moderate antibacterial activity against oxacillin-resistant *S. aureus* [86]. In another study, an antibiotic-loaded carboxymethylcellulose hydrogel was mixed into a crosslinked nano-electrospun fiber mat made of enzymatic poly-3-caprolactone nanofibers grafted with poly (gallic acid). The drug carrier was carboxymethylcellulose hydrogel, and the electrospun polymers provided mechanical support in a porous three-dimensional (3D) structure. The resulting composites were hemocompatible and non-cytotoxic (epithelial cells). Clindamycin was released using a Fickian diffusion model, and antibacterial tests revealed activity against *S. aureus* (moderate activity) [93]. In addition to the properties mentioned above, injectable hydrogels have the ability to fill wound sites. Qu et al. created injectable hydrogel dressings by combining oxidized hyaluronic acid-graft-aniline tetramer and N-carboxyethyl chitosan with Schiff base bond formation under physiological conditions. Aniline has conductivity and antioxidant properties, which aid in the proliferation of electrical stimuli sensitive cells and free radical scavenging, thereby accelerating wound healing. The addition of in situ encapsulated amoxicillin resulted in a Fickian diffusion model and demonstrated strong antibacterial activity against *S. aureus* and *E. coli*. Moreover, in vivo experiments using a full-thickness skin defect model showed better therapeutic effect of the developed hydrogels than the commercial film dressing Tegaderm [94].

A summary of the antimicrobial hydrogel carriers is depicted in Table 8.2.

8.3.4 **Other antimicrobial agents**

Nanoparticles and antibiotics frequently exhibit several issues, including insolubility, systemic toxicity, and extending the healing process of a wound. Furthermore, antibiotic overuse is dangerously increasing microorganism resistance [95−97]. To overcome these drawbacks, new antimicrobial agents can be incorporated into composites for wound treatment. There is literature on the use of antimicrobial hydrogels based on polysaccharides loaded with natural organic compounds, salts, or mixtures of these components, such as calcium alginate, modified starch, and hyaluronic acid derivatives [98−100]. Because of its inherent antimicrobial activity, chitosan is the most widely used polysaccharide in the production of dressings. Ren et al. created a hydrogel by crosslinking chitosan, genipin (crosslinker), and licorice polysaccharide. The presence of the crosslinker increased the stiffness of the composite and allowed it to degrade slowly. The bacteriostatic effect against *S. aureus* and *E. coli*, on the other hand, was achieved in the presence of licorice polysaccharide while maintaining the composite's biocompatibility, making it an excellent application in wound dressings [101]. Likewise, diazo resin crosslinked chitosan and cordycepin hydrogels, as well as chitosan or thiolated chitosan crosslinked with polyethylene glycol diacrylate and loaded with symmetric tryptophan-rich peptide, produced biocompatible composites and accelerated wound healing in vivo by 14 to 21 days [95,102,103]. As an alternative to conventional wound dressing application processes, injectable, and emulsion hydrogels have been evaluated [40]. According to this, Xuan and coworkers developed an injectable chitosan-silver hydrogel loaded with fibroblast growth factors [96]. After 17 days, the wound healing process ended due to the occurrence of a sustained release of growth factors and silver.

One of the topics that has been prioritized in the production of wound dressings is the use of natural compounds with biological properties to replace metallic nanoparticles. Some of the compounds used include curcumin, quercetin, dopamine, and matrine [104−106]. For example, curcumin has been incorporated into hydrogels produced through esterification reactions between lignin, polyethylene glycol, Gantrez S-97, and a polyacid (poly(methylvinyl ether-co-maleic acid)) [107]. Overall, the composites developed demonstrated antimicrobial activity against *S. aureus* and *Proteus mirabilis* (*P. mirabilis*).

Table 8.2 Summary of hydrogel dressings for release of antibiotics by diffusion.

Hydrogel composition	Functionalization	Diffusion model	Results	References
Hyaluronic acid/ carboxylated chitosan	Gentamicin sulfate (antibiotic)	n.d.	Drug release (39.91%, 24 h); Biocompatibility (NIH-3T3 cells); Antibacterial activity against *P. aeruginosa* and *S. aureus* (inhibition zone: 12.5 and 14.8 mm).	[88]
Xyloglucan/polyvinyl alcohol	Ampicillin (antibiotic)	Fickian	Biocompatibility (A549 epithelial cells viability of 90%); Hemocompatibility; Antibacterial activity against *E. coli* (inhibition zone: 40 mm).	[89]
Agar/κ-Carrageenan/ montmorillonite	Lidocaine hydrochloride (analgesic) and chloramphenicol (antibiotic)	n.d.	Biocompatibility (M-63 cells viability of 104%); Drug release (91.66% and 86.46% (LDC, CLP), 210 min); Antibacterial activity against *E. coli* and *S. aureus* (inhibition zone: 29.7 and 29.3 mm, respectively).	[84]
Gelatin/diosgenin/ carboxylated nanocellulose	Neomycin (antibiotic)	n.d.	Drug release (25% to 18%, pH 5.5 and pH 7.4, 15 min, 100% in 24 h); Biocompatibility (human dermal fibroblast cells viability of 80.1%); Antibacterial activity against *E. coli* and *S. aureus* (inhibition zone: 10.8 and 14.2 mm, respectively).	[90]
Carboxymethyl cellulose/keratin	Clindamycin (antibiotic)	Fickian	Drug release (13.2% in 4 h and 50% in 7 days.); Hemocompatibility (protein adsorption, 36.85 mg/g, clotting time, 13 s); Biocompatibility (L929 mouse fibroblasts viability >90%); Antibacterial activity against *S. aureus* (78.66%).	[91]
Poly(vinyl alcohol/ sodium alginate	Minocycline (antibiotic) and phages	n.d.	Drug release (99.9%, 15 min); Hemocompatibility; Biocompatibility (SK-1 skin epithelial cells viability >94.25%); Antibacterial activity against methicillin resistant *S. aureus*, *K. pneumoniae*, and *P. aeruginosa*.	[92]
Silk sericin/poly(vinyl alcohol)	Gentamicin (antibiotic)	n.d.	Drug release (~80%, 10 h); Biocompatibility (NIH-3T3 and HEK-293 cells viability of 130 and 140%, respectively); Antibacterial activity against *E. coli*, *S. aureus* and *P. aeruginosa*.	[87]

(Continued)

Table 8.2 Summary of hydrogel dressings for release of antibiotics by diffusion. *Continued*

Hydrogel composition	Functionalization	Diffusion model	Results	References
Calcium alginate	Ciprofloxacin (antibiotic)	Fickian (0.005 and 0.025%), non Fickian (0.010%)	Drug release 0.025% ciprofloxacin with the fastest release rate (68.36%, 5 min) followed by sustained drug release; Biocompatibility (human adult keratinocytes cell viability >85%, 72 h); Antibacterial activity against *E. coli*, *S. aureus*, and *P. aeruginosa* (inhibition zone: 43.43, 34.67, and 40.00 mm, respectively).	[85]
Carboxylated poly (vinyl alcohol)/gelatin/ hyaluronic acid/gelatin	Ampicillin (antibiotic)	Zero order	Drug release (34.5%, 6 h and 65%, 7 days); Biocompatibility (L929 cells viability); Antibacterial activity against oxacillin sensitive *S. aureus* (inhibition zone: ~ 14 mm).	[86]
Nanofibers of enzymatic PCL grafted with poly (gallic acid) (PGAL)/ sodium carboxymethylcellulose, and CMC	Clindamycin (antibiotic)	Fickian	Hemocompatibility; Biocompatibility (viability of epithelial cells); Antibacterial activity against *S. aureus* (inhibition zone: ~ 14 mm).	[93]
Oxidized hyaluronic acid-graft-aniline tetramer/N-carboxyethyl chitosan	Amoxicillin (antibiotic)	Fickian	Antibacterial activity against *E. coli* and *S. aureus* (inhibition zone: ~ 30 and 45 mm, respectively).	[94]

Depending on the type of hydrogel, curcumin was released for up to four days. A fit to the Korsmeyer-Peppas model was found in the formulation with polyethylene glycol (molecular weight 14,000). Curcumin is released through diffusion along the hydrogel's polymeric matrix (Fickian diffusion). A similar process was discovered in the formulation of the antimicrobial peptide hydrogel loaded with glucose oxidase based on this release mechanism [108]. *S. aureus* was successfully inhibited by this formulation. Furthermore, when used in diabetic wounds, glucose oxidation catalysis of glucose into hydrogen peroxide has two significant advantages: (1) it reduces the concentration of glucose at the wound site; (2) it produces hydrogen peroxide, which acts as an antimicrobial agent. To create a composite with healing properties, a bacterial nanocellulose-based hydrogel was impregnated with coniferyl alcohol [109]. The coniferyl alcohol release fits the Korsmeyer-Peppas model in this case, and it is classified as quasi-Fickian diffusion. Again, the release profile is determined by the swelling of bacterial nanocellulose (74%−97%), and no erosion of the hydrogel was observed. Anomaly diffusion was observed during the in vitro release of salicylic acid from a polyvinyl alcohol-fish gelatin-based hydrogel, on the other hand [110]. The swelling process (~60% to 85%) influences the diffusion of the solvent into the polymer matrix of the hydrogel. Diffusional-release hydrogels have mechanical, physicochemical, and biological properties that make them useful for wound management. The main disadvantage identified was the initial burst release, which promotes an insufficient release in a short period of time (Table 8.3).

Table 8.3 Summary of hydrogels dressings for release of other antimicrobial agents by diffusion.

Hydrogel composition	Functionalization	Diffusion model	Results	References
Chitosan	Silver nitrate	Quasi-Fickian	Rapid release within the first 24 h, followed by controlled over 11 days; Antimicrobial activity against *E. coli* and *S. aureus*; Anti-inflammatory.	[96]
Chitosan	Cordycepin	n.d.	Initial burst release of 80% in 12 h; Biocompatibility; Antimicrobial activity against *E. coli* and *S. aureus*; Wound-healing promoter.	[95]
Starch	Graphene and salvia	n.d.	Weak antimicrobial activity against *E. coli* and *S. aureus* (inhibition zone: 7.0 mm 9.5 mm, respectively); Electrical conductivity.	[100]
Calcium alginate	Zinc ions	n.d.	Biocompatible; Antimicrobial activity against *S. aureus* and *E. coli* (96.2% and 95.7%, respectively).	[98]
N-halamine hydrogel from hyaluronic acid	Chloride ions	n.d.	Biocompatible (NIH-3T3 and C28/I2, 92% and 78%, respectively); Antimicrobial activity against *E. coli* and *S. aureus* (in vivo and in vitro); Wound healing promoter.	[99]
Chitosan-genipin	Licorice	n.d.	Biocompatible ($>74\%$); Bacteriostatic effect against *E. coli* and *S. aureus* (81% and 75%, respectively).	[101]
Chitosan	Diazo-resin	n.d.	Biocompatible; Weak antimicrobial activity against *E. coli* and *S. aureus* (inhibition zone: 2.2 and 3.5 mm, respectively); Wound healing promoter.	[102]
Polyethylene glycol diacrylate and chitosan or thiolated chitosan	Symmetric Trp-rich peptide	n.d.	Initial burst release followed by a sustained release for 20 days; Biocompatibility; Weak antimicrobial activity against *S. aureus* and *E. coli* (inhibition zone: 4.5 and 4.7 mm, respectively); Anti-inflammatory.	[103]
Acrylamide and gelatin	Cinnamon oil	Fickian	Initial burst release in first 18 h, followed by a control release for over than 20 days (85%); Antimicrobial activity against *S. aureus* and *E. coli*.	[40]
Polyvinyl alcohol	Propolis	Fickian	Initial burst release of 45% in 3 h, followed by control release, continues for 96 h; Moderate antimicrobial activity against *S. aureus*.	[104]

(Continued)

Table 8.3 Summary of hydrogels dressings for release of other antimicrobial agents by diffusion. *Continued*

Hydrogel composition	Functionalization	Diffusion model	Results	References
Gelatin and Polyethylene glycol	Curcumin and quercetin	Second-order	Controlled release, 85% release; Antimicrobial activity against planktonic methicillin-resistant *S. aureus* (MRSA).	[106]
Konjac and fish gelatin	Matrine	n.d.	Initial burst release of 90% 45 min; Hemocompatible; Moderate antimicrobial activity against *E. coli* and *S. aureus* (inhibition zone: 12.0 and 11.5 mm, respectively).	[105]
Lignin-based hydrogel	Curcumin	Fickian	Controlled release for 4 days; Antimicrobial activity against *S. aureus* and *P. mirabilis*.	[107]
Heptapeptide	Glucose oxidase	Fickian	Maximum release of 55% in 4 h; Biocompatibile ($> 80\%$); Antimicrobial activity against *S. aureus*.	[108]
Bacterial nanocellulose	Coniferyl alcohol	Quasi-Fickian	Initial burst release of 24%, total release of 46.2% after 72 h; Antimicrobials against *S. aureus*, *Listeria monocytogenes*, and *Salmonella typhimurium*.	[109]
Polyvinyl alcohol and fish gelatin	Salicylic acid	Anomalous diffusion	Initial burst release of 34% and 72% after 3 h; Moderate antimicrobial activity *S. aureus* and *E. coli* (17 mm).	[110]

8.4 Stimuli-responsive hydrogels

The most frequently studied hydrogel application for wound management has been found to be diffusional-release hydrogels. However, uncontrolled diffusional release caused by drug concentration differences can result in dangerous burst releases, usually resulting in adverse effects on patients at the outset and greatly reduced drug efficacy (antimicrobials, growth factors, anti-oxidants, and antiinflammatory agents) in the long run [111,112]. As a result, the development of "smart" hydrogels is in high demand. These "smart" textiles have one or more of the following characteristics: (1) on-demand drug release triggered by external stimuli; (2) controllable "on/off" release; (3) tunable release kinetics [113,114]. Stimuli-responsive hydrogels can be classified according to their stimuli (physical or chemical)-responsive mechanisms: light-responsive, temperature-responsive, magnetic-responsive, mechano-responsive, electric-responsive, pH-responsive, enzyme-responsive, redox-responsive, among others [115−119]. When used in infected wounds, stimuli-responsive hydrogel wound dressings can reversibly change their phases or stiffness, enabling functions such as injectability, sensing, controlled drug release, self-healing, or shape memory [120−122]. These stimuli-responsive hydrogels can be further

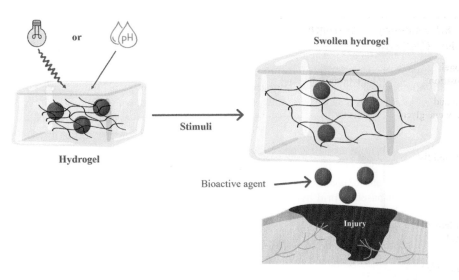

FIGURE 8.2

Light or pH stimuli-responsive antibacterial hydrogel.

classified based on how they react to various physical and chemical signals that may occur inside and outside the wound site: (1) external stimuli-responsive hydrogels, which include triggers such as light, temperature, electrical field, magnetic field, and so on; (2) internal stimuli-responsive hydrogels, which include triggers such as pH, enzymes, redox, and so on. The most extensively researched stimuli-responsive hydrogels are light-responsive and pH-responsive hydrogels and are depicted in Fig. 8.2. The composition, preparation methods, antimicrobial and other properties of light-responsive and pH-responsive hydrogels in wound treatment applications are depicted in the following sections.

8.4.1 Light-responsive hydrogels

Photoresponsive units found in light-responsive hydrogels include iron(III), zinc oxide, tungsten disulfide, gold nanorods, silver nanoparticles, trans-azobenzene, graphene oxide, and carbon nanotubes, among others. Different light signals, such as near infrared light, ultraviolet light, or visible light, can be converted into heat or reactive oxygen by these photo-responsive units. Heat or reactive oxygen species will produce photo-thermal or photo-dynamic responses, namely antimicrobial effects and reversible hydrogel phase modification [123,124]. For example, graphene oxide-based nanomaterials can locally raise the temperature when irradiated with near infrared light leading to photo-thermal bactericidal effect [125,126]. To achieve on-demand drug release, Rosselle et al. created a near infrared light-responsive and cefepime-loaded hydrophilic cryogel composed of butyl methacrylate and poly(ethylene glycol) methyl ether methacrylate incorporated with reduced graphene oxide [127]. When the reduced graphene oxide was exposed to near-infrared light, it heated the cryogel, causing it to swell and release the encapsulated antimicrobial drug. In ex vivo human skin model wound explants infected with *S. aureus*, these cefepime-loaded cryogels significantly reduced pro-inflammatory responses. Based on previous work, the researchers covalently conjugated maleimide-modified antimicrobial peptides onto furan-based cryogels using the Diels-Alder cycloaddition to reduce passive and burst antimicrobial peptide release and achieve on-demand peptide release [128]. The cryogel can effectively load either small molecular or biomacro-molecular drugs for controlled near infrared light-responsive antimicrobial action in the treatment of infected wounds, according to their research. Similarly, Zhang et al. incorporated reduced graphene oxide, molybdenum disulfide, and silver phosphate composites into a polyvinyl alcohol hydrogel to enhance reduced graphene oxide's photo-thermal conversion ability, achieving synergistic antibacterial activities under dual visible and near infrared light irradiation [129]. Furthermore, the hybrid hydrogels were endowed with improved mechanical property and swelling ratio with the developed composite, indicating great potential as antibacterial wound dressing. In another study, graphene oxide sheets were electrostatically modified with zinc oxide quantum dots to create a chitosan-based hybrid hydrogel. When exposed to near infrared irradiation, the composite hydrogel demonstrated synergistic photothermal and photodynamic antimicrobial effects, promoting wound healing in an in vivo wound infection model [130]. Other carbon materials, such as hollow carbon nanoparticles, were used as photosensitizers in the fabrication of light-responsive hydrogels, in addition to graphene oxide-based carbon material. Polyethylene glycol modified hollow carbon nanoparticles were co-embedded in a poly-2-dimethylaminoethyl methacrylate-based hydrogel with natural antibiotic aloe-emodin, exerting photo-dynamic and photo-thermal antimicrobial activity upon near infrared exposure. Furthermore, the aloe-emodin release from the gel had long-term therapeutic effects [131]. Furthermore, when compared to monotherapy, this aloe-emodin and carbon material co-loaded polymer hydrogel demonstrated the best anti-infection performance and promoted wound healing, indicating good clinical potential in wound management. Similarly, Liang et al. used carbon nanotubes as a photosensitizer in the development of multifunctional hydrogels. To summarize, composite hydrogels were created using chitosan, gelatin-grafted-dopamine, and polydopamine-coated carbon nanotubes as building blocks. The crosslinking was achieved through the oxidative coupling of catechol groups using a catalytic system based on hydrogen peroxide and horseradish peroxidase. Encapsulated doxycycline was also added to the polymeric network for effective and long-lasting antibacterial effects [132]. The hydrogel now has antibacterial (photo-thermal and chemical), adhesive, antioxidant, and conductive properties thanks to the combination of the aforementioned materials. In vitro and in the infected full-thickness mouse skin defect wound model, this complex composite demonstrated good therapeutic effect.

Because photosensitizers such as titanium dioxide, zinc oxide, copper(I) oxide, molybdenum disulfide, zinc phthalocyanine, and silver nanoparticles can generate reactive oxygen species when exposed to visible light, they are frequently used in the development of light-responsive hydrogels [133,134]. Wang et al. synthesized a polyvinyl alcohol hydrogel with silver-doped titanium dioxide nanoparticles. When exposed to visible light, the reactive oxygen species produced by this hydrogel were intended to combat multidrug-resistant bacteria in wound management [135]. Wounds infected with S. aureus by the hybrid hydrogel exhibited excellent antibacterial activity in both the in vitro and in vivo rat models within a short period of time. It is worth mentioning that the hybrid hydrogels presented accelerated wound healing in rat model compared with a traditional 3 M wound dressing. Similarly, in another work, silver phosphate and molybdenum disulfide photosensitizer composite were embedded in the polyvinyl alcohol-based hybrid hydrogel to exert synergistic photodynamic and photothermal bactericidal effects under coirradiation of visible (660 nm) and near infrared (808 nm) light. In comparison to pure hydrogel and commercial medical gauze, irradiated hydrogel accelerated wound healing in a rat wound infection model [136]. Bayat et al. used zinc phthalocyanine photosensitizer in conjunction with the amphiphilic polypeptide colistin to create chitosan-based hydrogels [137]. The addition of colistin increased not only the bioavailability and solubility of zinc phthalocyanine, but also its overall photo-bactericidal activity.

Overall, light-responsive hydrogels can reveal antimicrobial activity when exposed to specific light, providing on-demand therapeutic effects in the treatment of wound infections. In such cases, however, additional devices are required to trigger their activity. Furthermore, the action period is strongly influenced by the irradiation period. On the one hand, it can be viewed as a safe by design strategy. When dealing with infected wound management, it may be considered a deficient antimicrobial wound dressing that requires additional drug loading for sustained antimicrobial activity.

8.4.2 pH-responsive hydrogels

The pH of intact skin is commonly 4.5 to 5.0, and at the wound site it is slightly acidic (<7.0). The slightly acidic environment is attributed to the thriving metabolic activity of fibroblasts and keratinocytes during proliferation and angiogenesis [138]. Acidic substances commonly produced during aerobic bacteria metabolism are also expected to contribute to a pH drop at the wound site. However, the preferred metabolism at the wound site may be anaerobic, which is known to rapidly release polyamines and ammonia and may raise environmental pH if not quickly swept away by immune system components [139]. As a result, alkaline pH may be indicative of both bacterial activity and a faulty immune response at the wound site. Despite the historically alkaline pH of infected and chronic wounds, several pH-responsive hydrogels based on bacteria aerobic metabolism were developed to respond to a pH drop [140−142]. Khan et al. created a series of composite hydrogels from arabinoxylan and chitosan that were crosslinked with tetraethyl orthosilicate and anchored with reduced graphene oxide sheets. The hydrogels were then loaded with silver sulfadiazine, an antimicrobial agent. The planned goal was to achieve sustained and controlled wound disinfection [143]. Protonation of the alcoholic and carboxylic acid functional groups of arabinoxylan and chitosan in acidic conditions (pH ~ 6.4) reduced anion-anion repulsion, resulting in swelling of the composite hydrogel. The release of silver sulfadiazine is promoted during swelling, demonstrating a pH-responsive drug delivery feature. The composite hydrogels exhibited good antibacterial activity against several skin disease-causing pathogens, namely: *S. aureus*, *E. coli*, and *P. aeruginosa* and *Enterococcus faecalis*. Furthermore, it displayed excellent biocompatibility when incubated with mouse pre-osteoblast cells. Similarly, in the study of Mohamed et al., the pH-responsive carboxymethyl chitosan and carboxymethyl pullulan were employed with temperature sensitive polymer poly (N-isopropylacrylamide) as the crosslinker and zinc oxide nanoparticles as the antimicrobial agent to fabricate a hybrid nanogel [144]. The nanogel was then incorporated into a cellulosic fabric to create a pH and thermosensitive wound dressing with good antimicrobial activity against *S. aureus* and *E. coli*. Because it deforms or shrinks in acidic conditions, alginate could be used as a pH-responsive biopolymer [145]. Using a time-saving vacuum suction method, Shahriari-Khalaji et al. efficiently incorporated sodium alginate into the bacterial nanocellulose matrix, followed by crosslinking via immersion in separate solutions of various cations. In the full-thickness skin defect rat model, pH-responsive antibacterial activity and wound healing were achieved [146]. Researchers used pH-sensitive chemical bonds (such as imine, acylhydrazone, or carboxyl bonds) in the construction of pH-responsive hydrogels, in addition to different degradation behaviors of polysaccharides under different pH conditions. Guan et al., for example, used aldehyde hyaluronic acid and adipic acid dihydrazide graft hyaluronic acid as the main polymer building blocks and sisomicin sulfate as the antimicrobial agent in their study. Crosslinking occurred between the components via imine and acylhydrazone bonds, resulting in a hydrogel [147]. Under acid conditions, the composite hydrogel degraded rapidly and released and sisomicin sulfate, which exerted on-demand and sustained antibacterial activity, shortened the inflammation period, and promoted wound healing in the full-thickness mouse skin defect wound. Methacrylated gelatin was also polymerized with methacrylic acid and crosslinked with ethylene glycol dimethacrylate in another study. Gentamicin was added and covalently bonded via its primary amine group and the copolymer's carboxylic acid group, resulting in a polymer-drug conjugate with pH sensitivity [148]. This polymer-drug conjugate was then loaded with ampicillin and 2-amino guanidine for a synergistic antimicrobial effect as well as reactive oxygen species scavenging. In response to the pH shift, the designed hydrogel demonstrated good anti-inflammatory properties as well as prolonged, controlled, and synergistic antimicrobial activity.

Furthermore, researchers have combined several stimuli responsive mechanisms to create dual or multiple stimuli-responsive hydrogels for the treatment of infected wounds [149]. Zhao et al. created an injectable, anti-oxidative, near infrared and pH-responsive antimicrobial, tissue adhesive, and self-healing physical double-network hydrogel in their study. This hydrogel was created as a removable wound dressing for the treatment of multi-drug resistant bacterial infections [150]. In brief, the hydrogel was created by catechol-iron(III) coordination between the pre-polymer, iron (III), and quadrupole hydrogen bonding of synthon modified gelatin and poly(glycerol sebacate)-co-poly(ethylene glycol)-g-catechol prepolymer, UPy-hexamethylene diisocyanate synthon modified gelatin, and iron(III) chloride. It is worth noting that the coordination of prepolymer and iron (III) gave the hydrogel pH and near infrared responsiveness. Experiments in vivo confirmed their excellent antimicrobial activity against MRSA as well as good hemostasis to skin trauma. Furthermore, when compared to biomedical glue and surgical suture in the full-thickness skin defect model, the designed hydrogel achieved better wound closure and wound healing. This was attributed to inflammation regulation, accelerated collagen deposition, and angiogenesis promotion. Ma et al. created a thermo and pH-responsive hydrogel by simply assembling reversible thermosensitive hydroxypropyl chitin polymer, tannic acid as a crosslinker, and iron(III) chloride. The obtained chitin-based hydrogel gels spontaneously at physiological temperature ($\sim 37°C$), and the pyrogallol/catechol groups of tannic acid tended to deprotonate at physiological pH (~ 7.4), strengthening the complexation and slowing hydrogel release [151]. In the full-thickness mouse skin defect wound model, this composite hydrogel demonstrated extended bacterial infection control and accelerated wound healing effects, indicating its potential as a future bioactive wound dressing in clinical wound management. Unlike the previous study, which was thermo-sensitive for injectability and pH-sensitive for controlled drug release, Huang et al. used dual thermo- and pH-responsive mechanisms for controlled and prolonged release of silver ions in the presence of pathogens at the wound site. In summary, the hydrogel is created through hydrogen bonding and supramolecular complexation of carboxymethyl agarose and silver ions. The ionic interaction is disrupted under acidic and higher temperature conditions, resulting in increased release of silver ions from the hydrogel [152]. This composite hydrogel demonstrated improved antibacterial and anti-inflammatory properties, as well as excellent cytocompatibility and hemocompatibility. It significantly accelerated wound closure.

Regardless of the trigger, pH-responsive antimicrobial hydrogels can specifically release drugs at the wound infection site with a switch on/off property, significantly lowering the toxicity of uncontrolled drug release in general and increasing therapeutic efficacy in infected wound treatment. Hydrogels can achieve multifunction of on-demand antimicrobial action, anti-oxidation, anti-inflammation, tissue adhesion, self-healing, injectability, and easy removability, etc. when combined with other stimuli-responsive mechanisms, resulting in novel bioactive wound dressings for clinical wound management.

8.5 Sensor and imaging hydrogels for wound healing

It is critical to detect wound infections as soon as possible in order to provide adequate and effective treatment. The presence of pathogenic bacteria is a major contributor to wound healing impediment [153,154]. Wound evaluation in traditional clinical settings, on the other hand, is impractical, lacks objective basis, and may increase the risk of wound infection [155,156]. Thus, a technology that can interface with wounds, detect pathogenic bacteria, and wirelessly transmit data without interfering with treatment enables continuous and effective monitoring [38,157]. Furthermore, common components used in wound debridement are not entirely proven efficient in wound healing [158]. Zhang et al. created an integrated smart dressing that can monitor the wound microenvironment while not interfering with the healing process, presenting a promising strategy for intelligent wound care [159]. This three-layer integrated smart dressing contains a biomimetic nanofiber membrane, a microenvironment sensor, and an ultraviolet-crosslinked -cyclodextrin-containing gelatin methacryloyl hydrogel. Because the hydrogel has a similar structure to the ECM, it promotes the expression of vascular endothelial growth factors, which promotes neovascularization and wound healing. The wound microenvironment data collected by the integrated sensor chip were transmitted via Bluetooth low energy 4.0 antennae and displayed on a mobile device-specific application (Fig. 8.3). A flexible and battery-free sensor developed by Xiong et al. is able to respond selectively to the enzyme deoxyribonuclease which is secreted by pathogenic bacteria such as *S. aureus*, *P. aeruginosa*, and *Streptococcus pyogenes* [38]. In vitro experiments show that this engineered DNA hydrogel responds selectively to *S. aureus* concentrations near the clinical infection threshold (1×10^6 colony forming units gram^{-1} of viable tissue), and thus prior to visible manifestation of infection. The tunable dielectric changes caused by deoxyribonuclease are converted into a wireless signal detectable by a smartphone. Moreover, in vivo studies demonstrated the detection of clinically relevant amounts of *S. aureus* for 24 hour. These findings show that such a

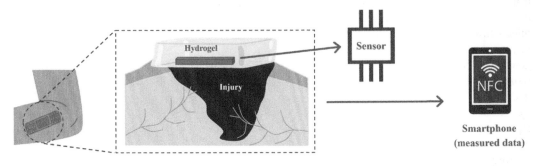

FIGURE 8.3

Sensor stimuli-responsive antibacterial hydrogel.

strategy for continuous infection monitoring has the potential to improve the management of surgical or chronic wounds.

Temperature and pH are two other physicochemical markers for wound infection. An increase in wound temperature could be an early sign of infection. Pang et al. developed a smart flexible electronics-integrated wound dressing based on this principle for early infection detection via real-time wound-temperature monitoring via an integrated sensor [157]. Furthermore, the mentioned smart wound dressing allowed for on-demand infection treatment by releasing antibiotics from the hydrogel via in situ ultraviolet irradiation. This wound dressing has a two-layer structure: an upper layer of flexible electronics encapsulated in polydimethylsiloxane that includes a temperature sensor and ultraviolet light-emitting diodes, and a lower layer of an ultraviolet-responsive antibacterial hydrogel. The temperature was continuously monitored and transmitted via Bluetooth by the integrated sensor. When the wound temperature exceeded a predetermined threshold, the integrated ultraviolet light-emitting diodes turned on, triggering antibiotic release in situ. However, pH monitoring of wound infections using electronic devices frequently necessitates the use of expensive and sophisticated technology [160]. Because of their lack of potential toxicity, indicator dyes, particularly natural dyes, are simple and cost-effective alternatives for pH monitoring [161–163]. Zepon et al. developed a pH-responsive hydrogel film as smart wound dressing based on κ-Carrageenan, locust bean gum, and cranberry extract for monitoring bacterial infections [162]. It responded well to the pH change caused by the basic compounds released from digested proteins during bacterial growth. In vitro studies with *S. aureus* and *P. aeruginosa* confirmed the color changes of the hydrogel film. The changes were visible with the naked eye, indicating that the obtained hydrogel film could be used as a visual system for monitoring bacterial wound infections (Fig. 8.4). Furthermore,

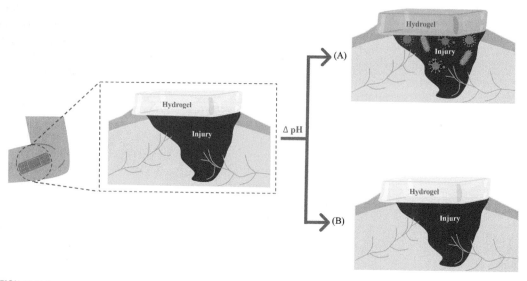

FIGURE 8.4

pH chromatic shift stimuli-responsive antibacterial hydrogel: (A) Unfavorable pH shift—infected injury; (B) Favorable pH variation—noninfected wound.

the presence of cranberry extract in the hydrogel film acts not only as a pH sensor, but also as a bacteria adhesion inhibitor on the hydrogel surface.

The use of antibacterial compounds as agents to promote wound healing by hastening wound closure is still unregulated, regardless of whether they are drugs, peptides, or nanoparticles. Chekini et al. reported the use of cellulose nanocrystals decorated with carbon dots composite hydrogel with strong iron(III) ion sequestration capability to overcome this issue [164]. The photoluminescence of the nanocolloidal hydrogel was quenched once the ionic iron was adsorbed onto the hydrogel surface, indicating the removal of iron(III) ions and the subsequent growth arrest of *E. coli*, *P. aeruginosa*, and *S. aureus*. Through three-dimensional printing, the wound dressings could be easily assembled.

Dynamic hydrogen and Schiff base crosslinking bonds were used to create an injectable, self-healing, and conductive chitosan-based hydrogel with inherent antibacterial properties [165]. The hydrogel's reversible pH responsiveness enables sol-gel conversion, promoting its degradation. Experiments in vivo confirmed the remarkable effect on wound healing. Furthermore, the conductive hydrogel could provide real-time analysis of the patient's medical data. A strain sensor with antibacterial properties was formed using a different conductive hydrogel based on a dual network of polyacrylamide and agarose, as well as tannic acid-borax complexes [166]. The hydrogel increased compressive stress by 58.14% when compared to the hydrogel composed of polyacrylamide and agar, allowing for more accurate measurements. Its transparency and excellent light transmission allow for wound assessment, avoiding the physical trauma and potential infection caused by conventional dressing removal. Furthermore, the antibacterial properties of tannic acid-borax reduce the risk of infection. Chai et al. created an adhesive hydrogel with up to seven times the adhesion capacity and excellent stretchability and resilience [167]. This hydrogel, which is made of poly(thioctic acid) and crosslinked with polydopamine, has anti-swelling and self-healing properties. Experiments confirmed the hydrogel's ability to accelerate wound healing and an increase in conductivity due to the addition of iron(III) ions, making it suitable for use as strain sensors. Jing et al. imagined a hydrogel for a variety of applications, including electronic skin, wound dressings, and wearable devices, inspired by the natural mussel adhesive mechanism [168]. Dopamine molecules were intercalated into talc by polydopamine-coated talc nanoflakes incorporated into a polyacrylamide hydrogel. The dispersion was improved and the catechol groups in the hydrogel were preserved after oxidation. The resulting hydrogel demonstrated exceptional stretchability, with over 1000% extension and a recovery rate of more than 99%. Strong adhesiveness to various substrates, including human skin, as well as rapid self-healing and mechanical property recovery without the need for external stimuli, were also demonstrated. Furthermore, the hydrogel demonstrated high sensitivity, with a gauge factor of 0.693 at 1000% strain, and was capable of monitoring human motions such as finger, knee, or elbow bending and deep breathing.

By introducing negatively charged clay nanosheets, Zhu et al. developed a smart ionic gelatin based polyacrylamide and clay hydrogel with high conductivity of 10.87 mS/cm [169]. The gel exhibited excellent self-healing properties, robust adhesion (interfacial toughness of up to 485 J/m^2 with pigskin), and multiple stimuli-responses driven by salt ions, pH, and stress. The hydrogel's applications include muscle movement sensors, wound pH monitoring, oxygen delivery, and drug delivery adjustment. A mechanically flexible, electroactive, and self-healing hydrogel was designed to perform both electrically-stimulated accelerated wound healing and motion sensing. Zheng et al. created a gelatin-based smart 3D-scaffold by combining water-dispersible conducting polymer complex, poly(3,4-ethylenedioxythiophene) polystyrene sulfonate, and multi-walled carbon nanotubes as functional building blocks [170]. Through precise electrical stimulation and a wearable motion sensing function at the wound injury area, the obtained hydrogel significantly promotes wound healing, with real-time monitoring of injury motion activities.

8.6 Concluding remarks and future perspectives

Because of their proficient antibacterial activity that employs novel agents and strategies to overcome their alarming emergence resistance, hydrogels represent a viable strategy to augment or replace traditional woven and non-woven antibacterial textiles. Furthermore, because of their structure and active ingredients, hydrogels effectively boost wound healing, lowering the likelihood of chronic wound development. The majority of developed hydrogels are focused on the release of their active agents, which is largely controlled by swelling. Various approaches have been used to control this important feature and to identify different triggers for swelling to begin and stop. A wound dressing with a tunable on-demand drug release feature can be used to control wound infections when they occur, and only when they occur. This significantly reduces the occurrence of bacterial resistance and potentially harmful side effects on patients caused by the active agents. Finally, remote wound dressing control may benefit not only the patient by reducing the risk of wound infections, pain, and complications, but it also represents an important feature to maximize the efficacy of treatment by health care professionals.

References

[1] M.G. Jeschke, M.E. Van Baar, M.A. Choudhry, K.K. Chung, N.S. Gibran, S. Logsetty, Burn injury, Nat. Rev. Dis. Prim. (2020) 6.
[2] M. Falcone, B. De Angelis, F. Pea, A. Scalise, S. Stefani, R. Tasinato, et al., Challenges in the management of chronic wound infections, J. Glob. Antimicrob. Resist. 26 (2021) 140−147.
[3] X. Song, P. Liu, X. Liu, Y. Wang, H. Wei, J. Zhang, et al., Dealing with MDR bacteria and biofilm in the post-antibiotic era: application of antimicrobial peptides-based nano-formulation, Mater. Sci. Eng. C. (2021) 128.
[4] J. Koehler, F.P. Brandl, A.M. Goepferich, Hydrogel wound dressings for bioactive treatment of acute and chronic wounds, Eur. Polym. J. 100 (2018) 1−11.
[5] M. Ovais, I. Ahmad, A.T. Khalil, S. Mukherjee, R. Javed, M. Ayaz, et al., Wound healing applications of biogenic colloidal silver and gold nanoparticles: recent trends and future prospects, Appl. Microbiol. Biotechnol. 102 (2018) 4305−4318.
[6] R.E. Jones, D.S. Foster, M.T. Longaker, Management of chronic wounds—2018, JAMA (2018) 320.
[7] M. Maaz Arif, S.M. Khan, N. Gull, T.A. Tabish, S. Zia, R. Ullah Khan, et al., Polymer-based biomaterials for chronic wound management: promises and challenges, Int. J. Pharmaceutics, (2021) 598.
[8] X. He, J. Xue, L. Shi, Y. Kong, Q. Zhan, Y. Sun, et al., Recent antioxidative nanomaterials toward wound dressing and disease treatment via ROS scavenging, Mater. Today Nano (2022) 17.
[9] H. Nosrati, M. Khodaei, Z. Alizadeh, M. Banitalebi-Dehkordi, Cationic, anionic and neutral polysaccharides for skin tissue engineering and wound healing applications, Int. J. Biol. Macromol. 192 (2021) 298−322.

[10] L. Chelkeba, T. Melaku, Epidemiology of staphylococci species and their antimicrobial-resistance among patients with wound infection in Ethiopia: a systematic review and meta-analysis, J. Glob. Antimicrob. Resist. (2021).

[11] S. Duan, R. Wu, Y.-H. Xiong, H.-M. Ren, C. Lei, Y.-Q. Zhao, et al., Multifunctional antimicrobial materials: from rational design to biomedical applications, Prog. Mater. Sci. (2022) 125.

[12] Q. Shi, X. Luo, Z. Huang, A.C. Midgley, B. Wang, R. Liu, et al., Cobalt-mediated multi-functional dressings promote bacteria-infected wound healing, Acta Biomater. 86 (2019) 465−479.

[13] B. Mai, M. Jia, S. Liu, Z. Sheng, M. Li, Y. Gao, et al., Smart hydrogel-based DVDMS/bFGF nanohybrids for antibacterial phototherapy with multiple damaging sites and accelerated wound healing, ACS Appl. Mater. Interfaces 12 (2020) 10156−10169.

[14] X. Zhang, W. Shu, Q. Yu, W. Qu, Y. Wang, R. Li, Functional biomaterials for treatment of chronic wound, Front. Bioeng. Biotechnol. (2020) 8.

[15] J. Ding, J. Zhang, J. Li, D. Li, C. Xiao, H. Xiao, et al., Electrospun polymer biomaterials, Prog. Polym. Sci. 90 (2019) 1−34.

[16] Y. Ding, Z. Sun, R. Shi, H. Cui, Y. Liu, H. Mao, et al., Integrated endotoxin adsorption and antibacterial properties of cationic polyurethane foams for wound healing, ACS Appl. Mater. Interfaces 11 (2018) 2860−2869.

[17] W. Wang, R. Narain, H. Zeng, Hydrogels, Polym. Sci. Nanotechnol (2020).

[18] Y. Fan, W. Wu, Y. Lei, C. Gaucher, S. Pei, J. Zhang, et al., Edaravone-loaded alginate-based nanocomposite hydrogel accelerated chronic wound healing in diabetic mice, Mar. Drugs (2019) 17.

[19] B.-D. Zheng, J. Ye, Y.-C. Yang, Y.-Y. Huang, M.-T. Xiao, Self-healing polysaccharide-based injectable hydrogels with antibacterial activity for wound healing, Carbohydr. Polym. (2022) 275.

[20] S. Li, S. Dong, W. Xu, S. Tu, L. Yan, C. Zhao, et al., Antibacterial hydrogels, Adv. Sci. (2018) 5.

[21] S. Tavakoli, A.S. Klar, Advanced hydrogels as wound dressings, Biomolecules (2020) 10.

[22] G. Tripodo, A. Trapani, A. Rosato, D.I. Franco, C. Tamma, R. Trapani, et al., Hydrogels for biomedical applications from glycol chitosan and PEG diglycidyl ether exhibit pro-angiogenic and antibacterial activity, Carbohydr. Polym. 198 (2018) 124−130.

[23] G. Liao, J. Hu, Z. Chen, R. Zhang, G. Wang, T. Kuang, Preparation, properties, and applications of graphene-based hydrogels, Front. Chem. (2018) 6.

[24] A. Moeini, P. Pedram, P. Makvandi, M. Malinconico, G. Gomez D'ayala, Wound healing and antimicrobial effect of active secondary metabolites in chitosan-based wound dressings: a review, Carbohydr. Polym. (2020) 233.

[25] W. Wang, K.-J. Lu, C.-H. Yu, Q.-L. Huang, Y.-Z. Du, Nano-drug delivery systems in wound treatment and skin regeneration, J. Nanobiotechnol. (2019) 17.

[26] R. Ahmad Raus, W.M.F. Wan Nawawi, R.R. Nasaruddin, Alginate and alginate composites for biomedical applications, Asian J. Pharm. Sci. 16 (2021) 280−306.

[27] A. Kirschning, N. Dibbert, G. Dräger, Chemical functionalization of polysaccharides-towards biocompatible hydrogels for biomedical applications, Chem. - A Eur. J. 24 (2018) 1231−1240.

[28] P. Rousselle, M. Montmasson, C. Garnier, Extracellular matrix contribution to skin wound re-epithelialization, Matrix Biol. 75−76 (2019) 12−26.

[29] A. Vijayan, P.P. James, C.K. Nanditha, G.S.V. Kumar, Multiple cargo deliveries of growth factors and antimicrobial peptide using biodegradable nanopolymer as a potential wound healing system, Int. J. Nanomed. 14 (2019) 2253−2263.

[30] Y.-H. Lee, Y.-L. Hong, T.-L. Wu, Novel silver and nanoparticle-encapsulated growth factor co-loaded chitosan composite hydrogel with sustained antimicrobility and promoted biological properties for diabetic wound healing, Mater. Sci. Eng. C. (2021) 118.

[31] M.J. Malone-Povolny, S.E. Maloney, M.H. Schoenfisch, Nitric oxide therapy for diabetic wound healing, Adv. Healthc. Mater. (2019) 8.

[32] J. Peng, H. Zhao, C. Tu, Z. Xu, L. Ye, L. Zhao, et al., In situ hydrogel dressing loaded with heparin and basic fibroblast growth factor for accelerating wound healing in rat, Mater. Sci. Eng. C. (2020) 116.

[33] Z. Xu, S. Han, Z. Gu, J. Wu, Advances and impact of antioxidant hydrogel in chronic wound healing, Adv. Healthc. Mater. (2020) 9.

[34] J. Wang, X.-Y. Chen, Y. Zhao, Y. Yang, W. Wang, C. Wu, et al., pH-switchable antimicrobial nanofiber networks of hydrogel eradicate biofilm and rescue stalled healing in chronic wounds, ACS Nano 13 (2019) 11686−11697.

[35] X. Zhao, H. Wu, B. Guo, R. Dong, Y. Qiu, P.X. Ma, Antibacterial anti-oxidant electroactive injectable hydrogel as self-healing wound dressing with hemostasis and adhesiveness for cutaneous wound healing, Biomaterials 122 (2017) 34−47.

[36] S. Hu, Y. Zhi, S. Shan, Y. Ni, Research progress of smart response composite hydrogels based on nanocellulose, Carbohydr. Polym. (2022) 275.

[37] S. Peers, A. Montembault, C. Ladavière, Chitosan hydrogels incorporating colloids for sustained drug delivery, Carbohydr. Polym. (2022) 275.

[38] Z. Xiong, S. Achavananthadith, S. Lian, L.E. Madden, Z.X. Ong, W. Chua, et al., A wireless and battery-free wound infection sensor based on DNA hydrogel, Sci. Adv. (2021) 7.

[39] A. Francesko, P. Petkova, T. Tzanov, Hydrogel dressings for advanced wound management, Curr. Med. Chem. 25 (2019) 5782−5797.

[40] J. Wang, Y. Li, Y. Gao, Z. Xie, M. Zhou, Y. He, et al., Cinnamon oil-loaded composite emulsion hydrogels with antibacterial activity prepared using concentrated emulsion templates, Ind. Crop. Prod. 112 (2018) 281−289.

[41] R. Rebelo, J. Padrão, M.M. Fernandes, S. Carvalho, M. Henriques, A. Zille, et al., Aging effect on functionalized silver-based nanocoating braided coronary stents, Coatings (2020) 10.

[42] J. Padrão, S. Ribeiro, S. Lanceros-Méndez, L.R. Rodrigues, F. Dourado, Effect of bacterial nanocellulose binding on the bactericidal activity of bovine lactoferrin, Heliyon (2020) 6.

[43] A. Basu, K. Heitz, M. Strømme, K. Welch, N. Ferraz, Ion-crosslinked wood-derived nanocellulose hydrogels with tunable antibacterial properties: candidate materials for advanced wound care applications, Carbohydr. Polym. 181 (2018) 345−350.

[44] K. Wang, J. Wang, L. Li, L. Xu, N. Feng, Y. Wang, et al., Novel nonreleasing antibacterial hydrogel dressing by a one-pot method, ACS Biomater. Sci. Eng. 6 (2020) 1259−1268.

[45] X. Tang, X. Gu, Y. Wang, X. Chen, J. Ling, Y. Yang, Stable antibacterial polysaccharide-based hydrogels as tissue adhesives for wound healing, RSC Adv. 10 (2020) 17280−17287.

[46] L. Tallet, V. Gribova, L. Ploux, N.E. Vrana, P. Lavalle, New smart antimicrobial hydrogels, nanomaterials, and coatings: earlier action, more specific, better dosing? Adv. Healthc. Mater. (2020) 10.

[47] M.L. Bruschi, Mathematical models of drug release, in: M.L. BRUSCHI (Ed.), Strategies to Modify the Drug Release from Pharmaceutical Systems, Woodhead Publishing, 2015.

[48] C. Alves, A. Ribeiro, E. Pinto, J. Santos, G. Soares, Exploring Z-Tyr-Phe-OH-based hydrogels loaded with curcumin for the development of fressings for wound healing, J. Drug. Deliv. Sci. Technol. (2022) 73.

[49] J. Padrão, S. Gonçalves, J.P. Silva, V. Sencadas, S. Lanceros-Méndez, A.C. Pinheiro, et al., Bacterial cellulose-lactoferrin as an antimicrobial edible packaging, Food Hydrocoll. 58 (2016) 126−140.

[50] M. Vigata, C. Meinert, D.W. Hutmacher, N. Bock, Hydrogels as drug delivery systems: a review of current characterization and evaluation techniques, Pharmaceutics (2020) 12.

[51] P.L. Ritger, N.A. Peppas, A simple equation for description of solute release I. Fickian and non-fickian release from non-swellable devices in the form of slabs, spheres, cylinders or discs, J. Control. Rel. 5 (1987) 23−36.

[52] J. Siepmann, Modeling of drug release from delivery systems based on hydroxypropyl methylcellulose (HPMC), Adv. Drug. Deliv. Rev. 48 (2001) 139−157.

[53] A.I. Ribeiro, A.M. Dias, A. Zille, Synergistic effects between metal nanoparticles and commercial anti-microbial agents: a review, ACS Appl. Nano Mater. 5 (2022) 3030–3064.

[54] P.V. Baptista, M.P. Mccusker, A. Carvalho, D.A. Ferreira, N.M. Mohan, M. Martins, et al., Nano-strategies to fight multidrug resistant bacteria—"a battle of the titans.", Front. Microbiol. (2018) 9.

[55] T.C. Dakal, A. Kumar, R.S. Majumdar, V. Yadav, Mechanistic basis of antimicrobial actions of silver nanoparticles, Front. Microbiol. (2016) 7.

[56] P. Le Thi, Y. Lee, T.T. Hoang Thi, K.M. Park, K.D. Park, Catechol-rich gelatin hydrogels in situ hybri-dizations with silver nanoparticle for enhanced antibacterial activity, Mater. Sci. Eng. C. 92 (2018) 52–60.

[57] N.S.V. Capanema, A.A.P. Mansur, S.M. Carvalho, L.L. Mansur, C.P. Ramos, A.P. Lage, et al., Physicochemical properties and antimicrobial activity of biocompatible carboxymethylcellulose-silver nanoparticle hybrids for wound dressing and epidermal repair, J. Appl. Polym. Sci. (2018) 135.

[58] K.M. Zepon, M.S. Marques, M.M. Da Silva Paula, F.D.P. Morisso, L.A. Kanis, Facile, green and scal-able method to produce carrageenan-based hydrogel containing in situ synthesized AgNPs for applica-tion as wound dressing, Int. J. Biol. Macromol. 113 (2018) 51–58.

[59] Y. Zhao, Z. Li, S. Song, K. Yang, H. Liu, Z. Yang, et al., Skin-inspired antibacterial conductive hydro-gels for epidermal sensors and diabetic foot wound dressings, Adv. Funct. Mater. (2019) 29.

[60] Y. Jiang, J. Huang, X. Wu, Y. Ren, Z. Li, J. Ren, Controlled release of silver ions from AgNPs using a hydrogel based on konjac glucomannan and chitosan for infected wounds, Int. J. Biol. Macromol. 149 (2020) 148–157.

[61] A. Kumar, H. Kaur, Sprayed in-situ synthesis of polyvinyl alcohol/chitosan loaded silver nanocomposite hydrogel for improved antibacterial effects, Int. J. Biol. Macromol. 145 (2020) 950–964.

[62] L. Gou, M. Xiang, X. Ni, Development of wound therapy in nursing care of infants by using injectable gelatin-cellulose composite hydrogel incorporated with silver nanoparticles, Mater. Lett. (2020) 277.

[63] A. Gupta, S.M. Briffa, S. Swingler, H. Gibson, V. Kannappan, G. Adamus, et al., Synthesis of silver nanoparticles using curcumin-cyclodextrins loaded into bacterial cellulose-based hydrogels for wound dressing applications, Biomacromolecules 21 (2020) 1802–1811.

[64] M. Tan, Y. Choi, J. Kim, J.-H. Kim, K. Fromm, Polyaspartamide functionalized catechol-based hydro-gels embedded with silver nanoparticles for antimicrobial properties, Polymers (2018) 10.

[65] J. Chalitangkoon, M. Wongkittisin, P. Monvisade, Silver loaded hydroxyethylacryl chitosan/sodium algi-nate hydrogel films for controlled drug release wound dressings, Int. J. Biol. Macromol. 159 (2020) 194–203.

[66] K. Chen, F. Wang, S. Liu, X. Wu, L. Xu, D. Zhang, In situ reduction of silver nanoparticles by sodium alginate to obtain silver-loaded composite wound dressing with enhanced mechanical and antimicrobial property, Int. J. Biol. Macromol. 148 (2020) 501–509.

[67] Y. Wang, R. Xie, Q. Li, F. Dai, G. Lan, S. Shang, et al., A self-adapting hydrogel based on chitosan/oxi-dized konjac glucomannan/AgNPs for repairing irregular wounds, Biomater. Sci. 8 (2020) 1910–1922.

[68] N. Masood, R. Ahmed, M. Tariq, Z. Ahmed, M.S. Masoud, I. Ali, et al., Silver nanoparticle impregnated chitosan-PEG hydrogel enhances wound healing in diabetes induced rabbits, Int. J. Pharm. 559 (2019) 23–36.

[69] G.G. De Lima, D.W.F. De Lima, M.J.A. De Oliveira, A.B. Lugão, M.T.S. Alcântara, D.M. Devine, et al., Synthesis and in vivo behavior of PVP/CMC/agar hydrogel membranes impregnated with silver nanoparticles for wound healing applications, ACS Appl. Bio Mater. 1 (2018) 1842–1852.

[70] M.T.S. Alcântara, N. Lincopan, P.M. Santos, P.A. Ramirez, A.J.C. Brant, H.G. Riella, et al., Simultaneous hydrogel crosslinking and silver nanoparticle formation by using ionizing radiation to obtain antimicrobial hydrogels, Radiat. Phys. Chem. (2020) 169.

[71] E.E. Khozemy, S.M. Nasef, G.A. Mahmoud, Synthesis and characterization of antimicrobial nanocomposite hydrogel based on wheat flour and poly (vinyl alcohol) using γ-irradiation, Adv. Polym. Technol. 37 (2018) 3252−3261.

[72] S. Kumaraswamy, S.L. Patil, S.H. Mallaiah, In vitro biocompatibility evaluation of radiolytically synthesized silver/polyvinyl hydrogel nanocomposites for wound dressing applications, J. Bioact. Compat. Polym. 35 (2020) 435−450.

[73] J. Baukum, J. Pranjan, A. Kaolaor, P. Chuysinuan, O. Suwantong, P. Supaphol, The potential use of cross-linked alginate/gelatin hydrogels containing silver nanoparticles for wound dressing applications, Polym. Bull. 77 (2019) 2679−2695.

[74] X. Xiao, Y. Zhu, J. Liao, T. Wang, W. Sun, Z. Tong, High-efficient and synergetic antibacterial nanocomposite hydrogel with quaternized chitosan/Ag nanoparticles prepared by one-pot UV photochemical synthesis, Biopolymers (2020) 111.

[75] J. Yang, Y. Chen, L. Zhao, Z. Feng, K. Peng, A. Wei, et al., Preparation of a chitosan/carboxymethyl chitosan/AgNPs polyelectrolyte composite physical hydrogel with self-healing ability, antibacterial properties, and good biosafety simultaneously, and its application as a wound dressing, Compos. Part. B: Eng. (2020) 197.

[76] A. Ounkaew, P. Kasemsiri, K. Jetsrisuparb, H. Uyama, Y.-I. Hsu, T. Boonmars, et al., Synthesis of nanocomposite hydrogel based carboxymethyl starch/polyvinyl alcohol/nanosilver for biomedical materials, Carbohydr. Polym. (2020) 248.

[77] A. Ponco, H. Helmiyati, Hydrogel of carboxymethyl cellulose and polyvinyl alcohol modified by CuNPs as antibacterial in wound dressing, in: Proceedings of the 5th International Symposium on Current Progress in Mathematics and Sciences (Iscpms2019), 2020.

[78] N.N. Mahmoud, S. Hikmat, D. Abu Ghith, M. Hajeer, L. Hamadneh, D. Qattan, et al., Gold nanoparticles loaded into polymeric hydrogel for wound healing in rats: effect of nanoparticles' shape and surface modification, Int. J. Pharm. 565 (2019) 174−186.

[79] J. Kai, Z. Xuesong, Preparation, characterization, and cytotoxicity evaluation of zinc oxide−bacterial cellulose−chitosan hydrogels for antibacterial dressing, Macromol. Chem. Phys. (2020) 221.

[80] R. Eivazzadeh-Keihan, F. Khalili, H.A.M. Aliabadi, A. Maleki, H. Madanchi, E.Z. Ziabari, et al., Alginate hydrogel-polyvinyl alcohol/silk fibroin/magnesium hydroxide nanorods: a novel scaffold with biological and antibacterial activity and improved mechanical properties, Int. J. Biol. Macromol. 162 (2020) 1959−1971.

[81] A. Ahmed, M.B.K. Niazi, Z. Jahan, T. Ahmad, A. Hussain, E. Pervaiz, et al., In-vitro and in-vivo study of superabsorbent PVA/Starch/g-C_3N_4/Ag@TiO_2 NPs hydrogel membranes for wound dressing, Eur. Polym. J. (2020) 130.

[82] X.-X. Li, J.-Y. Dong, Y.-H. Li, J. Zhong, H. Yu, Q.-Q. Yu, et al., Fabrication of Ag−ZnO@ carboxymethyl cellulose/K-carrageenan/graphene oxide/konjac glucomannan hydrogel for effective wound dressing in nursing care for diabetic foot ulcers, Appl. Nanosci. 10 (2019) 729−738.

[83] A.I. Raafat, N.M. El-Sawy, N.A. Badawy, E.A. Mousa, A.M. Mohamed, Radiation fabrication of Xanthan-based wound dressing hydrogels embedded ZnO nanoparticles: in vitro evaluation, Int. J. Biol. Macromol. 118 (2018) 1892−1902.

[84] T.G. Polat, O. Duman, S. Tunç, Agar/κ-carrageenan/montmorillonite nanocomposite hydrogels for wound dressing applications, Int. J. Biol. Macromol. 164 (2020) 4591−4602.

[85] A. Ahmed, G. Getti, J. Boateng, Ciprofloxacin-loaded calcium alginate wafers prepared by freeze-drying technique for potential healing of chronic diabetic foot ulcers, Drug. Deliv. Transl. Res. 8 (2017) 1751−1768.

[86] E. Tamahkar, B. Özkahraman, A.K. Süloğlu, N. Idil, I. Perçin, A novel multilayer hydrogel wound dressing for antibiotic release, J. Drug. Deliv. Sci. Technol. (2020) 58.

[87] G. Tao, Y. Wang, R. Cai, H. Chang, K. Song, H. Zuo, et al., Design and performance of sericin/poly (vinyl alcohol) hydrogel as a drug delivery carrier for potential wound dressing application, Mater. Sci. Eng. C. 101 (2019) 341−351.

[88] S. Huang, H.-J. Chen, Y.-P. Deng, X.-H. You, Q.-H. Fang, M. Lin, Preparation of novel stable microbicidal hydrogel films as potential wound dressing, Polym. Degrad. Stab. (2020) 181.

[89] P. Picone, M.A. Sabatino, A. Ajovalasit, D. Giacomazza, C. Dispenza, M.D. Carlo, Biocompatibility, hemocompatibility and antimicrobial properties of xyloglucan-based hydrogel film for wound healing application, Int. J. Biol. Macromol. 121 (2019) 784−795.

[90] S. Ilkar Erdagi, F. Asabuwa Ngwabebhoh, U. Yildiz, Genipin crosslinked gelatin-diosgenin-nanocellulose hydrogels for potential wound dressing and healing applications, Int. J. Biol. Macromol. 149 (2020) 651−663.

[91] S. Sadeghi, J. Nourmohammadi, A. Ghaee, N. Soleimani, Carboxymethyl cellulose-human hair keratin hydrogel with controlled clindamycin release as antibacterial wound dressing, Int. J. Biol. Macromol. 147 (2020) 1239−1247.

[92] P. Kaur, V.S. Gondil, S. Chhibber, A novel wound dressing consisting of PVA-SA hybrid hydrogel membrane for topical delivery of bacteriophages and antibiotics, Int. J. Pharm. (2019) 572.

[93] A. Romero-Montero, P. Labra-Vázquez, L.J. Del Valle, J. Puiggalí, R. García-Arrazola, C. Montiel, et al., Development of an antimicrobial and antioxidant hydrogel/nano-electrospun wound dressing, RSC Adv. 10 (2020) 30508−30518.

[94] J. Qu, X. Zhao, Y. Liang, Y. Xu, P.X. Ma, B. Guo, Degradable conductive injectable hydrogels as novel antibacterial, anti-oxidant wound dressings for wound healing, Chem. Eng. J. 362 (2019) 548−560.

[95] R. Song, J. Zheng, Y. Liu, Y. Tan, Z. Yang, X. Song, et al., A natural cordycepin/chitosan complex hydrogel with outstanding self-healable and wound healing properties, Int. J. Biol. Macromol. 134 (2019) 91−99.

[96] X. Xuan, Y. Zhou, A. Chen, S. Zheng, Y. An, H. He, et al., Silver crosslinked injectable bFGF-eluting supramolecular hydrogels speed up infected wound healing, J. Mater. Chem. B 8 (2020) 1359−1370.

[97] Y. Yu, Z. Yang, S. Ren, Y. Gao, L. Zheng, Multifunctional hydrogel based on ionic liquid with anti-bacterial performance, J. Mol. Liq. (2020) 299.

[98] M. Zhang, S. Chen, L. Zhong, B. Wang, H. Wang, F. Hong, Zn2 + -loaded TOBC nanofiber-reinforced biomimetic calcium alginate hydrogel for antibacterial wound dressing, Int. J. Biol. Macromol. 143 (2020) 235−242.

[99] W. Chen, Y. Zhu, Z. Zhang, Y. Gao, W. Liu, Q. Borjihan, et al., Engineering a multifunctional N-halamine-based antibacterial hydrogel using a super-convenient strategy for infected skin defect therapy, Chem. Eng. J. (2020) 379.

[100] K. González, C. García-Astrain, A. Santamaria-Echart, L. Ugarte, L. Avérous, A. Eceiza, et al., Starch/graphene hydrogels via click chemistry with relevant electrical and antibacterial properties, Carbohydr. Polym. 202 (2018) 372−381.

[101] H. Ren, L. Wang, H. Bao, Y. Xia, D. Xu, W. Zhang, et al., Improving the antibacterial property of chitosan hydrogel wound dressing with licorice polysaccharide, J. Renew. Mater. 8 (2020) 1343−1355.

[102] L. Gao, H. Zhang, B. Yu, W. Li, F. Gao, K. Zhang, et al., Chitosan composite hydrogels cross-linked by multifunctional diazo resin as antibacterial dressings for improved wound healing, J. Biomed. Mater. Res. Part. A 108 (2020) 1890−1898.

[103] L. Huang, Z. Zhu, D. Wu, W. Gan, S. Zhu, W. Li, et al., Antibacterial poly (ethylene glycol) diacrylate/chitosan hydrogels enhance mechanical adhesiveness and promote skin regeneration, Carbohydr. Polym. (2019) 225.

[104] F. Zeighampour, F. Alihosseini, M. Morshed, A.A. Rahimi, Comparison of prolonged antibacterial activity and release profile of propolis-incorporated PVA nanofibrous mat, microfibrous mat, and film, J. Appl. Polym. Sci. (2018) 135.

[105] L. Zhou, T. Xu, J. Yan, X. Li, Y. Xie, H. Chen, Fabrication and characterization of matrine-loaded konjac glucomannan/fish gelatin composite hydrogel as antimicrobial wound dressing, Food Hydrocoll. (2020) 104.

[106] M. di Luca, M. Curcio, E. Valli, G. Cirillo, F. Voli, M.E. Butini, et al., Combining antioxidant hydrogels with self-assembled microparticles for multifunctional wound dressings, J. Mater. Chem. B 7 (2019) 4361−4370.

[107] E. Larrañeta, M. Imízcoz, J.X. Toh, N.J. Irwin, A. Ripolin, A. Perminova, et al., Synthesis and characterization of lignin hydrogels for potential applications as drug eluting antimicrobial coatings for medical materials, ACS Sustain. Chem. Eng. 6 (2018) 9037−9046.

[108] Y. Zhao, X. Du, L. Jiang, H. Luo, F. Wang, J. Wang, et al., Glucose oxidase-loaded antimicrobial peptide hydrogels: potential dressings for diabetic wound, J. Nanosci. Nanotechnol. 20 (2020) 2087−2094.

[109] D. Zmejkoski, D. Spasojević, I. Orlovska, N. Kozyrovska, M. Soković, J. Glamočlija, et al., Bacterial cellulose-lignin composite hydrogel as a promising agent in chronic wound healing, Int. J. Biol. Macromol. 118 (2018) 494−503.

[110] T. Ren, J. Gan, L. Zhou, H. Chen, Physically crosslinked hydrogels based on poly (vinyl alcohol) and fish gelatin for wound dressing application: fabrication and characterization, Polymers (2020) 12.

[111] N. Bhattarai, J. Gunn, M. Zhang, Chitosan-based hydrogels for controlled, localized drug delivery, Adv. Drug. Deliv. Rev. 62 (2010) 83−99.

[112] Z. Yi, Y. Zhang, S. Kootala, J. Hilborn, D.A. Ossipov, Hydrogel patterning by diffusion through the matrix and subsequent light-triggered chemical immobilization, ACS Appl. Mater. Interfaces 7 (2015) 1194−1206.

[113] M. Vázquez-González, I. Willner, Stimuli-responsive biomolecule-based hydrogels and their applications, Angew. Chem. Int. Ed. 59 (2020) 15342−15377.

[114] C.-G. Wu, X. Wang, Y.-F. Shi, B.-C. Wang, W. Xue, Y. Zhang, Transforming sustained release into on-demand release: self-healing guanosine−borate supramolecular hydrogels with multiple responsiveness for Acyclovir delivery, Biomater. Sci. 8 (2020) 6190−6203.

[115] J. Wang, L. Wang, Y. Gao, Z. Zhang, X. Huang, T. Han, et al., Synergistic therapy of celecoxib-loaded magnetism-responsive hydrogel for tendon tissue injuries, Front. Bioeng. Biotechnol. (2020) 8.

[116] V. Castelletto, C.J.C. Edwards-Gayle, I.W. Hamley, G. Barrett, J. Seitsonen, J. Ruokolainen, Peptide-stabilized emulsions and gels from an arginine-rich surfactant-like peptide with antimicrobial activity, ACS Appl. Mater. Interfaces 11 (2019) 9893−9903.

[117] K. Fang, R. Wang, H. Zhang, L. Zhou, T. Xu, Y. Xiao, et al., Mechano-responsive, tough, and antibacterial zwitterionic hydrogels with controllable drug release for wound healing applications, ACS Appl. Mater. Interfaces 12 (2020) 52307−52318.

[118] P. Makvandi, G.W. Ali, F. Della Sala, W.I. Abdel-Fattah, A. Borzacchiello, Biosynthesis and characterization of antibacterial thermosensitive hydrogels based on corn silk extract, hyaluronic acid and nanosilver for potential wound healing, Carbohydr. Polym. (2019) 223.

[119] V. Pawar, M. Dhanka, R. Srivastava, Cefuroxime conjugated chitosan hydrogel for treatment of wound infections, Colloids Surf. B: Biointerfaces 173 (2019) 776−787.

[120] N. Rezaei, H.G. Hamidabadi, S. Khosravimelal, M. Zahiri, Z.A. Ahovan, M.N. Bojnordi, et al., Antimicrobial peptides-loaded smart chitosan hydrogel: release behavior and antibacterial potential against antibiotic resistant clinical isolates, Int. J. Biol. Macromol. 164 (2020) 855−862.

[121] Z. Deng, Y. Guo, X. Zhao, P.X. Ma, B. Guo, Multifunctional stimuli-responsive hydrogels with self-healing, high conductivity, and rapid recovery through host−guest interactions, Chem. Mater. 30 (2018) 1729−1742.

[122] A.J.R. Amaral, G. Pasparakis, Stimuli responsive self-healing polymers: gels, elastomers and membranes, Polym. Chem. 8 (2017) 6464−6484.

[123] Y. Takashima, S. Hatanaka, M. Otsubo, M. Nakahata, T. Kakuta, A. Hashidzume, et al., Expansion−contraction of photoresponsive artificial muscle regulated by host−guest interactions, Nat. Commun. (2012) 3.

[124] N. Yang, M. Zhu, G. Xu, N. Liu, C. Yu, A near-infrared light-responsive multifunctional nanocomposite hydrogel for efficient and synergistic antibacterial wound therapy and healing promotion, J. Mater. Chem. B 8 (2020) 3908−3917.

[125] H. Zhang, S. Zheng, C. Chen, D. Zhang, A graphene hybrid supramolecular hydrogel with high stretchability, self-healable and photothermally responsive properties for wound healing, RSC Adv. 11 (2021) 6367−6373.

[126] S. Huang, H. Liu, K. Liao, Q. Hu, R. Guo, K. Deng, Functionalized GO nanovehicles with nitric oxide release and photothermal activity-based hydrogels for bacteria-infected wound healing, ACS Appl. Mater. Interfaces (2020).

[127] L. Rosselle, A.R. Cantelmo, A. Barras, N. Skandrani, M. Pastore, D. Aydin, et al., An 'on-demand' photothermal antibiotic release cryogel patch: evaluation of efficacy on an ex vivo model for skin wound infection, Biomater. Sci. 8 (2020) 5911−5919.

[128] L. Chambre, L. Rosselle, A. Barras, D. Aydin, A. Loczechin, S. Gunbay, et al., Photothermally active cryogel devices for effective release of antimicrobial peptides: on-demand treatment of infections, ACS Appl. Mater. Interfaces 12 (2020) 56805−56814.

[129] C. Zhang, J. Wang, R. Chi, J. Shi, Y. Yang, X. Zhang, Reduced graphene oxide loaded with MoS_2 and Ag_3PO_4 nanoparticles/PVA interpenetrating hydrogels for improved mechanical and antibacterial properties, Mater. Des. (2019) 183.

[130] Y. Liang, M. Wang, Z. Zhang, G. Ren, Y. Liu, S. Wu, et al., Facile synthesis of ZnO QDs@GO-CS hydrogel for synergetic antibacterial applications and enhanced wound healing, Chem. Eng. J. (2019) 378.

[131] J. Xi, Q. Wu, Z. Xu, Y. Wang, B. Zhu, L. Fan, et al., Aloe-emodin/carbon nanoparticle hybrid gels with light-induced and long-term antibacterial activity, ACS Biomater. Sci. Eng. 4 (2018) 4391−4400.

[132] Y. Liang, X. Zhao, T. Hu, Y. Han, B. Guo, Mussel-inspired, antibacterial, conductive, antioxidant, injectable composite hydrogel wound dressing to promote the regeneration of infected skin, J. Colloid Interface Sci. 556 (2019) 514−528.

[133] W.T. Wang, T. Zheng, B.L. Sheng, T.C. Zhou, Q.C. Zhang, F. Wu, et al., Functionalization of polyvinyl alcohol composite film wrapped in a(m)-ZnO@CuO@Au nanoparticles for antibacterial application and wound healing, Appl. Mater. Today 17 (2019) 36−44.

[134] A.R. Karimi, A. Khodadadi, Mechanically robust 3D nanostructure chitosan-based hydrogels with autonomic self-healing properties, ACS Appl. Mater. Interfaces 8 (2016) 27254−27263.

[135] J. Wang, C. Zhang, Y. Yang, A. Fan, R. Chi, J. Shi, et al., Poly (vinyl alcohol) (PVA) hydrogel incorporated with Ag/TiO_2 for rapid sterilization by photoinspired radical oxygen species and promotion of wound healing, Appl. Surf. Sci. 494 (2019) 708−720.

[136] X. Zhang, C. Zhang, Y. Yang, H. Zhang, X. Huang, R. Hang, et al., Light-assisted rapid sterilization by a hydrogel incorporated with Ag_3PO_4/MoS_2 composites for efficient wound disinfection, Chem. Eng. J. 374 (2019) 596−604.

[137] F. Bayat, A.R. Karimi, Design of photodynamic chitosan hydrogels bearing phthalocyanine-colistin conjugate as an antibacterial agent, Int. J. Biol. Macromol. 129 (2019) 927−935.

[138] S.L. Percival, S. Mccarty, J.A. Hunt, E.J. Woods, The effects of pH on wound healing, biofilms, and antimicrobial efficacy, Wound Repair. Regen. 22 (2014) 174−186.

[139] D.G. Metcalf, M. Haalboom, P.G. Bowler, C. Gamerith, E. Sigl, A. Heinzle, et al., Elevated wound fluid pH correlates with increased risk of wound infection, Wound Med. (2019) 26.

[140] M. Sakthivel, D.S. Franklin, S. Sudarsan, G. Chitra, T.B. Sridharan, S. Guhanathan, Gold nanoparticles embedded itaconic acid based hydrogels, SN Appl. Sci. (2019) 1.

[141] Y. Jiang, Y. Wang, Q. Li, C. Yu, W. Chu, Natural polymer-based stimuli-responsive hydrogels, Curr. Med. Chem. 27 (2020) 2631−2657.

[142] H.T.P. Anh, C.-M. Huang, C.-J. Huang, Intelligent metal-phenolic metallogels as dressings for infected wounds, Sci. Rep. (2019) 9.

[143] M.U.A. Khan, S. Haider, M.A. Raza, S.A. Shah, S.I.A. Razak, M.R.A. Kadir, et al., Smart and pH-sensitive rGO/Arabinoxylan/chitosan composite for wound dressing: in-vitro drug delivery, antibacterial activity, and biological activities, Int. J. Biol. Macromol. 192 (2021) 820−831.

[144] A.L. Mohamed, A.G. Hassabo, Composite material based on pullulan/silane/ZnO-NPs as pH, thermo-sensitive and antibacterial agent for cellulosic fabrics, Adv. Nat. Sci.: Nanosci. Nanotechnol. (2018) 9.

[145] J.-J. Chuang, Y.-Y. Huang, S.-H. Lo, T.-F. Hsu, W.-Y. Huang, S.-L. Huang, et al., Effects of pH on the shape of alginate particles and its release behavior, Int. J. Polym. Sci. 2017 (2017) 1−9.

[146] M. Shahriari-Khalaji, S. Hong, G. Hu, Y. Ji, F.F. Hong, Bacterial nanocellulose-enhanced alginate double-network hydrogels cross-linked with six metal cations for antibacterial wound dressing, Polymers (2020) 12.

[147] S. Guan, Y. Li, C. Cheng, X. Gao, X. Gu, X. Han, et al., Manufacture of pH- and HAase-responsive hydrogels with on-demand and continuous antibacterial activity for full-thickness wound healing, Int. J. Biol. Macromol. 164 (2020) 2418−2431.

[148] T.S. Anirudhan, A.M. Mohan, Novel pH sensitive dual drug loaded-gelatin methacrylate/methacrylic acid hydrogel for the controlled release of antibiotics, Int. J. Biol. Macromol. 110 (2018) 167−178.

[149] W. Cheng, Y. Chen, L. Teng, B. Lu, L. Ren, Y. Wang, Antimicrobial colloidal hydrogels assembled by graphene oxide and thermo-sensitive nanogels for cell encapsulation, J. Colloid Interface Sci. 513 (2018) 314−323.

[150] X. Zhao, Y. Liang, Y. Huang, J. He, Y. Han, B. Guo, Physical double-network hydrogel adhesives with rapid shape adaptability, fast self-healing, antioxidant and NIR/pH stimulus-responsiveness for multi-drug-resistant bacterial infection and removable wound dressing, Adv. Funct. Mater. (2020) 30.

[151] M. Ma, Y. Zhong, X. Jiang, Thermosensitive and pH-responsive tannin-containing hydroxypropyl chitin hydrogel with long-lasting antibacterial activity for wound healing, Carbohydr. Polym. (2020) 236.

[152] W.C. Huang, R. Ying, W. Wang, Y. Guo, Y. He, X. Mo, et al., A macroporous hydrogel dressing with enhanced antibacterial and anti-inflammatory capabilities for accelerated wound healing, Adv. Funct. Mater. (2020) 30.

[153] P.G. Bowler, B.I. Duerden, D.G. Armstrong, Wound microbiology and associated approaches to wound management, Clin. Microbiol. Rev. 14 (2001) 244−269.

[154] V. Falanga, Wound healing and its impairment in the diabetic foot, Lancet 366 (2005) 1736−1743.

[155] J.G. Powers, C. Higham, K. Broussard, T.J. Phillips, Wound healing and treating wounds, J. Am. Acad. Dermatol. 74 (2016) 607−625.

[156] M. Collier, H. Hollinworth, Pain and tissue trauma during dressing change, Nurs. Stand. 14 (2000) 71−73.

[157] Q. Pang, D. Lou, S. Li, G. Wang, B. Qiao, S. Dong, et al., Smart flexible electronics-integrated wound dressing for real-time monitoring and on-demand treatment of infected wounds, Adv. Sci. (2020) 7.

[158] R.G. Wilkins, M. Unverdorben, Wound cleaning and wound healing, Adv. Skin. Wound Care 26 (2013) 160−163.

[159] Y. Zhang, T. Li, C. Zhao, J. Li, R. Huang, Q. Zhang, et al., An integrated smart sensor dressing for real-time wound microenvironment monitoring and promoting angiogenesis and wound healing, Front. Cell Dev. Biol. (2021) 9.

[160] M. Omidi, A. Yadegari, L. Tayebi, Wound dressing application of pH-sensitive carbon dots/chitosan hydrogel, RSC Adv. 7 (2017) 10638−10649.

[161] T.C. Wallace, M.M. Giusti, Anthocyanins, Adv. Nutr. 6 (2015) 620−622.

[162] K.M. Zepon, M.M. Martins, M.S. Marques, J.M. Heckler, F. Dal Pont Morisso, M.G. Moreira, et al., Smart wound dressing based on κ−carrageenan/locust bean gum/cranberry extract for monitoring bacterial infections, Carbohydr. Polym. 206 (2019) 362−370.

[163] T.R. Dargaville, B.L. Farrugia, J.A. Broadbent, S. Pace, Z. Upton, N.H. Voelcker, Sensors and imaging for wound healing: a review, Biosens. Bioelectron. 41 (2013) 30−42.

[164] M. Chekini, E. Krivoshapkina, L. Shkodenko, E. Koshel, M. Shestovskaya, M. Dukhinova, et al., Nanocolloidal hydrogel with sensing and antibacterial activities governed by iron ion sequestration, Chem. Mater. 32 (2020) 10066−10075.

[165] L. Fan, Z. He, X. Peng, J. Xie, F. Su, D.-X. Wei, et al., Injectable, intrinsically antibacterial conductive hydrogels with self-healing and pH stimulus responsiveness for epidermal sensors and wound healing, ACS Appl. Mater. Interfaces 13 (2021) 53541−53552.

[166] H. Lei, J. Zhao, X. Ma, H. Li, D. Fan, Antibacterial dual network hydrogels for sensing and human health monitoring, Adv. Healthc. Mater. (2021) 10.

[167] C. Chai, Y. Guo, Z. Huang, Z. Zhang, S. Yang, W. Li, et al., Antiswelling and durable adhesion biodegradable hydrogels for tissue repairs and strain sensors, Langmuir 36 (2020) 10448−10459.

[168] X. Jing, H.-Y. Mi, Y.-J. Lin, E. Enriquez, X.-F. Peng, L.-S. Turng, Highly stretchable and biocompatible strain sensors based on mussel-inspired super-adhesive self-healing hydrogels for human motion monitoring, ACS Appl. Mater. Interfaces 10 (2018) 20897−20909.

[169] Y. Zhu, L. Lin, Y. Chen, Y. Song, W. Lu, Y. Guo, A self-healing, robust adhesion, multiple stimuli-response hydrogel for flexible sensors, Soft Matter 16 (2020) 2238−2248.

[170] M. Zheng, X. Wang, O. Yue, M. Hou, H. Zhang, S. Beyer, et al., Skin-inspired gelatin-based flexible bio-electronic hydrogel for wound healing promotion and motion sensing, Biomaterials (2021) 276.

Medicated wound dressings

9

Bhingaradiya Nutan[1] and Arvind K. Singh Chandel[2]

[1]*Department of Biosciences and Bioengineering, Indian Institute of Technology Bombay, Mumbai, Maharashtra, India* [2]*Center for Disease Biology and Integrative Medicine, Graduate School of Medicine, The University of Tokyo, Bunkyo-Ku, Tokyo, Japan*

9.1 Introduction

Skin is the largest organ in the body, with an area of 2 m^2, that acts as a self-renewable barrier that protects internal tissues and organs from the external environment by forming a layer that separates the organism's internal environment from the outside world [1]. It performs a variety of physiological functions, including temperature regulation, protection of the body and inner organs from microbial attack, and support of immune and sensory functions. The skin's highly ordered architecture, which consists primarily of two tissue compartments, the Epidermis and Dermis, performs its critical bidirectional barrier function. The epidermis is a stratified layer of epithelial keratinocytes that form the outermost layer of the skin and are connected to one another via cell-cell adhesion molecules. By forming tight junctions (adherens junctions and desmosomes) with epidermal keratinocytes, this cell-cell communication provides mechanical stability and protects the organism from water loss. Through cell-matrix connections, stem cells in the epidermis' basal layer bind to the basement membrane. The Basement Membrane is a thin, rigid layer that separates the epidermis and dermis (Epidermal-Dermal junction). It is made up of cells and matrix components from both layers, and it is necessary for the mechanical stability of the skin. The dermis is a layer composed primarily of extracellular matrix (ECM) and sparsely of fibroblasts. Collagen (primarily Collagen Type I and III) is the primary ECM contributor to the dermis, conferring elasticity and stretchability as well as mechanical and tensile strength to the skin via collagen crosslinks [2]. The dermis is divided into two sections: papillary dermis and reticular dermis. The papillary dermis has a dense fibroblast network but a sparse collagen and elastin network. The basal reticular dermis, on the other hand, is densely packed with a collagen network. Maintaining the composition of skin architecture is critical for organism survival and well-being, as an organized epidermis confers mechanical strength, prevents water loss, and serves as a barrier function, protecting the skin from various external assaults and trauma. The dermis' structural compartmentalization is required for contractile properties, tissue homeostasis, and immune response regulation during wound insults. This highly ordered structure protects the skin from mechanical stress, chemical stress, and pathological stress. Any disruption to the skin's architecture and integrity is referred to as a wound, which initiates a series of signaling cascades and immune responses that, in turn, initiate the wound healing process.

Antimicrobial Dressings. DOI: https://doi.org/10.1016/B978-0-323-95074-9.00005-1

A wound is any disruption to the skin's integrity, tissue architecture, or the mucous membrane beneath the skin. It is also defined as a compromise of the skin's barrier function caused by a loss of physiological function of epithelial tissue as a result of physical/chemical/radiological damage or an underlying disease [3]. Skin is a self-renewing structure that is constantly subjected to wear and tear due to interaction with the external environment. Dead cells are sloughed off and replaced with new cells, and similarly, the skin heals after wound assaults. The extent and rate of healing are determined by the extent of damage in the skin compartments as well as the cause of the wound or injury. Wounds caused by external traumas, such as physical and chemical injuries, heal quickly and are referred to as acute wounds, whereas wounds caused by underlying chronic pathological and physiological disorders or diseases, such as ulcers and diabetes mellitus, take longer to heal [4].

Wound healing is a complex and highly orchestrated biological phenomenon that requires the participation of various mediators such as cytokines and growth factors, ECM components such as collagen, and cells from various parts of the body such as macrophages, fibroblasts, and platelets to repair the tissue, heal the wound, and restore tissue homeostasis. The stages of wound healing are well documented and have been the subject of extensive research over the years. Hemostasis, inflammatory and debridement, proliferative, and maturation are the four major stages of wound healing (Fig. 9.1) [5,6].

Platelets play an important and versatile role during the hemostatic stage. Platelet alpha and dense granules contain a variety of growth factors as well as cell receptors required for hemostasis. A blood clot forms after an injury to prevent vascular fluid loss and damage. Platelet receptors bind to ECM components such as fibronectin, collagen, and von Willebrand factor, which then adhere to endothelial tissue, resulting in the formation of thrombin, which in turn activates platelets via conformational change and the release of constituents of alpha and dense granules. Coagulation factors released from the granules initiate a coagulation cascade that results in the formation of insoluble fibrin clots. Several factors influence the formation and spread of a clot. The blood clot protects

Normal Wound-healing Process

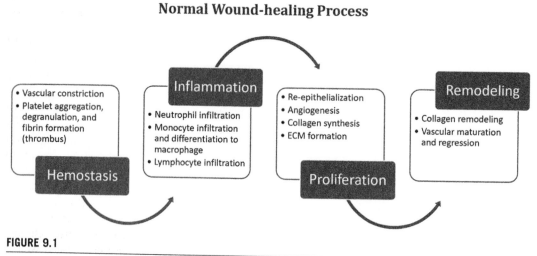

FIGURE 9.1

Normal hemostasis and wound healing process.

against bacterial invasion and serves as a platform for immune cells and factors to participate in the wound healing cascade. Platelet secretions also contain growth factors that stimulate keratinocytes and fibroblasts, which play an important role in wound healing later on [7].

Necrotic and injured cells release damage-associated molecular patterns and pathogen-associated molecular patterns in the Inflammation stage, which bind to resident T-cells and circulating macrophages, damage response elements, and promote inflammatory signaling pathways. These signaling effectors, particularly proinflammatory cytokines, act as signals (chemical signals) for neutrophil and macrophage migration. Other mediators released during inflammatory events, such as selectins, aid in recruitment by promoting vasodilation, which allows leukocytes to migrate (extravasation and diapedesis) [8]. When neutrophils migrate to the site of injury, they initiate Inflammatory signaling pathways and effectors, particularly the NF-KB pathway, which is required for Neutrophil trapping. Through phagocytosis, antimicrobial peptides (extracellular trap), and proteolytic enzymes, they remove dead, necrotic cells and engulf and destroy pathogens. After migrating to the injured site, monocytes differentiate into macrophages, scavenge apoptotic neutrophils, facilitate phagocytosis, and regulate inflammatory responses [7].

Keratinocytes, macrophages, and fibroblasts perform resurfacing of new epithelial tissue, wound contraction, ECM deposition, and angiogenesis during the proliferative stage. Fibroblasts and Keratinocytes are activated after injury in response to mechanical stress, pathogenic stress, and the release of cytokines in the inflammatory phase by macrophages. Keratinocytes go through an epithelial to mesenchymal transition and develop a more migratory phenotype, migrating laterally across the wound to re-epithelialize the epidermal layer, and the new Keratinocytes begin depositing ECM proteins. Fibroblasts at the site of injury and those derived from EMT respond to mediators released by platelets, macrophages, and vascular endothelium, causing the fibrin-rich matrix to be replaced by more pinkish granulation tissue. These fibroblasts differentiate into myofibroblasts, which are in charge of wound contraction. Angiogenesis is required for the formation of new blood vessels via various mediators such as VEGF, and wound macrophages release MMPs, which degrade the fibrin clot matrix and aid in the migration of newly formed blood vessels. In this phase, in addition to re-epithelialization, matrix deposition, and angiogenesis, nerve fiber regeneration (innervation) occurs [9].

The remodeling phase is another name for the maturation phase. Matrix remodeling and ECM deposition occur at almost all stages, but collagen reorganization occurs at this stage. During the preceding stages, the wound is thick and dominated by collagen III, while the uninjured tissue is dominated by collagen I. In this phase, collagen III is replaced by collagen I, and the wound healing process is terminated. As the healing process progresses, it is critical to keep the wound area pathogen-free, moist, and nutrient-rich. While the innate immune system responds to microbial attack via T-cells, macrophages, and phagocytosis, deep wounds such as burns and chronic wounds caused by diseases such as diabetes mellitus and diabetic foot ulcers take a long time to heal due to a variety of causative factors [7]. For example, the wound environment should be rich in growth factors and micronutrients to heal it, but these conditions also attract microorganisms (wound bioburden), and when they invade, they produce various factors such as inflammatory cytokines, which cause excessive inflammation, and MMPs, which can degrade the ECM and slow the rate of healing even further. So wound management is critical because thousands of people suffer from chronic wounds and disabilities in healing as a result of external factors that can cause decelerated wound healing and, if left untreated and under-managed, can lead to death. So wound dressing is any material that is applied to a wound to protect it from microorganisms, keep it moist, absorb excess fluid and exudates, and ensure the wound healing process progresses. Cotton gauze and honey have traditionally been used as dressings due to their fluid absorption and Biocompatibility, but in modern society, wound care and management is a clinically significant subject, and wound dressing has evolved to limit the disadvantages (excessive fluid absorption, dryness, and non-replaceable) associated with those dressings [3].

Medicated wound dressings are a type of wound dressing that has therapeutic value in addition to its normal function. In a chronic wound with bacterial invasion, for example, where the host's antimicrobial function/defense is compromised or diminished, the application of an antimicrobial ingredient-loaded dressing allows for accelerated wound healing in comparison to a normal dressing where the dressing material facilitates the normal wound healing process and the antimicrobial agent takes care of the invading microorganisms. While these medicated wound dressings can be divided into multiple sub-classes based on the type of dressing material used, such as hydrogel, fibers, and hydrocolloids, etc., we have broadly classified them based on the function of the therapeutic compound (such as antimicrobial, growth factors, or enzymes) added to the dressing, and this chapter deals with these medicated wound dressings.

9.2 General wound healing mechanism

Constrictions of peripheral blood vessels are released at the time of injury in an attempt to minimize bleeding into soft tissues. The platelet is the key cell responsible for this function, which causes the body to form a clot to prevent further bleeding. Platelets clump together in order for the blood vessel to complete the clotting process. Platelets also release important cytokines, such as platelet-derived growth factor, which gathers cells for later stages of healing [10]. Following hemostasis, the inflammatory phase of wound healing begins. The local signs and symptoms that occur during the inflammatory phase are swelling increased fluid perfusion of blood redness release of epinephrine histamine heat histamine response and the inflammatory phase is characterized by a host of cells leukocytes and macrophages infiltrating the wound sites. Hemostasis regulates bleeding [11]. Leukocytes, particularly polymorphonuclear neutrophils, destroy any foreign body or bacteria that is present. About four days after the injury, macrophages work to destroy bacteria and clean the wound of cellular debris, and macrophages replace the leukocytes. Following that, the macrophage produces a slew of cytokines and growth factors that act as chemoattractants to other cells required for tissue repair. It is also thought that the macrophage attracts contractual cells to the wound to encourage wound contraction vasodilation with resultant edema, heat, and redness are the result of factors secreted by the macrophage. The other leukocytes present at the wound site in response to the inflammatory process. The inflammatory phase of wound healing's objectives are to clean debris and bacteria and to prevent infection. Scar tissue formation is divided into three stages. contraction of granulation epithelialization Aspects of wound healing at the cellular level [12].

Granulation tissue is produced in an open wound, resulting in red beefy shiny tissue with a granular appearance. This tissue is composed of fibroblasts and neutrophils, and as this type of tissue proliferates, fibroblasts stimulate the production of collagen. As the wound site fills with granulation tissue, the margins contract or pull together, decreasing the size of the wound. The extent of contraction is dependent on the mobility of the surrounding tissue during epithelialization [13].

Cells move and divide from the wound margins in the last stage of the phase, where they eventually touch. Because of this, the maturation phase of sealing the wound epithelialization can only take place in the presence of living vascular tissue epithelial cells and will not migrate across a dry surface or necrotic tissue. The collagen fibers remodel, mature, and gain tensile strength as they age. Rearranging and redistributing the collagen fibers, proteoglycans, and fibronectin causes the scar to become less cellular and more tensile [14]. However, because its tensile strength is lower than that of uninjured skin, this tissue will always be vulnerable to breakdown. Collagen synthesis begins with the fibroblasts. Secrets of the fibroblast Procollagen fiber growth is a complicated process; macrophages and platelets are important growth factors in procollagen development. Procollagen fibers develop into collagen fibrils. The fibrils are then linked together to form a very strong ropelike collagen fiber; approximately 10,000 fibrils are interconnected within a single collagen fiber; research is currently being conducted to determine the chemical details associated with collagen [15].

9.3 Integra

9.3.1 Materials for wound care

Paraffin gauze is a type of wound contact layer. This is an illustration of a paraffin gauze dressing. These dressings have a limited placement wound management because they can trap extra day under the dressing due to the heavy paraffin load, which can cause maceration. Additionally, the goals component of this dressing can become adhered to the wound bed [16].

However, it is quite conformable to areas such as digits because the dressing can stick to itself quite nicely and they do not require changing daily emulsion gauze dressing. There are several brands available; unlike paraffin gauze dressings, these dressings effectively transfer exit at the perforations into a secondary dressing. They have a silky feel to them, have a low risk of adherence to the wound bed, and do not need to be changed every day. They also come in a variety of sizes, including some that are quite large for dressing areas such as circumferential wounds on the legs. One of the main disadvantages is that they do not have a high tack to the surrounding skin or the dressing itself, making them less conformable to digits [17].

Additional data transfer through the perforations into a secondary silicone sheet dressing is made possible by silicone sheet wound dressing. They make some very soft and conformable, and they help the dressing stay in place. These are great dressings for treating category one and two skin tears because they support the wound and surrounding tissue in addition to holding the flaps in place. The dressing can be left in place for up to 14 days with additional changes of the secondary dressing in between in the traditional dressing section. The practitioner can see through the dressing to examine the wound. For various levels of exudate, there are numerous different types of pads. For low doses of X-ray, these types of dressings are low absorption pads, and they are not recommended for moderate to high exudate. They may also be known as island dressings or non-adherent dressings. But keep in mind that these can stick to an open wound. Direct contact with open wounds is generally avoided. For modest amounts of additional date, it frequently uses those as a secondary dressing over contact layers over our absorption pads. Despite their variable composition, they are typically not recommended for dry wounds. They typically consist of cellulose and cotton fiber. Again, if applied directly to an open wound, the contact layer on these can adhere. As a result, they are typically used as a second dressing. Depending on the need for high levels of absorption, this may be applied over a specialized dressing such as a capillary action or antimicrobial dressing or over a traditional dressing like a wound contact layer. It may also be applied over a basic modern dressing like an alginate or jelling fiber. We have high-level action absorbent pads. These are typically made of cellulose and super absorbent polymers, which can absorb and retain large amounts of fluid again, and are not recommended for wounds with low levels of exudate or dry wounds. These are typically applied over a conventional wound contact layer as a secondary dressing. Understanding these dressings before using them is important because some of them can initially appear to be not very absorbent. Examples of basic modern dressings include an alginate or jelling fiber or a specialized dressing like a capillary action dressing or an antimicrobial dressing. But because of the makeup of the dressing and the absorption capacity of the polymers contained in it, this stressing can absorb and hold up to 100 mL of action [18].

In addition to conventional wound dressings, there are other kinds. Modern dressings are also available today, and they are particularly effective. As a result, you should consult your doctor about the best wound dressing. There is typically a difference between dry and moist wound care. Dry wound care entails applying a dry dressing to the wound. Blood and wound fluid are absorbed by it. The so-called exudate guards against contamination, and the most well-known dry wound dressing is probably mechanical injury compress [19].

It could be fixed with a plaster or a fixation bandage and tape. Apply the fixation bandage loosely if you plan to add compression bandages. The soft outer material of a self-adhesive wound dressing with a wound pad is sandwiched between a dry, netlike wound contact layer that quickly conducts fluid into the wound and keeps the wound and dressing from adhering to one another [20].

These materials are impregnated with a neutral, active-substance-free ointment compound for the treatment of dry wounds. The overlying compress ensures that the absorption of exudate of both pads is fixed together with an elastic bandage, and they act as an intermediate layer to prevent the dressing from sticking to the wound. As the name implies, moist wound care creates a climate that is known to encourage healing. In addition, excess exudate is absorbed in both wound and wound environments protected by hydrogel-coated foam dressings that primarily provide a moist wound environment that aids in wound healing. Additionally, they are offered with a self-adhesive edge for fixation [21,22] (Fig. 9.2).

In order to fix super absorbent and silicone wound dressings, no dressing is required. These products are highly absorbent, trapping wound exudate and preventing leakage even when used under a compression bandage. Because of the silicone coating pad, which does not adhere to the wound. It only adheres to the skin gently, and dressings can be removed gently with rinsing. Absorption effects this unique dressing type that continuously absorbs and retains sterile ringer solution in the wound dressing for three days. This rinsing absorption effect results in a long-lasting cleansing effect; fastening is accomplished, for example, with transparent foil bandage tips. Self-adhesive wound dressings are a good choice if dressing changes are rare; however, if the dressing is changed frequently, a wound dressing with a fixing bandage and, if necessary, a flexible tubular bandage that adapts to the contours of the body is recommended for additional support.

Traditional dressings mainly comprise lint, cotton wool, and synthetic and natural gauzes and bandages with various degrees of absorbency that cause injury and incomplete tissue growth after removal.

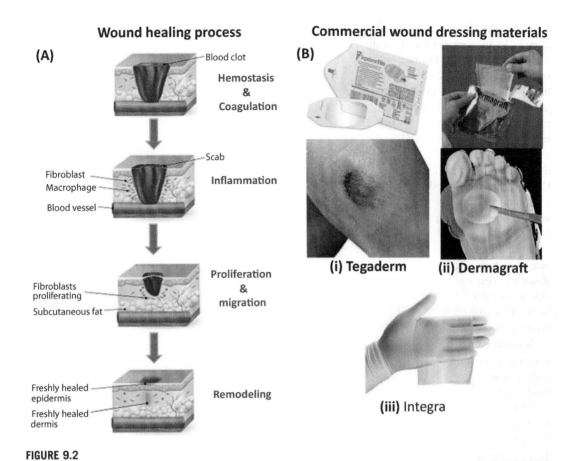

Wound healing process

(A)

- Blood clot — **Hemostasis & Coagulation**
- Scab — **Inflammation**
  - Fibroblast
  - Macrophage
  - Blood vessel
- **Proliferation & migration**
  - Fibroblasts proliferating
  - Subcutaneous fat
- **Remodeling**
  - Freshly healed epidermis
  - Freshly healed dermis

Commercial wound dressing materials

(B)

(i) Tegaderm   **(ii) Dermagraft**

(iii) Integra

FIGURE 9.2

(A) Illustration of the wound healing process (B) Digital images of commercial wound care materials (i) Tegaderm, (ii) Demagraft, (iii) Integra (images are adopted from respective websites).

Therefore, biodegradable wound dressings are based on bioactive materials that induce wound healing. The global advanced wound dressing market size was valued at USD 6.6 billion in 2020 and is expected to expand at a compound annual growth rate (CAGR) of 5.5% from 2021 to 2028 to the deposition of ECM are desirable [23,24].

9.3.2 Comparison

- Tegaderm polyurethane partial adherence to wound bed
- Integra Type I bovine collagen/chondroitin-6-sulfate/silicon high incidence of infection
- Opsite, a synthetic dressing substitute polyurethane incomplete adherence to wound bed
- Dermagraft polyglactin mesh/fibroblast full-thickness diabetic foot ulcers, allergic reactions
- Biobrane synthetic dressing substitute silicon/nylon mesh/peptides derived from porcine type I collagen allergic reactions

9.3.3 Polymeric biomaterials for wound healing

A suitable material has always been required to cover the wound to prevent infection in order to ensure effective wound healing. In the past, wound dressings made of biopolymers such as animal fats, plant fibers, and honey pastes were common. Cotton wool, lint, and gauzes were first used as dressings a few years later. Their primary function was to keep the wound dry by allowing wound exudates to evaporate and preventing bacterial invasion from the surrounding environment. When designing a bandage for wound healing that provides an alternative to the previous wound-healing environment, it is critical to consider the dressing's physical and chemical properties. To aid in the healing process, an active wound dressing regulates the biochemical state of the wound. No wound dressing is ideal, but the minimum requirements of rapid healing, patient affordability, esthetics, and infection prevention must be met during wound management.

9.3.4 Types of dressings used for wound healing

According to a widely accepted theory, moist wound dressings hasten wound healing. This hypothesis was tested in the winter on young domestic pigs, with moist wound dressings causing faster epithelialization. Scab formation was also thought to be preventative in moist wound conditions [25] (Fig. 9.3, Table 9.1).

FIGURE 9.3

Different types of wound healing materials.

Table 9.1 List of polymeric dressing, brand name and use.

S. no.	Type of polymeric dressing	Brand name	Use for
1	Polymeric foam	Flexzan	Chronic wounds
		Biopatch	
		Crafoams	Burns
		Biatain	Mohs surgery and wounds
		Cutinova	Laser resurfacing wounds
2	Polymeric hydrogels	Cultinova gel	Chemotherapy peels
		Biolex	
		TegaGel	
		2nd skin Flexderm	Ulcers
		Dry dressing	Laser resurfacing
3	Polymeric alginates	AlgiSite	Thickness burns
		AlginSan	
		Sorbsan	Surgical wounds
		Kaltostat	High exudate wounds
		Omiderm	Chronic ulcer
4	Polymeric hydrocolloids	Idosorb	Chronic ulcer
		Debrisan	Burns
		Sorbex	Average thickness wounds
		Duoderm	

9.3.5 Antimicrobial loaded Dressings

Prolonged skin healing in chronic wounds is caused primarily by the infection of microorganisms present in the immediate environment. The balance and barrier between the skin and microflora are damaged as the skin establishes a bi-directional barrier that protects the subcutaneous tissue from an infection that has been damaged due to a wound, and the microflora can invade the tissue, which prolongs skin healing from the estimated time of 8–12 weeks. Bacterial infections cause three-fourths of chronic wound infections. Gram-positive organisms such as Staphylococcus aureus and Escherichia coli predominate in both the early and late stages of the chronic wound. Pseudomonas species that are Gram-negative are more common in later stages, perfuse deeply into the wound, and cause significant tissue damage to the tissue microenvironment. Such skin and tissue infections have a high mortality rate. While the dressing promotes wound healing, some agents are required to prevent bacterial or microbial colonization and invasion. As a result, research groups all over the world began to investigate antimicrobial dressings, and microbial abundance in chronic wounds became a topic of interest to a variety of groups. Any pharmacological or therapeutic agent that combats or kills microorganisms is referred to as an antimicrobial. Antimicrobial Dressings are a type of medicated wound dressing that has an antimicrobial effect (therapeutic or medicinal use) due to the presence of an antimicrobial agent. Although there have been reports of fungal infections, antimicrobial dressings are discussed here as anti-bacterial dressings because multiple bacterial species from the mucus microflora and external environment contribute to the majority of infections in chronic wounds. Antibiotics, inorganic metal-based antiseptics (broad-spectrum anti-microbial activity), and natural products are examples of broad classifications. These antimicrobial agents are commonly loaded into wound dressings such as hydrogels, electro spun fibrous scaffolds, hydrocolloids, foams, and sponges.

9.3.5.1 Antibiotics

Antibiotics are substances that interfere with the function and structure of bacteria by inhibiting metabolic pathways, killing them and halting the invasion. Beta-lactams and glycopeptides, quino-lones, aminoglycosides, tetracyclines, and sulfonamides are the different types. Ciprofloxacin, ceftriaxone, tetracycline, and amoxicillin are the most commonly used antibiotics. Many microor-ganisms have developed antibiotic resistance as a result of overuse, and most antibiotics adminis-tered to the wound subcutaneously have an anti-metabolic effect, limiting their use as a direct agent in wound dressings. B-lactam antibiotics are a type of antibiotic that works by interfering with bac-terial cell wall synthesis by inhibiting the cross-linking of peptidoglycan subunits, which are impor-tant components of the bacterial cell wall. A deformed cell wall generates mechanical and osmotic tension, causing the cell wall to burst. Quinolones are a type of antibiotic that works by inhibiting nucleic acid synthesis and blocking transcription. They bind to bacterial topoisomerase II and IV, preventing mRNA transcript synthesis. Antibiotics are loaded using various methods. The most common method is to combine the antibiotic with a polymer matrix, such as a film, sponge, liquid, or gel [26]. Antibiotic-loaded Nano formulations or nanofibers have been used to load the drug [27]. Antibiotic-loaded dressing material is a potential material for wound dressing [28]. To load drugs, the electrospun or co-electrospun method is used. When a drug and a polymer are soluble in different solvents, the emulsion electrospinning method comes in handy [29]. different type of wound dressing materials with antibacterial activity has been reported as described in Table. 9.2.

9.4 Nanoparticles and nanomaterials

9.4.1 Silver nanoparticles (AgNPs)

Silver nanoparticles (AgNPs) is having wound healing, anticancer, antiangiogenesis, and biosensing property and hence have been used widely in biomedical applications [30,31]. Since the nineteenth century, silver-based formulations have been used to treat ulcers due to their antibacterial activity. Silver nitrate and silver sulfadiazine have been used to treat chronic wound dressing [32]. However, silver ions are extremely toxic. Silver ions have now been replaced by AgNPs and nano-crystals, which are less toxic and more effective in treating infections at lower concentrations [33]. Wound dressings containing AgNPs, such as Acticoat with sustained nanoparticle release, have been used to promote wound healing. It also helps to reduce protein toxicity and denaturation in the wound [34]. The safety was demonstrated by in vitro as well as in-vivo analysis [35]. The AgNPs were also synthesized by the use of phytochemicals. Biosynthesized AgNPs have also been reported by Chai et al. It has shown good wound healing after 17 days of application [36]. AgNPs synthesized by using biopolymers or phytochemicals have shown good antibacterial and wound healing properties.

Table 9.2 Antibacterial wound dressing their product name, composition, and indication.

S. no.	Product name	Composition		Indications
		Polymer	Antibacterial agent	
1	3M Kerracel Ag gelling fiber dressing	Carboxymethyl cellulose (CMC)	Ag^{3+}	Chronic and acute wounds
2	3M promogran prisma	Oxidized regenerated cellulose, collagen	Silver	Pressure injuries, diabetic ulcers, venous ulcers
3	3M Silvercel	G (guluronic acid) alginate and CMC	Silver	Moderate to heavily exuding partial- and full-thickness wounds
4	AQUACEL Ag SURGICAL	Polyurethane film	Ionic silver	Traumatic and elective post-operative wounds
5	BIAKŌS	Poloxamer 407	Polyaminopropyl biguanide	Burns, partial- and full-thickness wounds, large-surface-area wounds
6	ColActive plus	Collagen, CMC	Ionic silver	Venous ulcers, donor and graft sites, abrasions, and lacerations
7	Covalon IV clear	Silicone adhesive	Chlorhexidine and silver	Cover and protect insertion sites
8	Cutimed sorbact	–	Removes bacteria using a physical mode of action	Post-excision of fistulas and abscesses and dermal fungal infections.
9	DermaBlue + foam	Polyurethane/polyether foam	Methylene blue, gentian violet and silver sodium zirconium phosphate	Moderately to heavily exuding, partial- to full-thickness wounds
10	Hydrofera Blue CLASSIC	–	Methylene blue Gentian violet	Local management of wounds such as pressure injuries
11	IoPlex	I-Plexomer	Iodine	Infected ulcers and wounds
12	PolyMem MAX silver	–	Silver	Every stage of healing
13	Suprasorb	–	Silver	Moderately to highly exuding acute or chronic wounds
14	UrgoTul	Lipidocolloid	Silver	Acute and chronic wounds

9.4.2 Gold nanoparticles

Stabilized gold nanoparticles (AuNPs) have been synthesized and used widely in different biomedical applications like angiogenesis, cancer treatment, gene delivery, arthritis, drug delivery, and biosensing [37]. AuNPs have now also been used as tissue adhesive and wound healing applications [38]. AuNP composite chitosan-based adhesive materials showed good adhesion properties compared to chitosan alone [39]. MatriDerm, a commercially available nanocomposite sponge, has demonstrated good wound healing compared to collagen alone [40]. Gold nanodots prepared by the use of cyclic lipopeptide surfactin has been reported by Chen et al. for their antibacterial and wound healing property [41]. 101 Laser-activated AuNP-based sutures, thermos-responsive and IR light-sensitive wound dressings have been used as tissue closer and dressing [42]. AuNPs have shown good antibacterial, anti-inflammatory, and hemostatic activity.

9.4.3 Graphene oxides

Because of the ease and versatility of surface modification with active ingredients, biomolecules, and antibodies, graphene oxide has been used in a variety of biomedical applications such as imaging, drug delivery, and biosensing. Antibacterial properties of graphene oxides (GO) composites have recently been reported. Diabetic wounds have been treated using isabgol-based GO nanocomposite scaffolds with cultured fibroblasts [43]. An in-vivo study on diabetic rats revealed faster wound healing. The in-vivo study revealed increased collagen concentration and wound shrinkage, indicating its potential use as a low-cost wound dressing material for clinical use. For antibacterial activity, a nanocomposite prepared using a hybrid nanostructure containing graphene oxide and AgNPs was used [44]. In vivo, polycaprolactone-based hybrid nanocomposites demonstrated significant antibacterial and wound regeneration activity. Antibacterial activity of GO and curcumin-loaded nanocomposite scaffolds against gram-negative and gram-positive bacteria has been demonstrated. GO has been used in wound healing applications due to its versatility and ease of modulation of physicochemical properties.

9.4.4 Terbium nanoparticles

Terbium hydroxide nanoparticles have been used in different biomedical applications including angiogenesis with enhancement in cell proliferation, viability, and cell migration [45]. During the in-vivo study, researchers observed the formation of new blood vessels, the generation of intracellular reactive oxygen species (ROS), and the generation of angiogenesis signal molecules. Terbium-hydroxide nanospheres and nanorods were found to have significant proangiogenic activity in zebra embryonic primary cells and transgenic zebra fish models after administration [46]. Potential toxicity studies and clinical trials are required to further use terbium nanoparticles as a potential wound healing agent (Fig. 9.4).

9.4.5 Copper nanoparticles (CuNPs)

For the healing of the wound, angiogenesis is the key step in the generation of new blood vessels and the proliferation and migration of endothelial cells [47]. Copper is essential for angiogenesis and wound healing regulation. In an in-vivo study with rats, biosynthesized (using *Pseudomonas aeruginosa*) copper nanoparticles (CuNPs) were found to promote wound healing more than copper [48]. Wound dressings containing copper oxide have been shown to induce vascular endothelial growth factor, placental growth factor, and hypoxia-inducible growth factor in diabetic mice, resulting in wound healing [49]. In the rat model, chitosan nanocomposites loaded with CuNPs demonstrated antibacterial and antifungal activity, as well as angiogenesis-aided wound healing via induction of vascular endothelial growth factor [50]. As a result, nanocomposite copper-based materials have shown promising potential as wound healing agents, as well as antibacterial and antifungal activity.

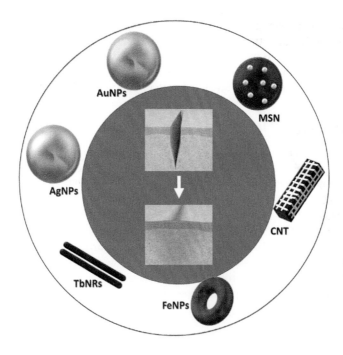

FIGURE 9.4

Application of different inorganic nanoparticles in wound healing.

9.4.6 Zinc nanoparticles

Zinc is a trace element that is essential for enzymes, cofactors, immune responses, and the central nervous system. A nanocomposite material containing zinc nanoparticles (ZnNPs) has been shown to significantly improve wound healing, cell migration, and cell adhesion. Polycaprolactone conjugated ZnNPs demonstrated good wound healing properties in an in-vivo study in Guinea Pigs by increasing ROS [51]. ZnNPs have also been used in skin replacement and wound healing processes. Biosynthesized ZnNPs derived from *Trianthema portulacastrum* demonstrated good dermal and epidermal permeability. In a rat model, the particles were found to have good wound healing ability via keratinocyte migration, reepithelization, collagen fiber deposition, and tissue granulationl [52]. Sodium alginate-based ZnNPs nanoconjugates have also displayed antioxidant and antiinflammatory activity [53]. Nanocomposite hydrogels containing ZnNPs with sustained release demonstrated antibacterial activity, increased cell migration and proliferation, and excellent wound healing properties. ZnNPs have antibacterial, anti-inflammatory, and wound healing properties.

9.4.7 Silica nanoparticles

Silica nanoparticles (SiNPs) are widely used in wound dressings. Fibroblast migration and proliferation are critical in wound healing. SiNPs have been shown to promote wound healing by converting silicic acid via fibroblast action [54]. In a rat model, collagen nanocomposites containing mupirocin and silica microspheres demonstrated good wound healing and antibacterial activity via a synergistic effect [55]. Sustained antibacterial drug release and rehydration via collagen use enhance the healing effect. As a wound dressing material, collagen-based nanocomposite hydrogel loaded with SiNPs, rifamycin, and gentamycin has also been reported [56]. Thus, SiNPs show good wound healing properties through migration and epithelization of cells.

9.4.8 Titanium nanoparticles

Titanium along has been used in different biomedical applications. Nanocomposite bacterial cellulose conjugates have been used as a wound dressing [57]. Cellulose has been used to improve the dressings' water retention, porosity, and biocompatibility. In the burn wound model, the formulation demonstrated good antibacterial and wound healing activity by enhancing angiogenesis and fibroblast migration. Artificial nanocomposite skin with average biodegradability, good thickness, and low density has also been developed [58]. Anti-inflammatory and wound healing properties have been observed in the skin. Biosynthesized titanium nanoparticles (TiNPs) have environmental, cost-effective, and green synthesis advantages. TiNPs have been shown to promote wound healing by increasing collagen deposition, fibroblast formation, and macrophage accumulation [59]. In the rat model, a biosynthesized nanocomposite made from *Moringa oleifera* leaves demonstrated good antibacterial and wound healing activity [60].

9.4.9 Iron nanoparticles

The superparamagnetic iron oxide nanoparticles have been used in imaging as MRI contrast agents, diagnostic agents, and vehicles for drug delivery as cancer theranostic [61]. It is also reported to enhance the wound healing property of thrombin by enhancing the half-life of thrombin by conjugating with iron nanoparticles (FeNPs) [62]. Antibacterial activity of iron oxide nanoparticles and carbon nanotubes against both gram-positive and gram-negative bacteria has been demonstrated [63]. FeNPs has demonstrated promising wound healing activity.

References

[1] J.W. Patterson, Weedon's Skin Pathology E-Book, Elsevier Health Sciences, 2014.
[2] F.M. Watt, H. Fujiwara, Cold Spring Harb. Perspect. Biol 3 (2011) a005124.
[3] F. Hofstädter, Chirurg 66 (1995) 174.
[4] T. Kondo, Y. Ishida, Forensic Sci. Int. 203 (2010) 93.
[5] P. Beldon, Surgery 28 (2010) 409.
[6] S. al Guo, L.A. DiPietro, J. Dent. Res. 89 (2010) 219.
[7] G. Hosgood, Vet. Clin. Small Anim. Pract 36 (2006) 667.
[8] T.J. Koh, L.A. DiPietro, Expert Rev. Mol. Med (2011) 13.
[9] A. Young, C.-E. McNaught, Surgery 29 (2011) 475.
[10] R.J. de Mendonça, J. Coutinho-Netto, An. Bras. Dermatol 84 (2009) 257.
[11] G.S. Schultz, J.M. Davidson, R.S. Kirsner, P. Bornstein, I.M. Herman, Wound Repair. Regen 19 (2011) 134.
[12] S.V. Dryden, W.G. Shoemaker, J.H. Kim, Atlas Oral Maxillofac. Surg. Clin. North. Am 21 (2013) 37.

[13] C. Ahn, P. Mulligan, Adv. Skin Wound, Care 21 (2008) 227.

[14] D.A. Anaya, E.P. Dellinger, Surg. Infect. (Larchmt) 7 (2006) 473.

[15] M. Arnold, A. Barbul, Plast. Reconstr. Surg. 117 (2006) 42S.

[16] R. Chernoff, J. Am. Coll. Nutr 23 (2004) 627S.

[17] S.S. Gropper, J.L. Smith, Advanced Nutrition and Human Metabolism, Cengage Learning, 2012.

[18] R. Laurano, M. Boffito, G. Ciardelli, V. Chiono, Eng. Regen. 2022.

[19] A. Sood, M.S. Granick, N.L. Tomaselli, Adv. Wound Care 3 (2014) 511.

[20] T.A. Mustoe, K. O'shaughnessy, O. Kloeters, Plast. Reconstr. Surg. 117 (2006) 35S.

[21] L. Ovington, Ostomy Wound Manage. 49 (2003) 8.

[22] K. Chung, Grabb and Smith's Plastic Surgery, Lippincott Williams & Wilkins, 2019.

[23] S. Rajendran, Advanced Textiles for Wound Care, Woodhead Publishing, 2018.

[24] Grand View Research, Adhesion Barrier Market Size, Share &Trends Analysis Report By Product (Synthetic, Natural), By Formulation(Film/Mesh, Gel), By Application (Cardiovascular, NeurologicalSurgery), By Region, And Segment Forecasts, 2020−2027.

[25] M. Mir, M.N. Ali, A. Barakullah, A. Gulzar, M. Arshad, S. Fatima, et al., Prog. Biomater 7 (2018) 1.

[26] Z. Peles, M. Zilberman, Acta Biomater 8 (2012) 209.

[27] L. Fan, C. Cheng, Y. Qiao, F. Li, W. Li, H. Wu, et al., PLoS One 8 (2013) e66890.

[28] X. Xu, W. Zhong, S. Zhou, A. Trajtman, M. Alfa, J. Appl. Polym. Sci 118 (2010) 588.

[29] S.H. Ranganath, C.-H. Wang, Biomaterials 29 (2008) 2996.

[30] C. You, Q. Li, X. Wang, P. Wu, J.K. Ho, R. Jin, et al., Sci. Rep 7 (2017) 1.

[31] C.R. Patra, S. Mukherjee, R. Kotcherlakota, Nanomedicine 9 (2014) 1445.

[32] S.M. Bergin, P. Wraight, N. Dewapura, P. Greenberg, D. Campbell, P. Colman, Cochrane Database Syst. Rev., 2005.

[33] A. Adhya, J. Bain, O. Ray, A. Hazra, S. Adhikari, G. Dutta, et al., J. Basic Clin. Pharm 6 (2014) 29.

[34] K. Dunn, V. Edwards-Jones, Burns 30 (2004) S1.

[35] C. Rigo, L. Ferroni, I. Tocco, M. Roman, I. Munivrana, C. Gardin, et al., Int. J. Mol. Sci 14 (2013) 4817.

[36] J. Ai, E. Biazar, M. Jafarpour, M. Montazeri, A. Majdi, S. Aminifard, et al., Int. J. Nanomed. 6 (2011) 1117.

[37] S. Balakrishnan, S. Mukherjee, S. Das, F.A. Bhat, P. Raja Singh, C.R. Patra, et al., Cell Biochem. Funct 35 (2017) 217.

[38] J.-G. Leu, S.-A. Chen, H.-M. Chen, W.-M. Wu, C.-F. Hung, Y.-D. Yao, et al., Nanomed. Nanotechnol. Biol. Med 8 (2012) 767.

[39] L. Sun, S. Yi, Y. Wang, K. Pan, Q. Zhong, M. Zhang, Bioinspir. Biomim 9 (2013) 16005.

[40] O. Akturk, K. Kismet, A.C. Yasti, S. Kuru, M.E. Duymus, F. Kaya, et al., J. Biomater. Appl 31 (2016) 283.

[41] W.-Y. Chen, H.-Y. Chang, J.-K. Lu, Y.-C. Huang, S.G. Harroun, Y.-T. Tseng, et al., Adv. Funct. Mater 25 (2015) 7189.

[42] H.-C. Huang, C.R. Walker, A. Nanda, K. Rege, ACS Nano 7 (2013) 2988.

[43] P. Thangavel, R. Kannan, B. Ramachandran, G. Moorthy, L. Suguna, V. Muthuvijayan, J. Colloid Interface Sci 517 (2018) 251.

[44] S. Shahmoradi, H. Golzar, M. Hashemi, V. Mansouri, M. Omidi, F. Yazdian, et al., Nanotechnology 29 (2018) 475101.

[45] S.K. Nethi, A.K. Barui, V.S. Bollu, B.R. Rao, C.R. Patra, ACS Biomater. Sci. Eng 3 (2017) 3635.

[46] H. Zhao, O.J. Osborne, S. Lin, Z. Ji, R. Damoiseux, Y. Wang, et al., Small 12 (2016) 4404.

[47] S.A. Eming, T. Krieg, J.M. Davidson, J. Invest. Dermatol 127 (2007) 514.

[48] M. Tiwari, K. Narayanan, M.B. Thakar, H.V. Jagani, J. Venkata Rao, IET Nanobiotechnol. 8 (2014) 230.

[49] G. Borkow, J. Gabbay, R. Dardik, A.I. Eidelman, Y. Lavie, Y. Grunfeld, et al., Wound Repair Regen. 18 (2010) 266.

[50] A. Gopal, V. Kant, A. Gopalakrishnan, S.K. Tandan, D. Kumar, Eur. J. Pharmacol. 731 (2014) 8.

[51] R. Augustine, E.A. Dominic, I. Reju, B. Kaimal, N. Kalarikkal, S. Thomas, RSC Adv 4 (2014) 24777.

[52] E. Yadav, D. Singh, P. Yadav, A. Verma, RSC Adv 8 (2018) 21621.

[53] R. Raguvaran, B.K. Manuja, M. Chopra, R. Thakur, T. Anand, A. Kalia, et al., Int. J. Biol. Macromol 96 (2017) 185.

[54] S. Quignard, T. Coradin, J.J. Powell, R. Jugdaohsingh, Colloids Surf. B. Biointerfaces 155 (2017) 530.

[55] S. Perumal, S. kumar Ramadass, B. Madhan, Eur. J. Pharm. Sci. Off. J. Eur. Fed. Pharm. Sci. 52 (2014) 26.

[56] G.S. Alvarez, C. Hélary, A.M. Mebert, X. Wang, T. Coradin, M.F. Desimone, J. Mater. Chem. B 2 (2014) 4660.

[57] A. Khalid, H. Ullah, M. Ul-Islam, R. Khan, S. Khan, F. Ahmad, et al., RSC Adv 7 (2017) 47662.

[58] D. Archana, J. Dutta, P.K. Dutta, Int. J. Biol. Macromol 57 (2013) 193.

[59] R. Sankar, R. Dhivya, K.S. Shivashangari, V. Ravikumar, J. Mater. Sci. Mater. Med. 25 (2014) 1701.

[60] V. Sivaranjani, P. Philominathan, Wound Med 12 (2016) 1.

[61] C. Sun, J.S.H. Lee, M. Zhang, Adv. Drug. Deliv. Rev 60 (2008) 1252.

[62] M. Arakha, S. Pal, D. Samantarrai, T.K. Panigrahi, B.C. Mallick, K. Pramanik, et al., Sci. Rep 5 (2015) 14813.

[63] K.S. Khashan, G.M. Sulaiman, R. Mahdi, Artif. Cells Nanomed. Biotechnol. 45 (2017) 1699.

Clinical effectiveness of antimicrobial dressings

10

Ayush Gupta

Department of Microbiology, AIIMS Bhopal, Madhya Pradesh, India

10.1 Introduction

A wound is caused by a disruption in the normal architecture and function of the skin caused by a variety of insults such as surgery or trauma from chemical, physical, mechanical, and infectious insults [1]. The restoration of skin integrity is a complex series of events that includes the resolution of the insult and resulting inflammation, removal of damaged tissue, angiogenesis, regeneration of functional extracellular tissue matrix, wound contraction, re-epithelialization, differentiation, and remodeling. Furthermore, infection prevention is critical during the repair process [2]. If this sequence of events is not completed in an orderly and timely manner, the wound will become chronic.

Ancient civilizations understood that the best wound care consists of three steps: washing the wound, applying topical treatments, and bandaging [3,4]. They washed wounds with beer, boiled water, vinegar, or wine; made topical remedies from plants, animal products (honey), and minerals (clay and metals); and bandaged wounds with leaves, grasses, wool, or linen [5,6]. Labarraque proposed the first application of a synthetic chemical to a wound in 1823, but as a deodorant. Smith used Creosote, a by-product of wood tar distillation, in the treatment of venereal ulcers and fistulas in 1836. Around the same time, halogen-containing disinfectants such as hypochlorites and iodine began to find use in wound dressings. However, it was Joseph Lister who popularized the use of an antiseptic agent, phenol, in the dressing of surgical wounds in 1867 [7]. Currently, all of these agents are prohibited from being used on human tissues due to their toxicity and negative effects on wound healing.

The twentieth century saw the development of newer, more tolerable compounds. To name a few, they are povidone iodine, chlorhexidine, octenidine, and silver [1]. The dressing material changed dramatically over time as the importance of a moist environment during healing became clear, and dry gauze dressing gave way to semipermeable dressing [8]. A wide variety of dressing material, including paraffin gauze, polyurethanes, hydrocolloids, hydrogels, alginates, and foams, are in use since 80s. The incorporation of various antimicrobial agents, including antibiotics, into these materials has resulted in a plethora of antimicrobial wound dressing options in wound management [1].

Antimicrobial Dressings. DOI: https://doi.org/10.1016/B978-0-323-95074-9.00003-8

10.2 Types of wounds & factors affecting healing

Understanding the clinical effectiveness of various antimicrobial wound dressings requires knowledge of the numerous causes of wound development, factors influencing the healing process, and the role of the wound microbiome in the repair process. As previously stated, a wound is caused by a break in the continuity of skin. It is sometimes created on purpose, such as during surgery, but it is more often caused by various types of insults, such as mechanical (traumatic), thermal or chemical injuries (burns), and infections. Every wound goes through a complex repair process after it develops. Acute wounds are those that complete the repair process in a timely manner, whereas chronic wounds are those that do not [2]. Venous ulcers, ischemic ulcers, pressure sores, and diabetic foot ulcers are some of the common examples of chronic wounds [9].

A chronic wound is caused by a combination of factors that impede the healing process. They are typically divided into two categories: intrinsic and extrinsic. Ischemia, infection, the presence of necrotic tissue, and foreign bodies in the wound are examples of intrinsic or local factors that can affect wound healing. Diabetes mellitus, cancer, chronic disease (chronic renal failure), steroid use, radiation injury, and malnutrition are extrinsic or external factors that must be considered in the evaluation of a chronic wound. These factors may be acting independently or in conjunction with each other to impede the healing process. Clinicians must identify these factors and remove or correct them in order to successfully manage a chronic wound [10].

10.3 Wound microbiology

Human skin is colonized with a plethora of microorganisms mainly gram positives including *Corynebacterium* spp., *Staphylococcus* spp., *Propionibacterium* spp., *Bacillus* spp., non-pathogenic *Mycobacterium* spp., and yeasts (*Candida* spp.). Furthermore, bacteria that are normally present on these mucosal surfaces, such as gram-negative *Enterobacterales* in the perineum region and *Streptococcus* spp. on skin on the face and neck region, inhabit the skin near the mucosal orifices. When a wound forms, the normal flora on the skin populates the wound's surface first. Typically, gram-positive bacteria predominantly including coagulase-negative *Staphylococcus* spp. (CoNS), enter the wound first and as the wound becomes chronic, it starts getting populated by gram-negative bacilli such as *Escherichia coli*, *Klebsiella pneumoniae*, *Enterobacter* spp., *Proteus* spp., *Pseudomonas aeruginosa*, *Acinetobacter* spp. depending on the immunological status and surroundings of the patient [11]. The presence of anaerobes on the wound has traditionally been underestimated due to an overreliance on routine culture methods. *Peptostreptococcus* and *Finegoldia* are the most common gram-positive anaerobes in wounds, while *Prevotella*, *Porphyromonas*, and *Bacteroides* are the most common gram-negative anaerobes [12].

Culture independent techniques have shown that the wound microbiology is far from simple [13,14]. Though *Staphylococcus* and *Pseudomonas* are the most common genera in the chronic wound microbiota, there is a large diversity and dynamic range in bacterial species. Chronic wounds have more anaerobes, gram-negative rods, and gram-positive cocci than normal skin, with fewer commensals like Propionibacterium [15]. Nevertheless, the wound microbiome varies widely on the basis of wound etiology, location, status of diabetes in patient and antibiotic usage [16].

Majority of wounds are polymicrobial and bacteria within them are present in the form of biofilms [17]. Bacterial communities in biofilms exhibit a number of characteristics that make them difficult to treat, including slow antimicrobial penetration and ineffective host defense mechanisms, up-regulation of horizontal gene transfer in response to stress, anoxia, and the formation of persister cells [18].

Microorganism colonization of wounds is a universal phenomenon, and the mere presence of these microorganisms has no negative impact on the wound healing process. For a wound to heal properly, the bioburden of the wound and the host's immune system must remain in balance. If the balance shifts in favor of the microbes, the microorganisms (usually bacteria) multiply uncontrollably and invade tissues, resulting in a local infection that, if left untreated, can lead to systemic illness and, eventually, death [19]. Some authors have suggested that certain microbial species such as CoNS, *Lactobacillus* help to maintain the autochthonous antimicrobial self-defence of skin by preventing the colonization of pathogenic species [20−22]. Improved wound healing has been linked to increased microbial diversity and instability of the wound's microbial composition, rather than the bacterial load itself [16]. Nonetheless, our understanding of the role of the microbiome in wound repair is limited, and this is an area that needs to be explored further.

10.4 **What are antimicrobial dressings**

Antimicrobials are agents that inhibit microorganisms, such as disinfectants, antiseptics, and antibiotics. While the term "disinfectants" refers to chemical agents that are only applied to inanimate surfaces, the term "antiseptics" refers to chemical agents that can be applied directly to human tissues [19]. It is important to remember that the distinctions between these agents are subject to change as more information about their safety profile becomes available. In the 19th century, hypochlorites and phenols were considered antiseptics, but they are no longer applied directly to human skin or tissues. Alcohols, on the other hand, such as ethyl alcohol and isopropyl alcohol, can be used to kill microorganisms on both inanimate surfaces and human skin. Such chemical agents typically have broad-spectrum antimicrobial activity, and microbial resistance to them is uncommon. Antibiotics are chemical agents that occur naturally or are synthesized. They can act selectively and can be administered topically or systemically.

Antimicrobial dressings are wound dressings that contain an antiseptic agent or antibiotic that is slowly released into the wound to kill or inhibit the growth of microorganisms. It is to be distinguished from the topical application of an antibiotic followed by the application of a dressing to cover the wound. Common systemic antibiotics used in wound dressings include beta-lactams, aminoglycosides, fluoroquinolones, tetracyclines, sulfonamides, and glycopeptides [23−28]. Another type of antimicrobial dressing contains agents that are not traditionally used as antibiotics. Silver, iodine, chlorhexidine, octenidine, polyhexamtheyl-biguanide (PHMB), and natural antibacterial products such as honey, curcumin, Aloe vera, St. John's Wort-EO, certain essential oils, and chitosan are examples (CS). Dressings containing nanoparticles (NPs) of various chemicals such as silver (Ag), iron oxide (Fe_3O_4), titanium dioxide (TiO_2), and zinc oxide (ZNO) have also demonstrated potent antibacterial properties [29]. These types of wound dressings are discussed elsewhere in the book, and this chapter will discuss the studies that determined their clinical effectiveness in humans.

10.5 Dressings with antiseptic agents

Before delving into the published scientific literature on the clinical effectiveness of these antimicrobial dressings, it is important to understand that there will be a great deal of heterogeneity among the studies that have been conducted on the topic to date. The patient population studied, study design, interventions, comparator group, and outcomes assessed can all vary greatly.

A. Silver

Antimicrobial properties of silver have been well established since ages [30]. Silver is a powerful antimicrobial agent that works through a variety of mechanisms. Silver is inert in its metallic state, but when it comes into contact with moisture from the skin or fluid from a wound bed, it ionizes to Silver ions (Ag +). These silver ions are extremely reactive, causing structural changes in bacterial cell walls, cytoplasmic and nuclear membranes, and ultimately cell lysis [31]. Furthermore, by binding to and denaturing bacterial DNA and RNA, they inhibit protein synthesis and replication. Furthermore, their microbicidal activity is broad-spectrum, as they affect gram-positive, gram-negative, anaerobes, and fungi [32,33]. Silver is appealing for use in antimicrobial dressings because microorganisms have a low tendency to develop resistance to it due to its multi-modal mechanism of action and potency at low concentrations [34].

Silver is being used in topical forms of silver nitrate and silver sulfadiazine for prophylaxis & treatment of wound infections since 1960s [35]. They fell out of favor, however, due to a number of issues with silver preparations. Silver ions were ineffective against deep-seated bacteria, required multiple applications, and caused skin discoloration. Furthermore, there were concerns about the overuse of silver and the subsequent toxicity after absorption, as well as the emergence of resistance [36]. As a result of the discovery of other antiseptic agents, their use declined. After it was discovered that these agents interfere with healing by being toxic to host cells and causing resistance, there was renewed interest in silver-based wound therapy. The steady release of silver ions in the wound can be controlled by incorporating silver into antimicrobial dressings, reducing toxicity to human cells [35].

There are several antimicrobial dressings available that act as a reservoir for silver, allowing to control its release into the wound. Creams, cloths, hydrofibers, alginates, foams, pad and island dressings, hydrocolloids, barrier dressings, and other delivery vehicles for silver cations (Ag +) are examples. A detailed review is beyond the scope of this chapter, so one can refer to an article by Tomaselli et al. [36]. The form, description, function, antimicrobial sensitivity, and indications of these antimicrobial silver dressings differ. Tomaselli's article includes a list of various commercial products that have been evaluated in various studies. Over 30 years of published literature, the use of silver-containing antimicrobial wound dressings in various clinical conditions exists. These studies, however, are extremely heterogeneous due to differences in major parameters such as antimicrobial dressing type, study design, comparator, patient population used in the study, and assessed outcomes. Antimicrobial silver dressings have been extensively studied in burn patients [37−44]. Topical application of silver compounds, such as silver sulphadiazine (SSD) and silver nitrate (SN), has traditionally been the standard of care for burn wounds that do not require surgical intervention. In multiple randomized controlled trials, various nanocrystalline silver dressings were compared to topical silver applications in patients with partial thickness burns. Their use was associated with less pain [40,41,43] especially during removal [40], less overall treatment costs [37], less number of dressing changes [37] and reduced hospital stay [43]. However, the results have been conflicting with regards to wound healing rates and development of secondary infections. In a recently published RCT by Moreira et al., the authors concluded that they found no evidence of a difference between nanocrystalline silver (Acticoat) and 1% SSD dressings regarding efficacy and safety outcomes [37] whereas Wu et al., who used a nanocrystalline silver dressing from another manufacturer concluded that there was significant improvement in wound healing rate, healing time, and pigmentation fading away time in patients where nano-silver dressings were used [42]. In one of the earliest RCTs on nano-silver dressings, Tredget et al. found significantly less occurrence of secondary bacteremia and Sepsis in the silver-dressing (Acticoat) group [40]. With regards to use of other types of silver containing antimicrobial dressings in patients with burn wounds, they have shown to significantly reduce the time to healing and pain (Alginate, Askina Calcitrol Ag and Hydrofiber, Aquacel Ag) [38,39], time to re-epithelization (Aquacel Ag) [44], and overall cost of treatment (Aquacel Ag) [38,44].

Gravante et al. published a systematic review and meta-analysis in 2009 comparing the use of nano-silver based dressings with topical silver applications in burn wounds, and the author concluded that nano-silver dressing groups had a significantly lower incidence of infections compared to the SSD group, as well as lower costs and pain values [45]. Chaganti et al. published a systematic review and meta-analysis in 2019 comparing the use of foam-based silver dressings versus SSD in burn wounds, and the authors concluded that there is moderate quality evidence indicating that there is no significant difference in wound healing between silver-containing foam dressing and SSD dressing. However, foam has the added benefit of reducing pain during the early stages of treatment and potentially lowering infection rates [46]. Nherera et al. concluded in their meta-analysis on the use of various silver-based antimicrobial dressings in partial thickness burn wounds that the use of nanocrystalline silver was associated with a statistically significant reduction in length of stay when compared to silver-impregnated hydrofiber dressing and a shorter time to healing when compared to silver-impregnated foam dressing. Infection rates and surgical procedures did not differ statistically between nanocrystalline silver, silver-impregnated hydrofiber dressing, and silver-impregnated foam dressing. However, using the Monte Carlo simulation method, it was discovered that nanocrystalline silver was the most beneficial of the three types of dressings for all outcomes, including infection rates and surgical procedures [47].

The published literature on the use of silver-based antimicrobial dressings in the management of other wound types, such as venous leg wounds, diabetic foot, pressure ulcers, and other acute and chronic wounds, is extremely heterogeneous, with different types of comparators and a small number of patients [48–57]. Notably, in the VULCAN trial, a large trial conducted on more than 200 patients to evaluate the use of silver-based antimicrobial dressings in venous leg ulcer management, there was no difference in the proportion of ulcers healed at 12 weeks between the silver-based antimicrobial and non-silver low adherence dressings, and the use of silver-based antimicrobial dressings was more expensive. Finally, the authors concluded that there was insufficient evidence to support the routine use of silver-donating dressings beneath compression for venous ulcers [56]. The authors concluded that the performance of both antimicrobials was comparable in terms of overall healing rate and number of wounds healed in another large RCT of over 140 patients in each arm of nano-silver

(Acticoat) or cadexomer iodine (Iodosorb) based antimicrobial wound dressings. The use of silver dressings, on the other hand, was associated with a faster healing rate during the first two weeks of treatment, as well as in wounds that were larger, older, and had more exudate [57]. In contrast, Zhao et al. concluded in a recent meta-analysis that there was sufficient evidence that silver-containing dressings can accelerate the healing rate of chronic venous leg ulcers and improve their healing in a short period of time [58].

To summarize, there is reasonable evidence of certain benefits to using silver-based antimicrobial dressings instead of topical silver preparations in the management of burn wounds. These include less pain for patients, lower costs to the healthcare system, and dressing changes and secondary infection prevention. Many of these advantages stem from the fact that these dressings must be changed at least every third day, whereas topical silver-based dressings must be applied at least daily. However, the evidence for an overall healing effect on burn wounds is lacking. In terms of their use on other types of chronic wounds or ulcers, the evidence is limited due to the small number of studies conducted on the topic and the heterogeneous nature of these studies, and more research is required to determine their role in the management of such conditions.

B. Iodine-based Dressings

Iodine has been known as a potent antiseptic since 18th century even before the actual discovery of the elemental iodine [59]. Iodine is a member of the halogen family, and various formulations have been used for antibacterial purposes for over a century. Among these formulations, "iodine tincture," an alcohol formulation, and iodophors, a formulation in which iodine is temporarily bound to a carrier molecule, demonstrate potent antibacterial efficacy. Under the right conditions, iodine will dissociate from the carrier molecule and release free iodine in a controlled manner from iodophors. As wound dressings, three types of iodophors are available. As carriers, they use povidone iodine (liquid), cadexomer iodine (semi-solid), and polyvinyl alcohol (PVA)-based foam (solid) [59]. The main advantage of these iodophors is that they release iodine in the wound in a controlled manner. This slow release is also beneficial in terms of cytotoxicity and systemic absorption [60]. Additionally, it has been shown that iodophors are able to penetrate and disrupt biofilms [61,62]. Phillips et al. demonstrated that cadexomer iodine and time-release silver gels are the most effective antimicrobial agents and dressings in reducing mature biofilms [63].

Iodine based dressings were extensively evaluated during the 80s and 90s of the 20th century in which povidone iodine and cadexomer iodine (in ointment and powder form) were compared against the standard care, which ranged from normal saline to paraffin gauze dressings in different studies [64–70]. The majority of these studies looked at the effect of iodine-based dressings on chronic venous ulcers, but only a few looked at decubitus (pressure) or diabetic foot ulcers [64,69]. Many of these studies concluded that wounds which were treated with cadexomer iodine healed much faster, were less painful, had less debris, and less costlier than those managed with standard care [64–66,68–70]. Vermeulen et al. conducted a systematic review to determine the benefits and harms of iodine in wound care and concluded that iodine is an effective antiseptic agent that does not impair wound healing or demonstrate the purported harmful effects, particularly in chronic and burn wounds [71]. Notably, Miller et al. concluded that the performance of cadexomer iodine ointment and silver dressing (Acticoat) in leg ulcers was comparable in terms of overall healing rate and number of wounds healed in a large randomized trial of 281 patients. The use of silver compounds, on the other hand, was associated with a faster healing rate during the first two weeks of treatment, as well as in wounds that were larger, older, and had more exudate [72].

In the 21st century, few studies have evaluated the effect of commercial iodine releasing antimicrobial dressings. These include a Foam based dressing containing povidone-iodine (Betafoam) [73,74] and a hydrogel based liposomal dressing (Repithel) [75–77]. In an RCT on skin graft donors, Pak et al. discovered that Betafoam was superior to a hydrocellular dressing and petrolatum gauze in terms of time of epithelialization and complete epithelialization [73]. The superiority of Repithel was also noted against SSD in burn patients [77], paraffin gauze, and chlorhexidine in patients with meshed skin grafts [75,76]. A recent systematic review of the published literature on various types of antiseptic agents in the management of chronic wounds of varying etiology concluded that iodine has better wound healing completion than saline [78].

C. Chlorhexidine

Chlorhexidine is a popular antiseptic agent for the skin and mucous membranes. Its applications in healthcare range from hand hygiene to skin cleansing prior to blood sample collection or surgery, catheter site preparation, mouth rinse for ventilator-associated pneumonia prevention, dental caries prevention, and endodontic treatment [79–83]. It is an effective antimicrobial agent that is active against gram-positive, gram-negative, fungi, and viruses. Chlorhexidine's antimicrobial effect is dose-dependent, with bacteriostatic activity at low concentrations (0.02%–0.06%) and bactericidal activity at higher concentrations ($>0.12\%$) [84].

Despite their potent antimicrobial action, chlorhexidine-based antimicrobial dressings are not recommended for the treatment of chromic wounds due to their cytotoxic activity on a variety of human cells [85–87]. One such commercial dressing Bactigrass (containing 0.5% Chlorhexidine) was found inferior to the silver-coated dressing (Acticoat) and Nystatin for topical antifungal effect in a contaminated full-skin-thickness rat burn wound model [88]. A Montmorillonite-chitosan-chlorhexidine composite film was recently developed as a delivery system capable of releasing chlorhexidine in a localized and prolonged manner. At 1% chlorhexidine concentration, all prepared films demonstrated good antimicrobial and antibiofilm activities with no cytotoxicity [89].

Chlorhexidine-based dressings were found to be inferior to silver-containing hydrofiber dressings and dialkylcarbamoyl chloride (DACC) dressings in patients with partial thickness burns and skin graft donor sites, respectively, in a few comparative clinical studies [90,91]. Wasiak et al. concluded in a systematic review of various treatment options for partial thickness burns that there was insufficient evidence to draw any conclusions on the efficacy of chlorhexidine-impregnated paraffin gauze dressings and that their use may be toxic to regenerating epithelial cells and may delay healing in uninfected wounds [92]. To summarize, chlorhexidine is preferred as an antiseptic, but its use on infected or uninfected chronic wounds is not recommended due to its cytotoxicity.

D. Octenidine

Octenidine dihydrochloride is a cationic surface active compound that binds readily to negatively charged surfaces like microbial cell envelopes and eukaryotic cell membranes [93]. Salts of fatty acid glycerol phosphates in the cell membrane are thought to be the main binding partners for cationic antiseptics like chlorhexidine and polyhexanide. It acts as an antimicrobial agent by interacting with enzymatic systems and polysaccharides in microorganism cell walls and inducing leakages in the cytoplasmic membrane [94]. Octenidine has antimicrobial activity against a broad range of microorganisms, including fungi, viruses, and parasites. It is approximately 3−10 times more effective than chlorhexidine and has comparable cytotoxicity [95]. It has also shown to be highly effective against biofilms and exerts a sustained antimicrobial effect as it is not percutaneously absorbed [93]. All of these properties make it an excellent agent for use in antimicrobial wound dressings. Octenidine is commercially available in the form of an aqueous solution containing phenoxyethanol (Octenisept, Schülke & Mayr GmbH) or in combination with aliphatic alcohols, glycerol, and detergents that do not contain phenoxyethanol. It is approved as a medicinal substance for skin, mucous membrane, and wound antisepsis and is used as a topical application [93]. There are no commercially available wound dressings containing octenidine.

Topical application of Octenidine in management of chronic wounds has shown its antibacterial efficacy in various clinical studies [96,97]. Sopata et al. investigate the impact of octenidine dihydrochloride on the clinical condition and bacterial flora of neoplastic ulcers. After three weeks of treatment, the clinical condition of the ulcers improved, as evidenced by a decrease in necrosis, exudate levels, erythema, and oedema. Bacteriologically, the wounds were free of *Staphylococcus aureus*, *Staphylococcus epidermidis*, and *Proteus mirabilis* [96]. In a 4-week double-blind, randomized, controlled clinical study on chronic wounds, OPE significantly improved granulation compared to Ringer solution [98]. Furthermore, a randomized, double-blind, controlled clinical trial found that the octenidine hydrogel significantly reduced bacterial colonization of skin graft donor site wounds, but there was no significant difference in the time for complete epithelialization of skin graft donor sites between the OCT and placebo groups [99]. Octenidine dihydrochloride was found to be effective and useful as a first treatment for a month in a two-period treatment study in patients with venous leg ulcers, followed by either hydrocolloid or foam dressings. According to the authors, it improved the clinical condition of VLU by preparing the wound for future treatment with modern dressings [97].

Recently, there has been a surge of interest in using octenidine in antimicrobial dressings. Alkhatib et al. created a bacterial nano-cellulose-based drug delivery system that had a one-week retention time for octenidine, improved mechanical and antimicrobial properties, and a high biocompatibility [100]. Mattos et al. also investigated bacterial nano-cellulose as a carrier for octenidine and povidone-iodine delivery in wound dressings. They discovered that octenidine delivered into a solid matrix in a sustained and prolonged manner, whereas povidone-iodine released faster [101]. Pandian et al. created a chitosan-based flexible bandage that contained octenidine. The prepared antiseptic bandage demonstrated a synergistic effect with Chitosan, as well as excellent antimicrobial and antibiofilm activity against *S. aureus* and *Candida auris* [102]. To summarize, octenidine is a preferred antiseptic over chlorhexidine for the treatment of chronic wounds due to its lower cytotoxicity and broad antimicrobial spectrum. However, because there are no commercial wound dressings available, its use as an antimicrobial dressing in the management of chronic wounds remains experimental.

E. Polyhexanide

Polyhexamethylene biguanide hydrochloride/polyhexanide (PHMB) is a highly positively charged synthetic polymer that has been used in the cosmetics and disinfectant industries for decades as an antimicrobial. It has broad antimicrobial activity against gram-negative and gram-positive bacteria, as well as some fungi and protozoa [103]. The exact mechanism of action is unknown, but it is thought to work by displacing the divalent cations in the cell wall that connect the core lipopolysaccharide molecules on the cell surface, thereby maintaining the outer membrane's rigidity and stability [104]. It is also thought to work by adsorbing onto the surface of an acidic phosphatidylglycerol bilayer, disrupting it and increasing fluidity and permeability [105]. Despite having been shown to be slightly more effective than chlorhexidine, PHMB requires a very long exposure time [106]. As a result, its use in perioperative skin disinfection and hand hygiene is not advised [107]. Due to its status as lethal when inhaled, polyhexanide was prohibited in cosmetic products from 1 January 2015 by European Chemicals Agency.

Use of polyhexanide based antiseptics has been evaluated in few clinical studies for management of chronic wounds [108−113]. A prospective randomized controlled double-blind study in 50 acutely injured patients with bacterially contaminated soft tissue wounds found that PHMB-based dressings (Lavasept) reduced microorganisms on wound surfaces faster and significantly than ringer solution. The authors discovered no evidence of impaired wound healing in either group, and Lavasept's anti-inflammatory effect and tissue compatibility were rated significantly higher than Ringer solution's [112]. In an RCT of partial thickness burn patients, the group treated with polyhexanide/betaine gel had significantly lower pain scores than the group treated with silver sulfadiazine, despite no significant differences in healing times, infection rates, bacterial colonization rates, or treatment cost in either group [114].

PHMB incorporated biocellulose dressings have shown better efficacy in reducing inflammation and bacterial colonization in chronic wounds against normal saline [115], non-antimicrobial foam dressings [108] and silver dressings [110]. A single-center RCT in patients with partial-thickness burns discovered that PHMB-based dressing was associated with significantly faster pain relief, fewer dressing changes, better ease of use, and lower cost when compared to silver sulfadiazine cream. Despite the fact that there were no differences in the healing times of the burn wounds [116]. A common criticism leveled at PHMB-based clinical studies is that they are of poor quality because they have a small number of participants, are industry-sponsored, and are published in low-quality journals. The authors included only one RCT (Piatkowski et al.) with PHMB-based management in a systematic review to assess the effects and safety of various antiseptics for the treatment of burns in any care setting. Overall, they concluded that because most trials were of poor quality, they were not confident that most trials were free of bias [117]. The broadly same conclusions were made by these authors in pressure ulcers and venous leg ulcers [118,119]. The authors identified two RCTs involving PHMB in a recently completed systematic review to evaluate four common antiseptic agents, including PHMB, in chronic wound care complete healing. They came to the conclusion that there was insufficient evidence to suggest a difference in wound healing using octenidine or polyhexanide [120].

F. Honey

Honey has been known since ancient civilizations for its medicinal properties. Records from ancient Indian, Egyptian, Greeks, Chinese civilizations have described its use in wound dressings to promote wound healing [121]. The antimicrobial properties of Honey were noted by Van Ketel in the late 19th century. It was used for wound care by Russian and Chinese soldiers during the First World War [122]. However, with the availability of commercial products suitable for medical use, the use of honey in the treatment of skin wounds, burns, and ulcers has recently experienced a revival [123,124]. Honey can be obtained from a variety of sources and has varying bactericidal properties, but the majority of studies have used Manuka honey, which is derived from the tree genus *Leptospermum* and is native to New Zealand and Australia [124]. Since last decade, US Food & Drug Administration (FDA) has approved a variety of medicinal honey based products, a detailed list of which can be accessed in a recent article by Hossain et al. [125]. These products range from topical gels, ointments, and creams to a variety of dressings for use over-the-counter as well as under medical supervision for specific wounds. The majority of these items include Manuka honey in various concentrations and combinations.

Honey's antibacterial properties are attributed primarily to its high sugar content, which produces an osmotic effect that causes bacterial dehydration. Other factors that contribute to its antibacterial properties include low pH, hydrogen peroxide production, reactive oxygen species, and the presence of specific compounds such as methylglyoxal (MGO) [125]. Manuka honey has been found to be particularly high in MGO, which is responsible for its excellent antibacterial activity. Additionally, phenolic compounds, organic acids, enzymes, nutrients, and other minor components have anticancer, antiparasitic, antiviral, and antidiabetic activity [126]. MGO activity has been demonstrated to be particularly effective at inhibiting biofilms. Furthermore, honey may act as an immunomodulator through a variety of mechanisms. It has been shown to stimulate the production of inflammatory cytokines in low-inflammatory environments while suppressing the same cytokines during infection, including TNF-α and interleukin-β [127]. Furthermore, honey has been shown to aid in some stages of the wound healing process, altering the natural physiology of wound healing. It has been shown to reduce oedema and wound exudate, and its acidic pH can increase oxygen off-loading from hemoglobin, promoting healing. Honey also has the unique property of providing rapid autolytic debridement of wounds, stimulating granulation and epithelialization, and providing a moist environment that aids in scar formation [121].

Clinical efficacy of honey based treatment on burn wounds has been evaluated in few studies. In the 1990s, multiple randomized trials done by Subrahmanyam found that honey impregnated gauze dressing were associated with earlier healing as compared to polyurethane film [128], amniotic membrane [129], and silver sulfadiazine cream [130] in the treatment of partial thickness burns. In the management of partial thickness burns, Malik et al. discovered that topical application of honey (Langnese-commercially available natural honey) was associated with earlier healing and a lower number of positive cultures for *P. aeruginosa* [131]. In their randomized study of partial-thickness burns, Maghsoudi et al. concluded that honey-dressed wounds were associated with earlier subsidence of acute inflammatory changes, better infection control, and faster wound healing than mafenide acetate treated wounds [132]. The authors of a Cochrane database systematic review and meta-analysis on antiseptics in burn wounds concluded that there is moderate certainty evidence that burns treated with honey heal faster than burns treated with topical antibiotics. They also concluded that it is unknown whether infection rates in honey-treated burns differ from non-antimicrobial treatments (very low certainty evidence) [133]. Published literature on the use of honey in chronic wounds such as pressure ulcers, diabetic foot ulcers and venous occlusive ulcers consists of poorly designed clinical trials with very small number of subjects [134—139]. When compared to different comparators such as povidone iodine or saline dressings, the majority of these studies concluded that honey-based dressings had beneficial effects on wound healing. A systematic review in the Cochrane database compared the use of honey with alternative wound dressings and topical treatments on the healing of acute (e.g., burns, lacerations) and/or chronic (e.g., venous ulcers) wounds. Because of the heterogeneity of the patient populations and comparators studied, as well as the generally low quality of the evidence, the authors concluded that it is difficult to draw broad conclusions about the effects of honey as a topical treatment for wounds. They did, however, conclude that honey appears to heal partial thickness burns faster than conventional treatment and infected postoperative wounds faster than antiseptics and gauze [140]. Furthermore, a recent meta-analysis comparing the efficacy of honey and povidone iodine-based dressings on wound healing outcomes in 1236 participants from 12 studies found that honey-based dressings significantly reduced the mean duration of healing, length of hospital stay, and pain score [141].

G. Chitosan

One of the most promising carbohydrate polymer with antimicrobial properties is Chitosan. It is a natural cationic polysaccharide consisting of $(1 \rightarrow 4)$-2-amino-2-deoxy-β-d-glucan and is the partially- to fully-deacetylated form of chitin [142]. It is found in crustacean shells, fungi and algae cell walls, insect exoskeletons, and mollusc radulae and is the second most abundant polymer after cellulose. It is commercially produced after demineralization and deproteinization processes, primarily from crustacean shells, which are byproducts of the seafood processing industry [143].

Chitosan is useful in wound dressings for a variety of reasons, including its biocompatibility, biodegradability, low toxicity, antimicrobial, and antioxidant properties. However, it is used in wounds for its hemostatic properties. It is an excellent procoagulant and many commercial products have been in use for its hemostatic properties. A list of various commercial chitosan-based dressings can be accessed in a review article by Matica et al. [143].

Though the precise mechanism of its antimicrobial properties is unknown, it is thought to exert them via multiple mechanisms. One of the most commonly assumed mechanisms for its antimicrobial action is likely due to electrostatic interactions between the positively charged chitosan molecule and the negatively charged cell membrane, which results in changes in cell permeability and cell membrane lysis [144,145]. Another postulated mechanism is by inhibition of the protein synthesis by interaction of chitosan hydrolysis products with microbial DNA [144]. Furthermore, chitosan has a high chelating capacity for a variety of metal ions. Ca^{2+} and Mg^{2+} cations are found in the cell walls of both gram-positive and gram-negative bacteria, and they help to maintain cell wall integrity. In this regard, chitosan's chelating property may destabilize the bacterial cell, making it susceptible to other antibiotics [146].

Despite the fact that there are numerous commercial chitosan-based dressings on the market, they are primarily used to control bleeding during surgeries or acute wounds [147]. In many studies, chitosan-based dressing has been combined with other antimicrobial compounds and used as a carrier for antimicrobial compound delivery over time [148−150]. There is little published literature on the actual use of chitosan-based dressings in wound management, and no randomized controlled trial study design has been used. Stone et al. compared the effects of chitosan-based dressing and soft silicone dressing on healing at split skin graft donor sites ($n = 20$). They concluded that chitosan aided in rapid wound re-epithelialization and nerve regeneration within the vascular dermis [151]. A product evaluation study looked at KytoCel, an absorbent wound dressing that was used to treat 30 wounds in community care and 10 split-thickness skin-graft donor sites in acute care. The authors discovered that 90% of community-treated wounds had moderate to high levels of exudate, with the majority ($n = 19$) healing or improving during treatment, and all ten split-thickness skin graft donor sites healing during the evaluation [152]. Varon et al. evaluated Opticell Ag, a chitosan-based dressing with silver nanoparticles on 19 split-thickness skin graft patients. They found that pain decreased significantly between postoperative day one and days 10−14 and mean percentage of re-epithelialization on days 10−14 was 92% and 99% by one month [153].

H. *Aloe vera*

Aloe vera is a well-known herb for wound healing. It is commonly found in tropical regions and is related to cactus in the Liliaceae family. Fresh plant leaves are processed to produce a clear mucilaginous gel known as Aloe vera gel or mucilage, which is used in the treatment of skin lesions. It is mostly made up of water, which accounts for 99%−99.5% of the total. The remaining solid fraction contains soluble sugars, polysaccharides, lipids, lignins, vitamins, enzymes, and minerals [154]. This gel has shown to influence the wound healing owing to its anti-inflammatory and antimicrobial properties [155]. In addition it has shown to stimulate fibroblast proliferation, collagen synthesis, and angiogenesis [156,157].

Few studies have looked into the antimicrobial activity of aloe vera gel when applied topically. Wound management in guinea pigs with full-thickness burns was done using SSD, salicylic acid cream, aloe vera gel, or plain gauze dressing in an experimental study. Only aloe vera gel extract and SSD application were found to significantly reduce wound bacterial counts [158]. Similarly in a study on rats, it was shown that application of aloe cream significantly increase re-epithelialization in burn wounds as compared with SSD cream [159].

One of the earlier wound dressings incorporating aloe vera is an amorphous hydrogel dressing (Carrasyn Gel wound dressing, Carrington laboratories, Inc. Irving, TX) approved by USFDA for pressure ulcers. Thomas et al. conducted an RCT in which 30 patients were randomly assigned to either daily topical application of the hydrogel study dressing or a moist saline gauze dressing. The authors discovered that the hydrogel dressing was as effective as, but not better than, the moist saline gauze wound dressing in treating pressure ulcers [160]. In a human study, 30 patients with infected leg ulcers were treated with aloe vera dressings, while 30 sex-matched controls were treated with topical antibiotics. After 11 days, 28 of 30 bacteria showed no growth, compared to none in the control group [161]. Khorasani et al., in their RCT of 30 patients with partial thickness burn wounds, concluded that the rate of re-epithelialization and healing of the burn wounds was significantly faster in the sites treated with aloe than with topical SSD (16 vs 19 days) [162]. A Cochrane systematic review and meta-analysis published in 2012, however, concluded that there was a lack of high-quality clinical trial evidence to support the use of Aloe vera topical agents or Aloe vera dressings as treatments for acute and chronic wounds [163].

Since then, a few more trials have been carried out to assess the efficacy of aloe vera-based products in wound healing. The authors discovered that topical aloe vera gel significantly accelerated healing but did not show significant pain relief when compared to the placebo group in a double-blind RCT on 12 patients with 24 split-thickness skin graft donor sites [164]. The authors found no significant difference between the two treatment groups in terms of wound healing, pain, pruritus, and patient discomfort in a recently completed double-blind, placebo-controlled RCT comparing a topical skin ointment with natural ingredients (aloe vera, honey, and peppermint) versus petroleum jelly on patients with split-thickness skin graft donor sites. However, the ointment outperformed petroleum jelly in terms of wound erythema reduction and was associated with significantly higher treatment satisfaction [165].

Aloe vera gel has been used in a variety of newer wound dressing materials, including alginate, wound films, nano-fibers, and chitosan-based materials, either alone or in combination with other antimicrobial agents like tetracycline, silver nanoparticles, honey, and curcumin [166–171]. However, most of these studies are preclinical evaluations and a detailed review on them is beyond the scope of this chapter.

I. Other herbal products

There are several other products derived from plants which have shown antimicrobial properties and are being considered to be incorporated in dressing materials for use in wound healing. They include Curcumin, a product from plant *Curcuma longa* (common name: turmeric, Haldi in Hindi), *Calendula officinalis* (marigold), *Hypericum perforatum* (common name: St. John's wort), and essential oils from various plants such as Green tea tree oil (*Melaleuca alternifolia*), neem oil, *Zingiber cassumunar*, Oregano oil (*Origanum vulgare* L. ssp. Hirtum), clove (*Eugenia* spp.), black pepper, ginger essential oils, lavender oil, Ajwain essential oil (*Tachyspermum ammi*). All of these plant products have been incorporated into various types of newer dressing materials, but the majority of them are still in the preclinical stage and thus outside the scope of this chapter. A recent article by Ramalingam et al. can be referred to for a detailed review of the role of herbal products in wound healing [172].

There have been few clinical trials of these dressings in patients with surgical or chronic wounds. Wound cleansing with water-miscible tree tea oil failed to achieve its primary target of wound decolonization by methicillin resistant *S. aureus* (MRSA) in 10 patients in an uncontrolled case-series on the application of tea tree oil (*M. alternifolia*) on acute and chronic wounds of mixed etiology [173]. In a recently published RCT on 97 patients with dehisced surgical wounds with critical colonization/infection, the authors concluded that the oil-based dressings were as effective as silver-based dressings in terms of inflammation resolution, and their use was associated with significantly less pain [174].

REOXCARE (Histocell), a curcumin-based antioxidant dressing containing an absorbent matrix derived from Locust Bean Gum galactomannan and a hydration solution containing curcumin and N-acetylcysteine, was evaluated in a multicenter, prospective case-series study design. Thirty-one patients with difficult-to-heal wounds were recruited, and the authors discovered that 9 wounds completely healed and 22 wounds improved significantly after an 8-week follow-up [175]. A randomized control trial using the same product is underway (Trial registration no: NCT03934671. Registered on 2 May 2019) and results are awaited [176].

10.6 Conclusions

To summarize, wound repair is a complex and time-consuming process that necessitates addressing the primary insult, creating conducive conditions (local and systemic), and controlling wound bio-burden during the repair process. There are a number of synthetic and natural antimicrobial products whose application on a wound, either as a topical application or as an antimicrobial dressing, aids in the control of wound bioburden or infection while also contributing to the wound repair process due to their anti-inflammatory and auto-debridement properties. There is a wealth of published literature on specific products such as silver, iodine, octenidine, and honey, but the majority of these studies have focused on topical application. There are a number of published studies as part of antimicrobial dressing that deliver these antimicrobial products in a controlled manner; however, due to varying technical aspects of conducting them such as type of wound, number of patient population, study design, interventions, comparator group, and outcomes that are evaluated, there is a lot of heterogeneity amongst these studies. Furthermore, relying on the results of many published trials is difficult due to poor study designs, a small number of patients, or a conflict of interest due to commercial involvement. As a result, many systematic reviews have refrained from conducting meta-analyses due to study heterogeneity and are inconclusive.

In terms of individual antimicrobial dressings, only a few conclusions can be drawn. First, silver-based antimicrobial dressings should be preferred over topical silver preparations such as SSD in the management of burn wounds because they are less painful, less expensive to the healthcare system, and require fewer visits to the hospital. However, the evidence for an overall healing effect on burn wounds is lacking. Second, when compared to standard dressing methods such as paraffin or saline gauze dressings, iodine-based dressings have shown greater potential; however, no recommendations can be made regarding their superiority or inferiority over silver-based dressings for wound management of various types of wounds. Third, chlorhexidine-based dressings should be avoided on chronic wounds due to their cytotoxic effect, which may impair wound healing. Fourth, scientific evidence on the use of octenidine or PHMB-based dressings for acute or chronic wound management is inconclusive, and more high-quality research is required. Among natural products, the use of honey-based dressings has been associated with superior effects in terms of duration of healing, length of hospital stay, and pain compared to commonly used antiseptics and gauze dressings, but its use has never been compared to silver-based antimicrobial dressings. Chitosan, Aloe vera gel, and essential oil-based dressings have promising antimicrobial wound dressing potential, but published trials have a small number of patients and thus cannot be relied on.

References

[1] J.-Y. Maillard, G. Kampf, R. Cooper, Antimicrobial stewardship of antiseptics that are pertinent to wounds: the need for a united approach, JAC Antimicrob. Resist. (2021) 3. Available from: https://doi.org/10.1093/JACAMR/DLAB027.

[2] G.S. Lazarus, D.M. Cooper, D.R. Knighton, R.E. Percoraro, G. Rodeheaver, M.C. Robson, Definitions and guidelines for assessment of wounds and evaluation of healing, Wound Repair. Regen. 2 (1994) 165—170. Available from: https://doi.org/10.1046/j.1524-475X.1994.20305.x.

[3] J.B. Shah, The history of wound care, J. Am. Col. Certif. Wound Spec. 3 (2011) 65—66. Available from: https://doi.org/10.1016/J.JCWS.2012.04.002.

[4] R.D. Forrest, Early history of wound treatment, J. R. Soc. Med. 75 (1982) 198.

[5] G. Broughton, J.E. Janis, C.E. Attinger, A brief history of wound care, Plast. Reconstr. Surg. (2006) 117. Available from: https://doi.org/10.1097/01.PRS.0000225429.76355.DD.

[6] R.D. Forrest, Development of wound therapy from the Dark Ages to the present, J. R. Soc. Med. 75 (1982) 268.

[7] W.B. Hugo, A brief history of heat and chemical preservation and disinfection, J. Appl. Bacteriol. 71 (1991) 9−18. Available from: https://doi.org/10.1111/j.1365-2672.1991.tb04657.x.

[8] G.D. Winter, Formation of the scab and the rate of epithelization of superficial wounds in the skin of the young domestic pig, Nature 193 (1962) 293−294. Available from: https://doi.org/10.1038/193293A0.

[9] K. Izadi, P. Ganchi, Chronic wounds, Clin. Plast. Surg. 32 (2005) 209−222. Available from: https://doi.org/10.1016/J.CPS.2004.11.011.

[10] W.K. Stadelmann, A.G. Digenis, G.R. Tobin, Impediments to wound healing, Am. J. Surg. 176 (1998) 39S−47S. Available from: https://doi.org/10.1016/S0002-9610(98)00184-6.

[11] G. Daeschlein, Antimicrobial and antiseptic strategies in wound management, Int. Wound J. 10 (Suppl 1) (2013) 9−14. Available from: https://doi.org/10.1111/IWJ.12175.

[12] R.S. Howell-Jones, M.J. Wilson, K.E. Hill, A.J. Howard, P.E. Price, D.W. Thomas, A review of the microbiology, antibiotic usage and resistance in chronic skin wounds, J. Antimicrob. Chemother. 55 (2005) 143−149. Available from: https://doi.org/10.1093/JAC/DKH513.

[13] R.D. Wolcott, J.D. Hanson, E.J. Rees, L.D. Koenig, C.D. Phillips, R.A. Wolcott, et al., Analysis of the chronic wound microbiota of 2,963 patients by 16S rDNA pyrosequencing, Wound Repair. Regen. 24 (2016) 163−174. Available from: https://doi.org/10.1111/WRR.12370.

[14] S.E. Dowd, Y. Sun, P.R. Secor, D.D. Rhoads, B.M. Wolcott, G.A. James, et al., Survey of bacterial diversity in chronic wounds using Pyrosequencing, DGGE, and full ribosome shotgun sequencing, BMC Microbiol. (2008) 8. Available from: https://doi.org/10.1186/1471-2180-8-43.

[15] A. Han, J.M. Zenilman, J.H. Melendez, M.E. Shirtliff, A. Agostinho, G. James, et al., The importance of a multifaceted approach to characterizing the microbial flora of chronic wounds, Wound Repair. Regen. 19 (2011) 532−541. Available from: https://doi.org/10.1111/J.1524-475X.2011.00720.X.

[16] Z. Xu, H.C. Hsia, The impact of microbial communities on wound healing: a review, Ann. Plast. Surg. 81 (2018) 113−123. Available from: https://doi.org/10.1097/SAP.0000000000001450.

[17] J.W. Costerton, P.S. Stewart, E.P. Greenberg, Bacterial biofilms: a common cause of persistent infections, Science 284 (1999) 1318−1322. Available from: https://doi.org/10.1126/SCIENCE.284.5418.1318.

[18] P. Stoodley, K. Sauer, D.G. Davies, J.W. Costerton, Biofilms as complex differentiated communities, Annu. Rev. Microbiol. 56 (2002) 187−209. Available from: https://doi.org/10.1146/ANNUREV.MICRO.56.012302.160705.

[19] P. Vowden, K. Vowden, C. Keryln, antimicrobial-dressings-made-easy, Wounds Int. 2 (2011) 1−6.

[20] T. Nakatsuji, T.H. Chen, S. Narala, K.A. Chun, A.M. Two, T. Yun, et al., Antimicrobials from human skin commensal bacteria protect against Staphylococcus aureus and are deficient in atopic dermatitis, Sci. Transl. Med. (2017) 9. Available from: https://doi.org/10.1126/SCITRANSLMED.AAH4680.

[21] J.C. Valdéz, M.C. Peral, M. Rachid, M. Santana, G. Perdigón, Interference of Lactobacillus plantarum with Pseudomonas aeruginosa in vitro and in infected burns: the potential use of probiotics in wound treatment, Clin. Microbiol. Infect. 11 (2005) 472−479. Available from: https://doi.org/10.1111/J.1469-0691.2005.01142.X.

[22] S. Blanchet-Réthoré, V. Bourdès, A. Mercenier, C.H. Haddar, P.O. Verhoeven, P. Andres, Effect of a lotion containing the heat-treated probiotic strain Lactobacillus johnsonii NCC 533 on Staphylococcus aureus colonization in atopic dermatitis, Clin. Cosmet. Investig. Dermatol. 10 (2017) 249−257. Available from: https://doi.org/10.2147/CCID.S135529.

[23] M. Mohseni, A. Shamloo, Z. Aghababaei, M. Vossoughi, H. Moravvej, Antimicrobial wound dressing containing silver sulfadiazine with high biocompatibility: in vitro study, Artif. Organs 40 (2016) 765−773. Available from: https://doi.org/10.1111/AOR.12682.

[24] H.v Pawar, J. Tetteh, J.S. Boateng, Preparation, optimisation and characterisation of novel wound healing film dressings loaded with streptomycin and diclofenac, Colloids Surf. B Biointerfaces 102 (2013) 102−110. Available from: https://doi.org/10.1016/J.COLSURFB.2012.08.014.

[25] M. Sabitha, R. Sheeja, Preparation and characterization of ampicillin-incorporated electrospun polyurethane scaffolds for wound healing and infection control, Polym. Eng. Sci. 55 (2015) 541−548. Available from: https://doi.org/10.1002/PEN.23917.

[26] J. Zhu, W. Qiu, C. Yao, C. Wang, D. Wu, S. Pradeep, et al., Water-stable zirconium-based metal-organic frameworks armed polyvinyl alcohol nanofibrous membrane with enhanced antibacterial therapy for wound healing, J. Colloid Interface Sci. 603 (2021) 243−251. Available from: https://doi.org/10.1016/J.JCIS.2021.06.084.

[27] T. Cerchiara, A. Abruzzo, R.A. Ñahui Palomino, B. Vitali, R. de Rose, G. Chidichimo, et al., Spanish Broom (Spartium junceum L.) fibers impregnated with vancomycin-loaded chitosan nanoparticles as new antibacterial wound dressing: preparation, characterization and antibacterial activity, Eur. J. Pharm. Sci. 99 (2017) 105−112. Available from: https://doi.org/10.1016/J.EJPS.2016.11.028.

[28] N. Adhirajan, N. Shanmugasundaram, S. Shanmuganathan, M. Babu, Collagen-based wound dressing for doxycycline delivery: in-vivo evaluation in an infected excisional wound model in rats, J. Pharm. Pharmacol. 61 (2009) 1617−1623. Available from: https://doi.org/10.1211/JPP/61.12.0005.

[29] D. Simões, S.P. Miguel, M.P. Ribeiro, P. Coutinho, A.G. Mendonça, I.J. Correia, Recent advances on antimicrobial wound dressing: a review, Eur. J. Pharm. Biopharm. 127 (2018) 130−141. Available from: https://doi.org/10.1016/J.EJPB.2018.02.022.

[30] J.J. Castellano, S.M. Shafii, F. Ko, G. Donate, T.E. Wright, R.J. Mannari, et al., Comparative evaluation of silver-containing antimicrobial dressings and drugs, Int. Wound J. 4 (2007) 114−122. Available from: https://doi.org/10.1111/J.1742-481X.2007.00316.X.

[31] T.C. Dakal, A. Kumar, R.S. Majumdar, V. Yadav, Mechanistic basis of antimicrobial actions of silver nanoparticles, Front. Microbiol. (2016) 7. Available from: https://doi.org/10.3389/FMICB.2016.01831.

[32] H. Shi, J. Ding, C. Chen, Q. Yao, W. Zhang, Y. Fu, et al., Antimicrobial action of biocompatible silver microspheres and their role in the potential treatment of fungal keratitis, ACS Biomater. Sci. Eng. 7 (2021) 5090−5098. Available from: https://doi.org/10.1021/ACSBIOMATERIALS.1C00815.

[33] M. Konop, T. Damps, A. Misicka, L. Rudnicka, Certain aspects of silver and silver nanoparticles in wound care: a minireview, J. Nanomater. (2016) 2016. Available from: https://doi.org/10.1155/2016/7614753.

[34] S.L. Percival, P.G. Bowler, D. Russell, Bacterial resistance to silver in wound care, J. Hosp. Infect. 60 (2005) 1−7. Available from: https://doi.org/10.1016/J.JHIN.2004.11.014.

[35] A. May, Z. Kopecki, B. Carney, A. Cowin, Antimicrobial silver dressings: a review of emerging issues for modern wound care, ANZ. J. Surg. 92 (2022) 379−384. Available from: https://doi.org/10.1111/ANS.17382.

[36] N. Tomaselli, The role of topical silver preparations in wound healing, J. Wound, Ostomy Cont. Nurs. 33 (2006) 367−380. Available from: https://doi.org/10.1097/00152192-200607000-00004.

[37] S.S. Moreira, M.C. Camargo, R. de, Caetano, M.R. Alves, A. Itria, T.V. Pereira, et al., Efficacy and costs of nanocrystalline silver dressings versus 1% silver sulfadiazine dressings to treat burns in adults in the outpatient setting: a randomized clinical trial, Burns 48 (2022) 568−576. Available from: https://doi.org/10.1016/J.BURNS.2021.05.014.

[38] P. Muangman, C. Pundee, S. Opasanon, S. Muangman, A prospective, randomized trial of silver containing hydrofiber dressing versus 1% silver sulfadiazine for the treatment of partial thickness burns, Int. Wound J. 7 (2010) 271−276. Available from: https://doi.org/10.1111/j.1742-481X.2010.00690.x.

[39] S. Opasanon, P. Muangman, N. Namviriyachote, Clinical effectiveness of alginate silver dressing in outpatient management of partial-thickness burns. Int. Wound, J. 7 (2010) 467−471. Available from: https://doi.org/10.1111/j.1742-481X.2010.00718.x.

[40] E.E. Tredget, H.A. Shankowsky, A. Groeneveld, R. Burrell, A matched-pair, randomized study evaluating the efficacy and safety of acticoat silver-coated dressing for the treatment of burn wounds, J. Burn. Care Rehabil. 19 (1998) 531−537. Available from: https://doi.org/10.1097/00004630-199811000-00013.

[41] R.P. Varas, T. O'Keeffe, N. Namias, L.R. Pizano, O.D. Quintana, M.H. Tellachea, et al., A prospective, randomized trial of acticoat versus silver sulfadiazine in the treatment of partial-thickness burns: which method is less painful? J. Burn. Care Rehabil. 26 (2005) 344−347. Available from: https://doi.org/10.1097/01.BCR.0000170119.87879.CA.

[42] B. Wu, F. Zhang, W. Jiang, A. Zhao, Nanosilver dressing in treating deep II degree burn wound infection in patients with clinical studies, Comput. Math. Methods Med. (2021) 2021. Available from: https://doi.org/10.1155/2021/3171547.

[43] F. Abedini, A. Ahmadi, A. Yavari, V. Hosseini, S. Mousavi, Comparison of silver nylon wound dressing and silver sulfadiazine in partial burn wound therapy, Int. Wound J. 10 (2013) 573−578. Available from: https://doi.org/10.1111/J.1742-481X.2012.01024.X.

[44] D.M. Caruso, K.N. Foster, S.A. Blome-Eberwein, J.A. Twomey, D.N. Herndon, A. Luterman, et al., Randomized clinical study of hydrofiber dressing with silver or silver sulfadiazine in the management of partial-thickness burns, J. Burn. Care Res. 27 (2006) 298−309. Available from: https://doi.org/10.1097/01.BCR.0000216741.21433.66.

[45] G. Gravante, R. Caruso, R. Sorge, F. Nicoli, P. Gentile, V. Cervelli, Nanocrystalline silver: a systematic review of randomized trials conducted on burned patients and an evidence-based assessment of potential advantages over older silver formulations, Ann. Plastic Surg. 63 (2009) 201−205. Available from: https://doi.org/10.1097/SAP.0b013e3181893825.

[46] P. Chaganti, I. Gordon, J.H. Chao, S. Zehtabchi, A systematic review of foam dressings for partial thickness burns, Am. J. Emerg. Med. 37 (2019) 1184−1190. Available from: https://doi.org/10.1016/J.AJEM.2019.04.014.

[47] L. Nherera, P. Trueman, C. Roberts, L. Berg, Silver delivery approaches in the management of partial thickness burns: a systematic review and indirect treatment comparison, Wound Repair. Regen. 25 (2017) 707−721. Available from: https://doi.org/10.1111/wrr.12559.

[48] G. Mosti, A. Magliaro, V. Mattaliano, P. Picerni, N. Angelotti, Comparative study of two antimicrobial dressings in infected leg ulcers: a pilot study, J. Wound Care 24 (2015) 121−127. Available from: https://doi.org/10.12968/JOWC.2015.24.3.121.

[49] C. Trial, H. Darbas, J.P. Lavigne, A. Sotto, G. Simoneau, Y. Tillet, et al., Assessment of the antimicrobial effectiveness of a new silver alginate wound dressing: a RCT, J. Wound Care 19 (2010) 20−26. Available from: https://doi.org/10.12968/jowc.2010.19.1.46095.

[50] K.Y. Woo, P.M. Coutts, R. Gary Sibbald, A randomized controlled trial to evaluate an antimicrobial dressing with silver alginate powder for the management of chronic wounds exhibiting signs of critical colonization, Adv. Skin. Wound Care 25 (2012) 503−508. Available from: https://doi.org/10.1097/01.ASW.0000422628.63148.4B.

[51] H. Beele, F. Meuleneire, M. Nahuys, S.L. Percival, A prospective randomised open label study to evaluate the potential of a new silver alginate/carboxymethylcellulose antimicrobial wound dressing to promote wound healing, Int. Wound J. 7 (2010) 262−270. Available from: https://doi.org/10.1111/j.1742-481X.2010.00669.x.

[52] O. Chansanti, A. Mongkornwong, Treating hard-to-heal ulcers: biocellulose with nanosilver compared with silver sulfadiazine, J. Wound Care 29 (2020) S33−S37. Available from: https://doi.org/10.12968/JOWC.2020.29.SUP12.S33.

[53] A. Chuangsuwanich, O. Charnsanti, V. Lohsiriwat, C. Kangwanpoom, N. Thong-In, The efficacy of silver mesh dressing compared with silver sulfadiazine cream for the treatment of pressure ulcers, J. Med. Assoc. Thai 94 (2011) 559−565.

[54] G. Krasowski, A. Jawień, A. Tukiendorf, Z. Rybak, A. Junka, M. Olejniczak-Nowakowska, et al., A comparison of an antibacterial sandwich dressing vs dressing containing silver, Wound Repair. Regeneration 23 (2015) 525−530. Available from: https://doi.org/10.1111/WRR.12301.

[55] S. Meaume, D. Vallet, M.N. Morere, L. Téot, Evaluation of a silver-releasing hydroalginate dressing in chronic wounds with signs of local infection, J. Wound Care 14 (2005) 411−419. Available from: https://doi.org/10.12968/jowc.2005.14.9.26835.

[56] J.A. Michaels, B. Campbell, B. King, S.J. Palfreyman, P. Shackley, M. Stevenson, Randomized controlled trial and cost-effectiveness analysis of silver-donating antimicrobial dressings for venous leg ulcers (VULCAN trial), Br. J. Surg. 96 (2009) 1147−1156. Available from: https://doi.org/10.1002/bjs.6786.

[57] C.N. Miller, N. Newall, S.E. Kapp, G. Lewin, L. Karimi, K. Carville, et al., A randomized-controlled trial comparing cadexomer iodine and nanocrystalline silver on the healing of leg ulcers, Wound Repair. Regen. 18 (2010) 359−367. Available from: https://doi.org/10.1111/j.1524-475X.2010.00603.x.

[58] M. Zhao, D. Zhang, L. Tan, H. Huang, Silver dressings for the healing of venous leg ulcer: a meta-analysis and systematic review, Medicine 99 (2020) e22164. Available from: https://doi.org/10.1097/MD.0000000000022164.

[59] R.D. Wolcott, R.G. Cook, E. Johnson, C.E. Jones, J.P. Kennedy, R. Simman, et al., A review of iodine-based compounds, with a focus on biofilms: results of an expert panel, J. Wound Care 29 (2020) S38−S43. Available from: https://doi.org/10.12968/JOWC.2020.29.SUP7.S38.

[60] D.J. Leaper, P. Durani, Topical antimicrobial therapy of chronic wounds healing by secondary intention using iodine products, Int. Wound J. 5 (2008) 361−368. Available from: https://doi.org/10.1111/J.1742-481X.2007.00406.X.

[61] K.E. Hill, S. Malic, R. McKee, T. Rennison, K.G. Harding, D.W. Williams, et al., An in vitro model of chronic wound biofilms to test wound dressings and assess antimicrobial susceptibilities, J. Antimicrob. Chemother. 65 (2010) 1195−1206. Available from: https://doi.org/10.1093/JAC/DKQ105.

[62] K. Johani, M. Malone, S.O. Jensen, H.G. Dickson, I.B. Gosbell, H. Hu, et al., Evaluation of short exposure times of antimicrobial wound solutions against microbial biofilms: from in vitro to in vivo, J. Antimicrob. Chemother. 73 (2018) 494−502. Available from: https://doi.org/10.1093/JAC/DKX391.

[63] P.L. Phillips, Q. Yang, S. Davis, E.M. Sampson, J.I. Azeke, A. Hamad, et al., Antimicrobial dressing efficacy against mature Pseudomonas aeruginosa biofilm on porcine skin explants, Int. Wound J. 12 (2015) 469−483. Available from: https://doi.org/10.1111/IWJ.12142.

[64] S. Moberg, L. Hoffman, M.-L. Grennert, A. Holst, A randomized trial of cadexomer iodine in decubitus ulcers, J. Am. Geriatr. Soc. 31 (1983) 462−465. Available from: https://doi.org/10.1111/j.1532-5415.1983.tb05117.x.

[65] M.C. Ormiston, M.T.J. Seymour, G.E. Venn, R.I. Cohen, J.A. Fox, Controlled trial of Iodosorb in chronic venous ulcers, Br. Med. J. (Clin. Res. ed.) 291 (1985) 308−310. Available from: https://doi.org/10.1136/bmj.291.6491.308.

[66] E. Skog, B. Arnesjö, T. Troëng, J.E. Gjöres, L. Bergljung, J. Gundersen, et al., A randomized trial comparing cadexomer iodine and standard treatment in the out-patient management of chronic venous ulcers, Br. J. Dermatol. 109 (1983) 77−83. Available from: https://doi.org/10.1111/j.1365-2133.1983.tb03995.x.

[67] C. Hansson, L.M. Persson, B. Stenquist, P. Nordin, J. Roed-Petersen, W. Westerhof, et al., The effects of cadexomer iodine paste in the treatment of venous leg ulcers compared with hydrocolloid dressing and paraffin gauze dressing, Int. J. Dermatol. 37 (1998) 390−396. Available from: https://doi.org/10.1046/j.1365-4362.1998.00415.x.

[68] G.A. Holloway, K.H. Johansen, R.W. Barnes, Pierce, Multicenter trial of cadexomer iodine to treat venous stasis ulcer, West. J. Med. 151 (1989) 35−38.

[69] J. Apelqvist, G. Ragnarson Tennvall, Cavity foot ulcers in diabetic patients: a comparative study of cadexomer iodine ointment and standard treatment: an economic analysis alongside a clinical trial, Acta Dermato-Venereol. 76 (1996) 231−235. Available from: https://doi.org/10.2340/0001555576231235.

[70] H. Laudanska, B. Gustavson, In-patient treatment of chronic varicose venous ulcers. a randomized trial of cadexomer iodine versus standard dressings, J. Int. Med. Res. 16 (1988) 428−435. Available from: https://doi.org/10.1177/030006058801600604.

[71] H. Vermeulen, S.J. Westerbos, D.T. Ubbink, Benefit and harm of iodine in wound care: a systematic review, J. Hosp. Infect. 76 (2010) 191−199. Available from: https://doi.org/10.1016/J.JHIN.2010.04.026.

[72] C.N. Miller, N. Newall, S.E. Kapp, G. Lewin, L. Karimi, K. Carville, et al., A randomized-controlled trial comparing cadexomer iodine and nanocrystalline silver on the healing of leg ulcers, Wound Repair. Regen. 18 (2010) 359−367. Available from: https://doi.org/10.1111/j.1524-475X.2010.00603.x.

[73] C.S. Pak, D.H. Park, T.S. Oh, W.J. Lee, Y.J. Jun, K.A. Lee, et al., Comparison of the efficacy and safety of povidone-iodine foam dressing (Betafoam), hydrocellular foam dressing (Allevyn), and petrolatum gauze for split-thickness skin graft donor site dressing, Int. Wound J. 16 (2019) 379−386. Available from: https://doi.org/10.1111/iwj.13043.

[74] H.C. Gwak, S.H. Han, J. Lee, S. Park, K.S. Sung, H.J. Kim, et al., Efficacy of a povidone-iodine foam dressing (Betafoam) on diabetic foot ulcer, Int. Wound J. 17 (2020) 91−99. Available from: https://doi.org/10.1111/IWJ.13236.

[75] P.M. Vogt, J. Hauser, O. Robbach, B. Bosse, W. Fleischer, H.U. Steinau, et al., Polyvinyl pyrrolidone-iodine liposome hydrogel improves epithelialization by combining moisture and antisepis. A new concept in wound therapy, Wound Repair. Regen. 9 (2001) 116−122. Available from: https://doi.org/10.1046/J.1524-475X.2001.00116.X.

[76] P.M. Vogt, K. Reimer, J. Hauser, O. Roßbach, H.U. Steinau, B. Bosse, et al., PVP-iodine in hydrosomes and hydrogel−a novel concept in wound therapy leads to enhanced epithelialization and reduced loss of skin grafts, Burns 32 (2006) 698−705. Available from: https://doi.org/10.1016/J.BURNS.2006.01.007.

[77] H.H. Homann, O. Rosbach, W. Moll, P.M. Vogt, G. Germann, M. Hopp, et al., A liposome hydrogel with polyvinyl-pyrrolidone iodine in the local treatment of partial-thickness burn wounds, Ann. Plastic Surg. 59 (2007) 423−427. Available from: https://doi.org/10.1097/SAP.0b013e3180326fcf.

[78] K. Barrigah-Benissan, J. Ory, A. Sotto, F. Salipante, J.P. Lavigne, P. Loubet, Antiseptic agents for chronic wounds: a systematic review, Antibiotics (Basel) (2022) 11. Available from: https://doi.org/10.3390/ANTIBIOTICS11030350.

[79] B.S. Oriel, K.M.F. Itani, Surgical hand antisepsis and surgical site infections, Surg. Infect. 17 (2016) 632−644. Available from: https://doi.org/10.1089/sur.2016.085.

[80] P. Ramirez Galleymore, M. Gordón Sahuquillo, Antisepsis for blood culture extraction. Blood culture contamination rate, Med. Intensiva 43 (2019) 31−34. Available from: https://doi.org/10.1016/j.medin.2018.08.007.

[81] N. Buetti, J.F. Timsit, Management and prevention of central venous catheter-related infections in the ICU, Semin. Respir. Crit. Care Med. 40 (2019) 508−523. Available from: https://doi.org/10.1055/s-0039-1693705.

[82] M.A. Fuglestad, E.L. Tracey, J.A. Leinicke, Evidence-based prevention of surgical site infection, Surg. Clin. North. Am. 101 (2021) 951−966. Available from: https://doi.org/10.1016/j.suc.2021.05.027.

[83] B.P.F.A. Gomes, M.E. Vianna, A.A. Zaia, J.F.A. Almeida, F.J. Souza-Filho, C.C.R. Ferraz, Chlorhexidine in endodontics, Braz. Dent. J. 24 (2013) 89−102. Available from: https://doi.org/10.1590/0103-6440201302188.

[84] S. Jenkins, M. Addy, W. Wade, The mechanism of action of chlorhexidine. A study of plaque growth on enamel inserts in vivo, J. Clin. Periodontol. 15 (1988) 415−424. Available from: https://doi.org/10.1111/J.1600-051X.1988.TB01595.X.

[85] H. Babich, B.J. Wurzburger, Y.L. Rubin, M.C. Sinensky, L. Blau, An in vitro study on the cytotoxicity of chlorhexidine digluconate to human gingival cells, Cell Biol. Toxicol. 11 (1995) 79−88. Available from: https://doi.org/10.1007/BF00767493.

[86] J.P. Barrett, E. Raby, F. Wood, R. Coorey, J.P. Ramsay, G.A. Dykes, et al., An in vitro study into the antimicrobial and cytotoxic effect of Acticoat™ dressings supplemented with chlorhexidine, Burns (2022) 48. Available from: https://doi.org/10.1016/J.BURNS.2021.09.019.

[87] C.T. Cabral, M.H. Fernandes, In vitro comparison of chlorhexidine and povidone-iodine on the long-term proliferation and functional activity of human alveolar bone cells, Clin. Oral. Investig. 11 (2007) 155−164. Available from: https://doi.org/10.1007/S00784-006-0094-8.

[88] A. Acar, F. Uygur, H. Diktaş, R. Evinç, E. Ülkür, O. Öncül, et al., Comparison of silver-coated dressing (Acticoat®), chlorhexidine acetate 0.5% (Bactigrass®) and nystatin for topical antifungal effect in Candida albicans-contaminated, full-skin-thickness rat burn wounds, Burns 37 (2011) 882−885. Available from: https://doi.org/10.1016/J.BURNS.2011.01.024.

[89] V. Ambrogi, D. Pietrella, M. Nocchetti, S. Casagrande, V. Moretti, S. de Marco, et al., Montmorillonite-chitosan-chlorhexidine composite films with antibiofilm activity and improved cytotoxicity for wound dressing, J. Colloid Interface Sci. 491 (2017) 265−272. Available from: https://doi.org/10.1016/J.JCIS.2016.12.058.

[90] S.P. Cebeci, R. Acaroglu, Use of silver-containing hydrofiber and chlorhexidine-impregnated tulle gras dressings for second-degree burns, Adv. Skin. Wound Care (2019) 32. Available from: https://doi.org/10.1097/01.ASW.0000553598.12820.E7.

[91] J.W. Lee, S.H. Park, I.S. Suh, H.S. Jeong, A comparison between DACC with chlorhexidine acetate-soaked paraffin gauze and foam dressing for skin graft donor sites, J. Wound Care 27 (2018) 28−35. Available from: https://doi.org/10.12968/JOWC.2018.27.1.28.

[92] J. Wasiak, H. Cleland, Burns: dressings. BMJ Clin Evid 2015, 2015.

[93] N.O. Hübner, J. Siebert, A. Kramer, Octenidine dihydrochloride, a modern antiseptic for skin, mucous membranes and wounds, Skin. Pharmacol. Physiol. 23 (2010) 244−258. Available from: https://doi.org/10.1159/000314699.

[94] P. Gilbert, L.E. Moore, Cationic antiseptics: diversity of action under a common epithet, J. Appl. Microbiol. 99 (2005) 703−715. Available from: https://doi.org/10.1111/J.1365-2672.2005.02664.X.

[95] A. Kramer, V. Adrian, P. Rudolph, S. Wurster, H. Lippert, [Explant test with skin and peritoneum of the neonatal rat as a predictive test of tolerance of local anti-infective agents in wounds and body cavities], Der Chirurg; Z. fur alle Gebiete der operativen Medizen 69 (1998) 840−845. Available from: https://doi.org/10.1007/S001040050498.

[96] M. Sopata, M. Ciupińska, A. Głowacka, Z. Muszyński, E. Tomaszewska, Effect of Octenisept antiseptic on bioburden of neoplastic ulcers in patients with advanced cancer, J. Wound Care 17 (2008) 24−27. Available from: https://doi.org/10.12968/JOWC.2008.17.1.27975.

[97] M. Sopata, M. Kucharzewski, E. Tomaszewska, Antiseptic with modern wound dressings in the treatment of venous leg ulcers: clinical and microbiological aspects, J. Wound Care 25 (2016) 419−426. Available from: https://doi.org/10.12968/jowc.2016.25.8.419.

[98] W. Vanscheidt, K. Harding, L. Téot, J. Siebert, Effectiveness and tissue compatibility of a 12-week treatment of chronic venous leg ulcers with an octenidine based antiseptic−a randomized, double-blind controlled study, Int. Wound J. 9 (2012) 316−323. Available from: https://doi.org/10.1111/J.1742-481X.2011.00886.X.

[99] Eisenbeiß, F. Siemers, G. Amtsberg, P. Hinz, B. Hartmann, T. Kohlmann, et al., Prospective, double-blinded, randomised controlled trial assessing the effect of an Octenidine-based hydrogel on bacterial colonisation and epithelialization of skin graft wounds in burn patients, Int. J. Burn. Trauma. 2 (2012) 71.

[100] Y. Alkhatib, M. Dewaldt, S. Moritz, R. Nitzsche, D. Kralisch, D. Fischer, Controlled extended octenidine release from a bacterial nanocellulose/Poloxamer hybrid system, Eur. J. Pharm. Biopharm.: Off. J. Arbeitsgemeinschaft fur Pharmazeutische Verfahrenstechnik e.V 112 (2017) 164−176. Available from: https://doi.org/10.1016/J.EJPB.2016.11.025.

[101] I. Bernardelli de Mattos, S.P. Nischwitz, A.C. Tuca, F. Groeber-Becker, M. Funk, T. Birngruber, et al., Delivery of antiseptic solutions by a bacterial cellulose wound dressing: uptake, release and antibacterial efficacy of octenidine and povidone-iodine, Burns 46 (2020) 918−927. Available from: https://doi.org/10.1016/j.burns.2019.10.006.

[102] M. Pandian, V.A. Kumar, R. Jayakumar, Antiseptic chitosan bandage for preventing topical skin infections, Int. J. Biol. Macromol. 193 (2021) 1653−1658. Available from: https://doi.org/10.1016/j.ijbiomac.2021.11.002.

[103] S. Wessels, H. Ingmer, Modes of action of three disinfectant active substances: a review, Regul. Toxicol. Pharmacol. 67 (2013) 456−467. Available from: https://doi.org/10.1016/J.YRTPH.2013.09.006.

[104] H. Nikaido, M. Vaara, Molecular basis of bacterial outer membrane permeability, Microbiol. Rev. 49 (1985) 1−32. Available from: https://doi.org/10.1128/MR.49.1.1-32.1985.

[105] T. Ikeda, A. Ledwith, C.H. Bamford, R.A. Hann, Interaction of a polymeric biguanide biocide with phospholipid membranes, Biochim. Biophys. Acta 769 (1984) 57−66. Available from: https://doi.org/10.1016/0005-2736(84)90009-9.

[106] T. Koburger, N.O. Hübner, M. Braun, J. Siebert, A. Kramer, Standardized comparison of antiseptic efficacy of triclosan, PVP-iodine, octenidine dihydrochloride, polyhexanide and chlorhexidine digluconate, J. Antimicrob. Chemother. 65 (2010) 1712−1719. Available from: https://doi.org/10.1093/JAC/DKQ212.

[107] H. Fjeld, E. Lingaas, Polyhexanide - Safety and efficacy as an antiseptic, Tidsskr. den. Norske Laegeforening 136 (2016) 707−711. Available from: https://doi.org/10.4045/tidsskr.14.1041.

[108] R.G. Sibbald, P. Coutts, K.Y. Woo, Reduction of bacterial burden and pain in chronic wounds using a new polyhexamethylene biguanide antimicrobial foam dressing-clinical trial results, Adv. Skin. Wound Care 24 (2011) 78−84. Available from: https://doi.org/10.1097/01.asw.0000394027.82702.16.

[109] T. Wild, M. Bruckner, M. Payrich, C. Schwarz, T. Eberlein, A. Andriessen, Eradication of methicillin-resistant Staphylococcus aureus in pressure ulcers comparing a polyhexanide-containing cellulose dressing with polyhexanide swabs in a prospective randomized study, Adv. Skin. Wound Care 25 (2012) 17−22. Available from: https://doi.org/10.1097/01.ASW.0000410686.14363.ea.

[110] T. Eberlein, G. Haemmerle, M. Signer, U. GruberMoesenbacher, J. Traber, M. Mittlboeck, et al., Comparison of PHMB-containing dressing and silver dressings in patients with critically colonised or locally infected wounds, J. Wound Care 21 (14−6) (2012) 18−20. Available from: https://doi.org/10.12968/jowc.2012.21.1.12. 12.

[111] G. Elzinga, J. van Doorn, aM. Wiersema, R.J. Klicks, a Andriessen, J.G. Alblas, et al., Clinical evaluation of a PHMB-impregnated biocellulose dressing on paediatric lacerations, J. Wound Care 20 (2011) 280−284.

[112] W. Fabry, C. Trampenau, C. Bettag, A.E. Handschin, B. Lettgen, F.X. Huber, et al., Bacterial decontamination of surgical wounds treated with Lavasept®, Int. J. Hyg. Environ. Health 209 (2006) 567−573. Available from: https://doi.org/10.1016/j.ijheh.2006.03.008.

[113] E. Lenselink, A. Andriessen, A cohort study on the efficacy of a polyhexanide-containing biocellulose dressing in the treatment of biofilms in wounds, J. Wound Care 20 (2011) 534−539. Available from: https://doi.org/10.12968/jowc.2011.20.11.534.

[114] S. Wattanaploy, K. Chinaroonchai, N. Namviriyachote, P. Muangman, Randomized controlled trial of polyhexanide/betaine gel versus silver sulfadiazine for partial-thickness burn treatment, Int. J. Low. Extrem. Wounds 16 (2017) 45−50. Available from: https://doi.org/10.1177/1534734617690949.

[115] A. Bellingeri, F. Falciani, P. Traspedini, A. Moscatelli, A. Russo, G. Tino, et al., Effect of a wound cleansing solution on wound bed preparation and inflammation in chronic wounds: a single-blind RCT, J. Wound Care 25 (2016) 160−168. Available from: https://doi.org/10.12968/jowc.2016.25.3.160.

[116] A. Piatkowski, N. Drummer, A. Andriessen, D. Ulrich, N. Pallua, Randomized controlled single center study comparing a polyhexanide containing bio-cellulose dressing with silver sulfadiazine cream in partial-thickness dermal burns, Burns 37 (2011) 800−804. Available from: https://doi.org/10.1016/J.BURNS.2011.01.027.

[117] G. Norman, J. Christie, Z. Liu, M.J. Westby, J.M. Jefferies, T. Hudson, et al., Antiseptics for burns, Cochrane Database Syst. Rev. (2017) 7. Available from: https://doi.org/10.1002/14651858.CD011821.PUB2.

[118] G. Norman, J.C. Dumville, Z.E. Moore, J. Tanner, J. Christie, S. Goto, Antibiotics and antiseptics for pressure ulcers, Cochrane Database Syst. Rev. (2016) 4. Available from: https://doi.org/10.1002/14651858.CD011586.PUB2.

[119] N.E.M. McLain, Z.E.H. Moore, P. Avsar, Wound cleansing for treating venous leg ulcers, Cochrane Database Syst. Rev. (2021) 3. Available from: https://doi.org/10.1002/14651858.CD011675.PUB2.

[120] K. Barrigah-Benissan, J. Ory, A. Sotto, F. Salipante, J.P. Lavigne, P. Loubet, Antiseptic agents for chronic wounds: a systematic review, Antibiotics (Basel) (2022) 11. Available from: https://doi.org/10.3390/ANTIBIOTICS11030350.

[121] S.K. Saikaly, A. Khachemoune, Honey and wound healing: an update, Am. J. Clin. Dermatology 18 (2017) 237—251. Available from: https://doi.org/10.1007/s40257-016-0247-8.

[122] A. Oryan, E. Alemzadeh, A. Moshiri, Biological properties and therapeutic activities of honey in wound healing: a narrative review and meta-analysis, J. Tissue Viabil. 25 (2016) 98—118. Available from: https://doi.org/10.1016/J.JTV.2015.12.002.

[123] T. Rogalska, Healing the bee's knees—on honey and wound healing, JAMA Dermatol. 152 (2016) 275. Available from: https://doi.org/10.1001/JAMADERMATOL.2015.3692.

[124] D.A. Carter, S.E. Blair, N.N. Cokcetin, D. Bouzo, P. Brooks, R. Schothauer, et al., Therapeutic manuka honey: no longer so alternative, Front. Microbiol. (2016) 7. Available from: https://doi.org/10.3389/FMICB.2016.00569.

[125] M.L. Hossain, L.Y. Lim, K. Hammer, D. Hettiarachchi, C. Locher, Honey-based medicinal formulations: a critical review, Appl. Sci. 11 (2021) 5159. Available from: https://doi.org/10.3390/APP11115159. 2021, Vol. 11, Page 5159.

[126] M. Johnston, M. McBride, D. Dahiya, R. Owusu-Apenten, P. Singh Nigam, Antibacterial activity of Manuka honey and its components: an overview, AIMS Microbiol. 4 (2018) 655—664. Available from: https://doi.org/10.3934/MICROBIOL.2018.4.655.

[127] J. Majtan, Honey: an immunomodulator in wound healing, Wound Repair. Regen. 22 (2014) 187—192. Available from: https://doi.org/10.1111/WRR.12117.

[128] M. Subrahmanyam, Honey impregnated gauze versus polyurethane film (OpSiteR) in the treatment of burns - a prospective randomised study, Br. J. Plastic Surg. 46 (1993) 322—323. Available from: https://doi.org/10.1016/0007-1226(93)90012-Z.

[129] M. Subrahmanyam, Honey-impregnated gauze versus amniotic membrane in the treatment of burns, Burns 20 (1994) 331—333. Available from: https://doi.org/10.1016/0305-4179(94)90061-2.

[130] M. Subrahmanyam, A prospective randomised clinical and histological study of superficial burn wound healing with honey and silver sulfadiazine, Burns 24 (1998) 157—161. Available from: https://doi.org/10.1016/S0305-4179(97)00113-7.

[131] K.I. Malik, M.N. Malik, A. Aslam, Honey compared with silver sulphadiazine in the treatment of superficial partial-thickness burns. International Wound, Journal 7 (2010) 413—417. Available from: https://doi.org/10.1111/j.1742-481X.2010.00717.x.

[132] H. Maghsoudi, F. Salehi, M.K. Khosrowshahi, M. Baghaei, M. Nasirzadeh, R. Shams, Comparison between topical honey and mafenide acetate in treatment of burn wounds, Ann. Burn. Fire Disasters 24 (2011) 132—137.

[133] G. Norman, J. Christie, Z. Liu, M.J. Westby, J.M. Jefferies, T. Hudson, et al., Antiseptics for burns, Cochrane Database Syst. Rev. (2017) 2017. Available from: https://doi.org/10.1002/14651858.CD011821.pub2.

[134] B. Biglari, P.H. vd Linden, A. Simon, S. Aytac, H.J. Gerner, A. Moghaddam, Use of Medihoney as a non-surgical therapy for chronic pressure ulcers in patients with spinal cord injury, Spinal Cord. 50 (2012) 165—169. Available from: https://doi.org/10.1038/SC.2011.87.

[135] B. Biglari, P.H. vd Linden, A. Simon, S. Aytac, H.J. Gerner, A. Moghaddam, Use of Medihoney as a non-surgical therapy for chronic pressure ulcers in patients with spinal cord injury, Spinal Cord. 50 (2012) 165—169. Available from: https://doi.org/10.1038/sc.2011.87.

[136] S. Gulati, A. Qureshi, A. Srivastava, K. Kataria, P. Kumar, A.B. Ji, A prospective randomized study to compare the effectiveness of honey dressing vs. povidone iodine dressing in chronic wound healing, Indian. J. Surg. 76 (2014) 193−198. Available from: https://doi.org/10.1007/S12262-012-0682-6.

[137] L.C. Holland, J.M. Norris, Medical grade honey in the management of chronic venous leg ulcers, Int. J. Surg. 20 (2015) 17−20. Available from: https://doi.org/10.1016/J.IJSU.2015.05.048.

[138] A.N. Mphande, C. Killowe, S. Phalira, H.W. Jones, W.J. Harrison, Effects of honey and sugar dressings on wound healing, J. Wound Care 16 (2007) 317−319. Available from: https://doi.org/10.12968/jowc.2007.16.7.27053.

[139] A. Saha, S. Chattopadhyay, M. Azam, P. Sur, The role of honey in healing of bedsores in cancer patients, South. Asian J. Cancer 1 (2012) 66−71. Available from: https://doi.org/10.4103/2278-330X.103714.

[140] A.B. Jull, N. Cullum, J.C. Dumville, M.J. Westby, S. Deshpande, N. Walker, Honey as a topical treatment for wounds, Cochrane Database Syst. Rev. (2015) 2015. Available from: https://doi.org/10.1002/14651858.CD005083.pub4.

[141] F. Zhang, Z. Chen, F. Su, T. Zhang, Comparison of topical honey and povidone iodine-based dressings for wound healing: a systematic review and meta-analysis, J. Wound Care 30 (2021) S28−S36. Available from: https://doi.org/10.12968/JOWC.2021.30.SUP4.S28.

[142] M. Hosseinnejad, S.M. Jafari, Evaluation of different factors affecting antimicrobial properties of chitosan, Int. J. Biol. Macromol. 85 (2016) 467−475. Available from: https://doi.org/10.1016/J.IJBIOMAC.2016.01.022.

[143] M.A. Matica, F.L. Aachmann, A. Tøndervik, H. Sletta, V. Ostafe, Chitosan as a wound dressing starting material: antimicrobial properties and mode of action, Int. J. Mol. Sci. (2019) 20. Available from: https://doi.org/10.3390/IJMS20235889.

[144] D. Raafat, K. von Bargen, A. Haas, H.G. Sahl, Insights into the mode of action of chitosan as an antibacterial compound, Appl. Env. Microbiol. 74 (2008) 3764−3773. Available from: https://doi.org/10.1128/AEM.00453-08.

[145] F.R. Tamara, C. Lin, F.L. Mi, Y.C. Ho, Antibacterial effects of chitosan/cationic peptide nanoparticles, Nanomaterials (Basel) (2018) 8. Available from: https://doi.org/10.3390/NANO8020088.

[146] L.A. Clifton, M.W.A. Skoda, A.P. le Brun, F. Ciesielski, I. Kuzmenko, S.A. Holt, et al., Effect of divalent cation removal on the structure of gram-negative bacterial outer membrane models, Langmuir 31 (2015) 404−412. Available from: https://doi.org/10.1021/LA504407V.

[147] C.H. Wang, J.H. Cherng, C.C. Liu, T.J. Fang, Z.J. Hong, S.J. Chang, et al., Procoagulant and antimicrobial effects of chitosan in wound healing, Int. J. Mol. Sci. (2021) 22. Available from: https://doi.org/10.3390/IJMS22137067.

[148] H. Chen, X. Xing, H. Tan, Y. Jia, T. Zhou, Y. Chen, et al., Covalently antibacterial alginate-chitosan hydrogel dressing integrated gelatin microspheres containing tetracycline hydrochloride for wound healing, Mater. Sci. Eng. C. Mater Biol. Appl. 70 (2017) 287−295. Available from: https://doi.org/10.1016/J.MSEC.2016.08.086.

[149] Y. Ma, L. Xin, H. Tan, M. Fan, J. Li, Y. Jia, et al., Chitosan membrane dressings toughened by glycerol to load antibacterial drugs for wound healing, Mater. Sci. Eng. C. Mater Biol. Appl. 81 (2017) 522−531. Available from: https://doi.org/10.1016/J.MSEC.2017.08.052.

[150] S. Massand, F. Cheema, S. Brown, W.J. Davis, B. Burkey, P.M. Glat, The use of a chitosan dressing with silver in the management of paediatric burn wounds: a pilot study, J. Wound Care 26 (2017) S26−S30. Available from: https://doi.org/10.12968/JOWC.2017.26.SUP4.S26.

[151] C.A. Stone, H. Wright, V.S. Devaraj, T. Clarke, R. Powell, Healing at skin graft donor sites dressed with chitosan, Br. J. Plastic Surg. 53 (2000) 601−606. Available from: https://doi.org/10.1054/bjps.2000.3412.

[152] J. Stephen-Haynes, L. Toner, S. Jeffrey, Product evaluation of an absorbent, antimicrobial, haemostatic dressing, Br. J. Nurs. 27 (2018) S24−S30. Available from: https://doi.org/10.12968/BJON.2018.27.6.S24.

[153] D.E. Varon, J.D. Smith, D.R. Bharadia, N. Shafique, D. Sakthivel, E.G. Halvorson, et al., Use of a novel chitosan-based dressing on split-thickness skin graft donor sites: a pilot study, J. Wound Care 27 (2018) S12−S18. Available from: https://doi.org/10.12968/jowc.2018.27.Sup7.S12.

[154] R.F. Pereira, P.J. Bártolo, Traditional therapies for skin wound healing, Adv. Wound Care (N. Rochelle) 5 (2016) 208−229. Available from: https://doi.org/10.1089/WOUND.2013.0506.

[155] D. Vijayalakshmi, R. Dhandapani, S. Jayaveni, P.S. Jithendra, C. Rose, A.B. Mandal, In vitro anti inflammatory activity of Aloe vera by down regulation of MMP-9 in peripheral blood mononuclear cells, J. Ethnopharmacol. 141 (2012) 542−546. Available from: https://doi.org/10.1016/J.JEP.2012.02.040.

[156] A. Atiba, M. Nishimura, S. Kakinuma, T. Hiraoka, M. Goryo, Y. Shimada, et al., Aloe vera oral administration accelerates acute radiation-delayed wound healing by stimulating transforming growth factor-β and fibroblast growth factor production, Am. J. Surg. 201 (2011) 809−818. Available from: https://doi.org/10.1016/J.AMJSURG.2010.06.017.

[157] N. Takzare, M.J. Hosseini, G. Hasanzadeh, H. Mortazavi, A. Takzare, P. Habibi, Influence of Aloe vera gel on dermal wound healing process in rat, Toxicol. Mech. Methods 19 (2009) 73−77. Available from: https://doi.org/10.1080/15376510802442444.

[158] M. Rodríguez-Bigas, N.I. Cruz, A. Suárez, Comparative evaluation of aloe vera in the management of burn wounds in guinea pigs, Plast. Reconstr. Surg. 81 (1988) 386−389. Available from: https://doi.org/10.1097/00006534-198803000-00012.

[159] S.J. Hosseinimehr, G. Khorasani, M. Azadbakht, P. Zamani, M. Ghasemi, A. Ahmadi, Effect of aloe cream versus silver sulfadiazine for healing burn wounds in rats, Acta Dermatovenerologica Croatica 18 (2010) 2−7.

[160] D.R. Thomas, P.S. Goode, K. LaMaster, T. Tennyson, Acemannan hydrogel dressing versus saline dressing for pressure ulcers. A randomized, controlled trial, Adv. Wound Care 11 (1998) 273−276.

[161] A. Banu, B.C. Sathyanarayana, G. Chattannavar, Efficacy of fresh Aloe vera gel against multi-drug resistant bacteria in infected leg ulcers, Australas. Med. J. 5 (2012) 305−309. Available from: https://doi.org/10.4066/AMJ.2012.1301.

[162] G. Khorasani, S.J. Hosseinimehr, M. Azadbakht, A. Zamani, M.R. Mahdavi, Aloe versus silver sulfadiazine creams for second-degree burns: a randomized controlled study, Surg. Today 39 (2009) 587−591. Available from: https://doi.org/10.1007/S00595-008-3944-Y.

[163] A.D. Dat, F. Poon, K.B. Pham, J. Doust, Aloe vera for treating acute and chronic wounds, Cochrane Database Syst. Rev. (2012). Available from: https://doi.org/10.1002/14651858.CD008762.PUB2.

[164] C. Burusapat, M. Supawan, C. Pruksapong, A. Pitiseree, C. Suwantemee, Topical Aloe vera gel for accelerated wound healing of split-thickness skin graft donor sites: a double-blind, randomized, controlled trial and systematic review, Plast. Reconstr. Surg. 142 (2018) 217−226. Available from: https://doi.org/10.1097/PRS.0000000000004515.

[165] M.S. Abbasi, J. Rahmati, A.H. Ehsani, A. Takzare, A. Partoazar, N. Takzaree, Efficacy of a natural topical skin ointment for managing split-thickness skin graft donor sites: a pilot double-blind randomized controlled trial, Adv. Skin. Wound Care 33 (2020) 1−5. Available from: https://doi.org/10.1097/01.ASW.0000666916.00983.64.

[166] R. Barbosa, A. Villarreal, C. Rodriguez, H. de Leon, R. Gilkerson, K. Lozano, Aloe Vera extract-based composite nanofibers for wound dressing applications, Mater. Sci. Eng. C. (2021) 124. Available from: https://doi.org/10.1016/j.msec.2021.112061.

[167] M. Tummalapalli, M. Berthet, B. Verrier, B.L. Deopura, M.S. Alam, B. Gupta, Composite wound dressings of pectin and gelatin with aloe vera and curcumin as bioactive agents, Int. J. Biol. Macromol. 82 (2016) 104−113. Available from: https://doi.org/10.1016/j.ijbiomac.2015.10.087.

[168] J. Yin, L. Xu, Batch preparation of electrospun polycaprolactone/chitosan/aloe vera blended nanofiber membranes for novel wound dressing, Int. J. Biol. Macromol. 160 (2020) 352−363. Available from: https://doi.org/10.1016/j.ijbiomac.2020.05.211.

[169] Q. Zhang, M. Zhang, T. Wang, X. Chen, Q. Li, X. Zhao, Preparation of aloe polysaccharide/honey/PVA composite hydrogel: antibacterial activity and promoting wound healing, Int. J. Biol. Macromol. 211 (2022) 249−258. Available from: https://doi.org/10.1016/j.ijbiomac.2022.05.072.

[170] H. Ezhilarasu, R. Ramalingam, C. Dhand, R. Lakshminarayanan, A. Sadiq, C. Gandhimathi, et al., Biocompatible aloe vera and tetracycline hydrochloride loaded hybrid nanofibrous scaffolds for skin tissue engineering, Int. J. Mol. Sci. (2019) 20. Available from: https://doi.org/10.3390/ijms20205174.

[171] S. Anjum, A. Gupta, D. Sharma, D. Gautam, S. Bhan, A. Sharma, et al., Development of novel wound care systems based on nanosilver nanohydrogels of polymethacrylic acid with Aloe vera and curcumin, Mater. Sci. Eng. C. 64 (2016) 157−166. Available from: https://doi.org/10.1016/j.msec.2016.03.069.

[172] S. Ramalingam, M.J.N. Chandrasekar, M.J. Nanjan, Plant-based natural products for wound healing: a critical review, Curr. Drug. Res. Rev. 14 (2022) 37−60. Available from: https://doi.org/10.2174/2589977513666211005095613.

[173] M. Edmondson, N. Newall, K. Carville, J. Smith, Tv Riley, C.F. Carson, Uncontrolled, open-label, pilot study of tea tree (Melaleuca alternifolia) oil solution in the decolonisation of methicillin-resistant Staphylococcus aureus positive wounds and its influence on wound healing, Int. Wound J. 8 (2011) 375−384. Available from: https://doi.org/10.1111/J.1742-481X.2011.00801.X.

[174] R. Arena, M.G. Strazzeri, T. Bianchi, A. Peghetti, Y. Merli, D. Abbenante, et al., Hypericum and neem oil for dehisced post-surgical wounds: a randomised, controlled, single-blinded phase III study, J. Wound Care 31 (2022) 492−500. Available from: https://doi.org/10.12968/JOWC.2022.31.6.492.

[175] B. Castro, F.D. Bastida, T. Segovia, P.L. Casanova, J.J. Soldevilla, J. Verdú-Soriano, The use of an antioxidant dressing on hard-to-heal wounds: a multicentre, prospective case series, J. Wound Care 26 (2017) 742−750. Available from: https://doi.org/10.12968/JOWC.2017.26.12.742.

[176] I.M. Comino-Sanz, M.D. López-Franco, B. Castro, P.L. Pancorbo-Hidalgo, Antioxidant dressing therapy versus standard wound care in chronic wounds (the REOX study): study protocol for a randomized controlled trial, Trials (2020) 21. Available from: https://doi.org/10.1186/s13063-020-04445-5.

Future research directions of antimicrobial wound dressings

Deepinder Sharda[1], Komal Attri[1,2] and Diptiman Choudhury[1,2]

[1]*School of Chemistry and Biochemistry, Thapar Institute of Engineering and Technology, Patiala, Punjab, India*
[2]*BioX, Centre of Excellence for Emerging Materials, Thapar Institute of Engineering and Technology, Patiala, Punjab, India*

11.1 Introduction

Wound healing is a complex physiological process involving a series of sequential yet overlapping healing phases to maintain tissue integrity following trauma caused by any physical or chemical means [1]. According to a retrospective analysis conducted in 2018, nearly 8.2 million people worldwide are infected with wounds (with or without infections). The cost of healing for acute and chronic wounds ranges from \$28.1 billion to \$96.8 billion, with surgical wounds costing the most, followed by diabetic ulcers [2]. Chronic wounds are those that remain open for a month or more and do not heal according to the normal healing mechanism. Patients with diabetes, obesity, and other major health issues are the most affected by chronic wounds [2]. Wounds that heal uneventfully over time, on the other hand, are acute and are typically surgical wounds, abrasions, or burns to the outer skin layer.

The entire healing process is divided into different yet intersecting phases that begin with the inflammatory, proliferative, and terminate with the remodeling phase [3]. In inflammation, hemostasis, chemotaxis, enhanced vascular permeability, vasoconstriction, and blood coagulation occur. It happens by aggregation of thrombocytes and platelets in the fibrin network [3,4]. The fibrin network prevents further damage by enhancing wound closure, preventing infection from microorganisms, removing microbes and debris from the surface, and forming the matrix essential for cellular migration, which helps maintain skin integrity and fibroblast proliferation [5,6]. Further, the inflow of leukocytes into the damaged tissue and the excretion of lysosomal enzymes and reactive oxygen species by inflammatory cells [7]. Re-epithelialization and migration of epithelial cells from the wound's boundary to the center occur during the proliferative phase. It occurs as a result of specific cytokines acting on the wound to close it, followed by keratinocyte migration [8]. Endothelial cells migrate in the wound's extracellular matrix region, leading to angiogenesis [9]. Myofibroblasts initiate wound contraction during this phase. These are high in alpha-smooth muscle actin and cause the lesions' borders to move inward for contractive activity [8,10,11]. The last phase is remodeling, in which the lesion attains the maximum possible toughness and tensile strength by restructuring, degradation, and reformation of the extracellular matrix [12]. Remodeling causes less cellular and vascular scar tissue to form from granulation tissue, followed by an increase in collagen fiber concentration [13]. When

Antimicrobial Dressings. DOI: https://doi.org/10.1016/B978-0-323-95074-9.00007-5

the entire wound area is covered by keratinocytes, the formation of the new epidermal layer from the border to the inside begins, accompanied by matrix deposition and changes in its constituents [8,14]. The thickening and parallel arrangement of collagen fibers in the tissue is what gives the wound its increased tensile strength [15]. Emigration or apoptosis causes inflammatory cells, fibroblasts, and blood vessels to vanish, which results in the formation of scars with fewer cells [10]. The entire wound healing process occurs under normal conditions, as shown in Fig. 11.1. However, in some extreme conditions, this process is affected due to one or the other reason.

Healing is hampered by local factors such as necrosis, maceration, desiccation, pressure, trauma, edema, or bacterial colonization. Tissue damage, including necrosis and a delay in recovery, occurs as a result of microorganism proliferation, which starts a microbiological chain and results in an inflammatory response, resulting in infection [16]. The wound will enter into the chronic stage due to the prolonged inflammatory phase caused by the elevation in the release of pro-inflammatory cytokines followed by enhanced levels of proteases that severely damages the extracellular matrix [17]. After that, macrophages release occurs, which damages the tissue and enhances jpro-inflammatory signaling resulting in a prolonged M-phase [18]. There is a delay in the transition of M1 macrophages to M2 macrophages in infective or diabetic wounds, as shown in Fig. 11.2. A higher rate of mortality and morbidity is observed due to bacterial contamination of the wound site [19].

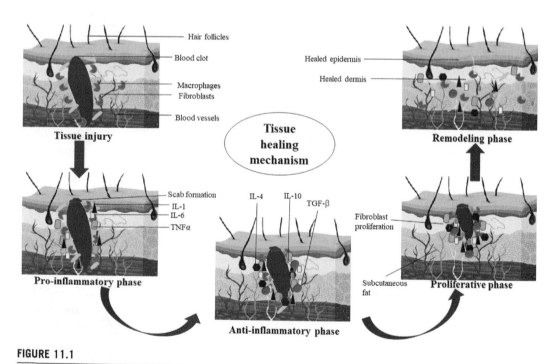

FIGURE 11.1

Wound healing mechanism followed in a normal wound. It consists of different healing stages, which begin after tissue injury. It includes the pro-inflammatory, anti-inflammatory, proliferative, and remodeling phases.

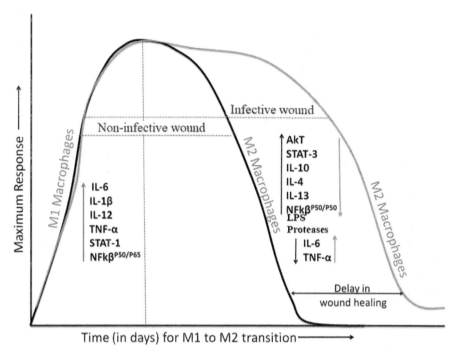

FIGURE 11.2

Role of cytokines in infective and non-infective wound healing. It exhibits a delay in recovery due to the persistent expression of M1 macrophages in infection, resulting in a delayed transition of M1 macrophages to M2 macrophages in the infective wound.

In open wounds caused due to burning or surgery [20], bacterial colonization occurs, which can be from the normal flora of the patient [21] or get in touch the patient from the contaminated water, fomites, or soil [22]. The significant bacteria involved include *Staphylococcus aureus*, *Enterococcus* spp., *Pseudomonas aeruginosa*, *Acinetobacter* spp., etc., as you can see in Fig. 11.3. The fungi involved include *Candida* spp., *Aspergillus* spp., etc. [23]. Gram-positive bacteria, such as *S taphylococcus aureus* and *Escherichia coli*, are the most common in skin tissue wounds. Gram-negative bacteria, such as Pseudomonas, cause damage to the injury later on by invading deep within the tissue or skin [17,24]. The microbes group together to form polymicrobial consortia, which aid in the formation of biofilms around the infected site. With the help of the glycocalyx covering, this biofilm protects pathogenic microbes from antimicrobial treatment and the body's immune system. It also secretes bacterial components that make phagocytic penetration difficult [25,26]. After biofilm formation, there is little or defective wound closure in which the tissue seems to be repaired but lacks normal functioning [27,28], as you can see in Fig. 11.4.

The bacterial lipopolysaccharides affect the wound healing process in several manners. They increase the time duration of the inflammatory phase by enhanced secretion of pro-inflammatory cytokines such as IL-6, IL-1β, leukemia inhibitory factor, CC-chemokines including

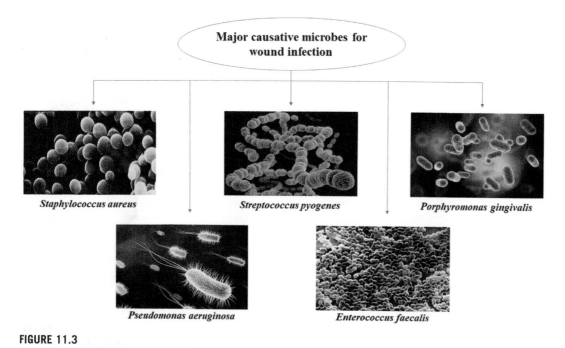

FIGURE 11.3

Significant causative microbes involved in the bacterial contamination of the wound and causes infection.

CCL2/MCP-1, CCL7/MCP-3, CCL3/MIP-1α, and CCL5/RANTES. However, there is no effect on CXC- chemokines. It enhances the infiltration of macrophages, collagen synthesis, and alteration in the secretion of various mediators crucial for the regeneration process of skin or tissue. There is also an upregulation in the secretion of different growth factors such as fibroblast growth factor 2 (FGF-2), vascular endothelial growth factors, and transforming growth factor-β1 [29]. Also, oxygen is essential for wound healing and preventing infection by activating different healing stages and initiating angiogenesis, proliferation, re-epithelialization, migration, and collagen synthesis [26]. In case of injury, temporary hypoxia and oxygen accelerate the healing due to fibroblast, macrophages, and keratinocytes' secretion of cytokines and growth factors [30]. Hypoxic conditions prevail in chronic wounds due to the decrease in oxygen levels from 30−50 to 5−20 mm/Hg^{-1} [31]. Diabetes also hinders proper healing by causing gangrene, amputation, infection, ulceration, or wound dehiscence [32]. Diabetic patients also have reduced vascular blood flow, which lowers the oxygen level in the infected area. It is suggested that interventional revascularization therapy can prevent this harm to patients [26,33]. All of these factors must be considered in order to achieve a smooth and efficient wound healing process. Different bacteria have external and internal characteristics that render them infectious to wounds.

For instance, *S. aureus* is a Gram-positive, spherical, facultative anaerobe. It is found in humans' mucous membrane and skin surface [34] and worsens the wound conditions. It does not

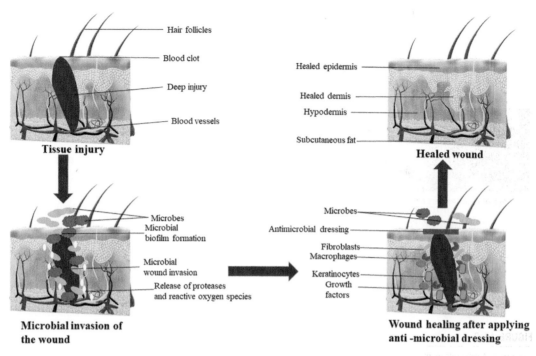

FIGURE 11.4

Role of antimicrobial dressings in wound healing. After tissue injury, the bacterial invasion of the wound occurs, which further forms the biofilm, thus preventing the antibiotics from acting on infection. Antimicrobial dressings can be applied on the infected portion to prevent the entry of microbes into the wound and lead to effective wound healing.

produce spores, is non-motile, and can readily adhere to the protein-rich chronic wound surface. Extracellular adherence protein is present, which prevents neovascularization and wound closure. It also reduces leukocyte-endothelial interactions by decreasing intercellular adhesion molecule-1 and NFκ β activation in leukocytes, capillary tube formation, and integrin-mediated endothelial-cell migration. [35]. Antibacterial membranes have been used to kill the bacteria, but their effectiveness has been limited by antibiotic resistance and a higher rate of infection.

Enterococcus spp. is a gram-positive bacteria of spherical shape present in the intestinal tract of humans and animals. It is responsible for many secondary and nosocomial infections [36]. It causes delayed healing because it participates in the formation of bacterial biofilms on the wound surface, which promotes survival by facilitating a defense mechanism against the host's immune system. Sortase enzymes, SrtA and SrtC, aid in polymerization and cell wall attachment via endocarditis and biofilm-associated pili. Its mechanism of action is still unknown, but one of the major factors could be the overuse of cephalosporins, which makes bacteria resistant to antibiotics [37].

P. aeruginosa is a gram-negative bacteria found in the deepest part of the wound. Biofilm formation, biofilm formation causes nosocomial infection and has high intrinsic and acquired resistance to antibiotics. Virulence factors such as elastase A, elastase B, protease IV, and alkaline proteases are released. They aid in cell adhesion by degrading the extracellular matrix and changing the signaling mechanism. It causes tissue damage and blood vessel invasion. It induces elastase in chronic ulcers, which causes immunoglobulin G and complement system elements to deteriorate, thereby increasing wound chronicity [38]. Because of their ability to form biofilm in a hostile environment, bacteria cause peripheral arterial disease in diabetics, leading to gangrene and limb amputation [39].

In solid media, *Acinetobacter* spp. is a gram-negative bacterium with smooth and mucoid colonies. These are antibiotic-resistant, and their mechanism varies depending on the species and geographical location. The presence of transposons, plasmids, or integrons with gene sets encoding resistance to multiple drugs is the cause of resistance. A large number of bacterial genes collaborate to regulate biofilm formation, making it a complex multifactorial process. Chaperon-usher pilus E, outer membrane protein A, a two-component system, and autoinducer synthase, which aids in adherence and biofilm formation, are among the important genes. Furthermore, it works by inhibiting the production of β-lactamases, switching the penicillin-binding protein, decreasing permeability, and increasing drug efflux. [40].

11.2 The areas that need a particular emphasis on antimicrobial activity

1. **Diabetes and diabetic foot ulcers**—Diabetes, a metabolic syndrome, is characterized by hyperglycemic conditions caused by irregular insulin release or IR receptor resistance. According to the WHO, 171 million people had diabetes in 2000, and that number is expected to double by 2030 [41]. Several complications are associated with diabetes, including nephropathy, neuropathy, retinopathy, atherosclerosis, and foot ulcers, being the most complicated [18]. A diabetic foot ulcer results from combined factors, such as trauma, peripheral neuropathy or vascular diseases, deformities in the foot, insufficient arterial network, or impaired resistance to infection [41,42]. Compared to patients without infection, having a foot infection doubles the risk of lower leg amputation [43]. The anatomy of the foot is unique due to structured compartments, tendons, sheaths, and neurovascular bundles, which increases the risk of infection [44]. To diagnose patients with foot ulcers, the secretion of pus from the wound is monitored, as well as other physical factors such as pain, edema, tenderness, and so on [45]. The primary causative microbes include *Staphylococcus*, *Streptococcus*, *P. aeruginosa*, and *Proteobacteria*, of which *S. aureus* alone is responsible for 43% of infections, followed by *P. aeruginosa* infecting 7%−33% of ulcers [45,46]. According to some studies, the interaction of these various microbes in biofilm results in the production of some virulent substances, such as proteases, collagenases, and a few short-chain fatty acids. Researchers must create a substance that has excellent antimicrobial activity in diabetic patients because these factors

collectively cause inflammation, obstruction in routine wound healing, and increased chronicity of the infection [47].

2. **Burn Wounds**—Burns are one of the most traumatic injuries. Patients with burns require immediate medical attention to avoid morbidity and mortality. Hemostasis begins in the body and immediately causes blood vessel contraction, retraction, and coagulation. There are three zones associated with burn wounds: the coagulation zone, the stasis zone, and the hyperemia zone [48,49]. Thermal injuries cause widespread skin loss, allowing bacteria to quickly infiltrate the affected body parts. The severity of infection is determined by the location of the skin, the thermal or chemical agent, and the length of exposure to the agent [48]. Burn wounds have ideal conditions for microbial growth and proliferation because of the avascular necrotic tissue and a protein-rich environment [50,51]. Burn wound infections are caused by a variety of gram-positive and gram-negative bacteria. Streptococcus pyogenes was the main pathogen infecting the burn wound area prior to the development of antibiotics [52]. *S. aureus*, which is found in sweat glands and hair follicles, colonizes the infected area within a few hours of injury if no antimicrobial dressing is applied and is the leading cause of infection in burn wounds since the discovery of antibiotics. [53]. Similarly, *P. aeruginosa* will form the biofilm within the initial 10 hours of injury, thus, preventing the action of antibiotics over the infected surface [54].

11.3 The need for different antimicrobial agents in wound healing

Researchers around the world are currently working on developing new antimicrobial agents for improved wound healing, with no interruptions in the normal cascade of phases. The dressings were created for a variety of reasons. The most serious is that bacteria are becoming resistant to various drugs [55]. Furthermore, due to poor blood circulation in the extremities of the lower body, oral and IV antibiotics become ineffective in elderly or diabetic patients. We require slightly higher inhibitory concentrations of antibiotics in topical dressings to protect the patient from the side effects of traditional antibiotics such as allergy, diarrhea, vomiting, headache, vaginosis, and so on [26]. Antiseptic agents in antimicrobial dressings act as a biocide, killing or inhibiting the growth of microorganisms in or around the wound area of tissue or skin. Several antiseptic products have been developed in recent years that are highly effective against microbial colonization while being less or less harmful to the body's healthy cells [56]. These antimicrobial substances are used to promote quick healing and prevent contamination of the infected area [57]. They can also be used to kill bacteria and provide a supportive environment for the growth of natural body flora, thereby promoting the host's defense abilities and general health conditions of the infected patient. As a result, bioburden reduction is critical for enabling the host's defense mechanism as well as improving the body's natural immunity. Finally, the cost of making dressings is less than that of taking oral or intravenous antibiotics [26]. Antimicrobial dressings are a cost-effective, efficient, and biocompatible method of removing bioburden that is widely used around the world. Various antimicrobial agents are used to functionalize wound dressings and will be briefly discussed in the following stanzas.

11.4 The different antimicrobial agents and their mode of action in wound healing

1. **Antibiotics as antibacterial agents**—Due to toxicity issues or little or no uptake by cells, only about 1% of the thousands of available antibiotics are used to produce wound dressings with antibacterial effects [58]. Glycopeptides, aminoglycosides, beta-lactams, tetracyclines, sulfonamides, and quinolones [59−64]. These antibiotics can either act on the bacterial surface or on the metabolic pathways that bacteria use to grow in four different ways. To begin, it can inhibit bacterial cell wall synthesis. This mechanism is influenced by beta-lactams and glycopeptides. They prevent the formation of peptide bonds via the reaction catalyzed by penicillin-binding proteins. It also prevents cross-linking between peptidoglycan units or simply inhibits peptidoglycan synthesis [65]. Due to increased osmotic pressure, it has a negative impact on the shape and structure of bacteria and causes bacterial lysis [58]. Secondly, certain antibiotics interfere with protein synthesis and can be divided into two different categories: 50S and 30S inhibitors. Aminoglycosides and tetracyclines act as a hindrance to access of aminoacyl-tRNAs to the ribosomes. It also acts as a 30S ribosome inhibitor to treat skin infections [66,67]. Thirdly, it is essential for preventing bacterial growth to block important metabolic pathways like folate, which is present in microbes but not in mammals [68]. Sulfonamides mimic the structure of folic acid and thus bind to bacterial enzymes, affecting DNA, RNA, and protein production as well as bacterial proliferation [66]. Finally, it inhibits nucleic acid synthesis by inhibiting either the replication or transcription processes. Antibiotics inhibit bacterial mRNA production by inhibiting polymerase activity. It also targets topoisomerase II and IV in bacteria [69]. However, the number of bacteria resistant to natural and synthetic antibiotics is growing at an alarming rate, necessitating the development of new therapeutic agents [70]. Making nanoparticles with antimicrobial agents for use in nanomedicine tools prevents bacterial invasion [71].

2. **Nanoparticles as antimicrobial agent**s—They have desirable properties such as bactericidal activity against a wide range of bacterial strains and the ability to reduce the harmful side effects of drugs on the human body. They do not cause microbial resistance, making them a better alternative to traditional antibiotics [72]. As a result, they are widely used to create new formulations that can overcome bacterial antibiotic resistance. Their mechanism of action differs significantly from that of antibiotics. They can act by producing reactive oxygen species (ROS), coming into direct contact with bacterial cell walls, or releasing toxic metal ions on the surface [58]. Negatively charged groups on the surface of bacteria, such as lipopolysaccharides in gram-negative bacteria and teichoic acid or peptidoglycan in gram-positive bacteria, interact with positively charged nanoparticles and are linked via hydrophobic interactions, receptor-ligand interactions, or Van der Waals forces. As a result, pores form on the bacterial surface, affecting the permeability of the bacterial wall and causing its disruption and loss of intracellular components [73]. Nanoparticles can also cross the cell wall, after which they can affect the metabolic pathway or target the mitochondria and causes its disruption, which leads to ROS generation [74]. The ability of nanoparticles to affect the proton efflux pumps causes variations in membrane surface charge and pH deregulation. The interaction of nanoparticles with bacterial DNA, lysosomes, enzymes, and ribosomes causes oxidative stress, electrolytic

imbalance, protein deactivation, and enzyme inhibition [75]. Despite all of these positive characteristics, nanoparticles are harmful to the human body due to their small size, geometry, extreme solubility, and chemical composition. These factors allow them to easily pass through various biological barriers, enter the human body, and disrupt normal biochemical pathways within cells [76]. To avoid harmful effects, it is necessary to thoroughly investigate the cytotoxicity profile of such nanoparticles. One method for avoiding toxicity is to obtain nanocarriers from natural sources such as plants [77].

3. **Polymers as antimicrobial agents**—Antibacterial applications of polymers with biocompatibility and biodegradability are growing in popularity [78]. They play a crucial role in designing functional wound dressings [79]. The polymers are categorized into natural polymers such as chitosan and synthetic polymers such as lignin or antibacterial peptides such as ε-poly-l-lysine [80,81]. Chitosan exhibits antimicrobial activity against gram-positive and gram-negative bacteria, viruses, algae, and fungi [82]. It operates through other mechanisms. To begin, it inhibits microbial growth by electrostatic interactions between negatively charged groups on the bacterial surface and positively charged groups in chitosan, which affect cell wall permeability and cause an imbalance in internal osmotic pressure. The interactions also cause intracellular component leakage by inducing the hydrolysis of peptidoglycans on the cell wall surface [83]. Secondly, it inhibits the absorption of nutrients and exchanges amongst cells by forming a polymeric network around the bacteria [84]. Third, it inhibits the production of toxic elements, thereby preventing microbial growth, by causing chelation of trace metals and oligo-elements, which are important in bacterial evolution [84].

4. **Natural products as antimicrobial dressings**—Nowadays, many naturally derived products are used to functionalize the wound dressings for better antimicrobial activity [85,86]. Honey, chitosan, and essential oils are among the most commonly used products because of their exceptional antimicrobial properties. Honey has been used as a wound-healing agent since time immemorial because of its antimicrobial activity and ability to promote re-epithelialization, wound contraction, angiogenesis, granulation, and other processes [87,88]. Honey's antimicrobial activity is due to its acidity, low water content, and presence of antimicrobial flavonoids, phenolic acids, and hydrogen peroxide [89]. Due to the presence of gluconic acid, which helps macrophages kill the bacteria and prevent the formation of biofilms, the solution is acidic [90]. Low water content and high osmolarity provide an unfavorable environment that prevents microbial growth and survival [91]. Additionally, the production of hydrogen peroxide inhibits the growth of bacteria by interfering with their cell walls, lipids, nucleic acids, and proteins [92]. Essential oils, which are plant-derived secondary metabolites, have antimicrobial properties primarily because of phenolic compounds like thymols and carvacrol [93,94]. They exhibit antimicrobial activity by increasing cell permeability and lysis after attacking the phospholipids and lipids found in the cell membrane and cell wall of bacteria. It results in cytoplasm leakage, a drop in pH, and the loss of major cellular activities such as protein synthesis, ATP biosynthesis, and DNA transcription. Furthermore, they cause bacterial cell contents to coagulate and nutrients to be transported across the cell membrane, disrupting normal cell membrane function [58]. In this way, natural compounds with excellent antimicrobial properties are being used to create efficient and biocompatible wound dressings.

5. **Graphene-based materials as an antimicrobial agent**—Since the discovery of the antibacterial properties of graphene oxide and reduced graphene oxide, there has been a surge in interest in graphene materials [95]. The initial research was focused on how the physicochemical properties of graphene materials affected the ability of microbes to proliferate [96]. The physicochemical characteristics and nanoscale structures of these compounds, such as their solubility, size, dispersion, and oxidative species, determine their antimicrobial activity [97]. The mechanism of killing bacteria by graphene materials is a combination of several factors, including physical puncturing, causing oxidative stress to the cellular membrane or components, limiting cell growth, blocking transport across the membrane, or destabilizing the membrane by inserting or extracting membrane elements. These factors work alone or in combination to destroy cell envelopes or inhibit cell growth [98,99].

Thus, a wide range of antimicrobial formulations was made using different agents ranging from inorganic salts to graphene-based materials, polymers, or natural products with effective antibacterial activity, as shown in Fig. 11.5.

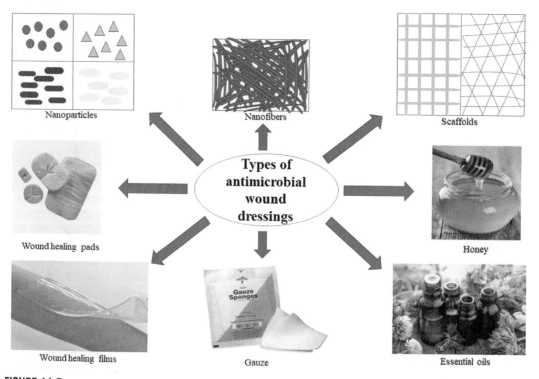

FIGURE 11.5

Various types of antimicrobial agents that can be used to get rid of microbial infections and includes nanoparticles, nanofibers, scaffolds, natural products, bandages, gauze, and films.

11.5 **The different methods under consideration for preventing wounds from microbes**

The infected area is cleaned of bacterial debris using a variety of techniques. The simplest method is to simply rinse the wound, whereas the complicated method is mechanically assisted lavage. The most effective and widely used treatment to get rid of any infection or contamination brought on by microbes is simple surgical debridement. For debris removal, a variety of techniques are used, including the application of sterile water at high pressure, saline, and low-temperature plasma. Other electrophysical techniques are also employed to remove debris, such as ultra-sonification and live maggot antisepsis [100,101]. For this purpose, negative pressure wound therapy is also frequently used. To effectively remove bacteria from the site without harming the body, antimicrobial wound dressings with specific physical features, such as a gradient from the dressing to the wound bed, can be used [102]. The most efficient methods are briefly discussed below.

1. **Negative pressure wound therapy**—In this technique, exudate or infectious materials are removed by applying sub-atmospheric pressure intermittently or continuously through tubes connected to open-cell dressings. The majority of patients exhibit obvious wound healing activity after applying pressure, with the exception of a few cases where no bacterial reduction or accelerated growth is seen [103,104]. Negative pressure therapy is used in conjunction with antiseptic installation in contaminated, infectious, or colonized wounds. During pressure therapy, it will collectively apply a topical antiseptic solution to the wound site, allow it to stay there for a predetermined amount of time, and then remove it [105].

2. **Supporting pathogen competition**—This technique makes use of the skin-supporting microbes that make up the physiological microflora of the skin and protect it. CoNS are dominant florae that meet the necessary requirements and function according to two different mechanisms. The first involves taking up residence in the area surrounding the wound, and the second involves the secretion of antimicrobial enzymes. Due to hygienic concerns or the infection's intervention, only pure antimicrobial treatment or agent is advised for this method, which is still in its experimental stage [102].

3. **Experimental methods to reduce the bioburden** include different techniques such as laser therapy, photodynamic therapy, and cold plasma therapy, which are still experimental. By applying microbicidal peptides produced in vitro or through genetic synthesis derived from human skin bacteria or by auto-transplanting live colonies to the wound site, these techniques involve the reinforcement of physiological flora. In order to maintain a sufficient distance between the wound site and healthy body parts, this multi-barrier system generates self-defense skills, activates living and physical barriers, and involves the appropriate hygienic conditions [102].

4. **Development of smart devices**—The rapid development of new materials, including bioelectronic devices, 3D printers, and innovative materials, is a result of technological advancements. These resources will facilitate continuous wound monitoring and on-demand care at the wound site, both of which are very beneficial [106]. In comparison to conventional antibacterial dressings, it will be more precise, prevent material waste, and be more effective. To track changes in the wound area, numerous wearable pH- or temperature-sensitive systems or sensors have been developed. As they differ greatly between healthy and infected skin, these

are essential for effective healing [107]. Additionally, they quickly and efficiently provide on-demand therapeutic opportunities by detecting the pH or temperature of an early infection in the wound [107].

5. **Development of new formulations**—The disadvantage of the herbal dressings is that they adhere strongly to the wound surface and are typically applied in layers. This can be avoided by using a variety of materials, such as paper, polypropylene, dacron, etc. To increase the material's flexibility or mechanical stiffness, thermoplastic should be combined with herbal dressings [108]. Collagen-based dressings can be used on stubborn wounds to speed up their healing because they have been shown to accelerate the formation of granular tissue, extracellular synthesis, maturation, and wound closure [109].

11.6 Future perspective of wound dressings

Antibacterial dressings are essential for the care of microbial infections as well as the healing of wounds. Traditional bandage administration comes with a variety of difficulties. The main issue with these dressings is the unabated release of antibacterial agents, which leads to prolonged and ineffective wound healing because of the emergence of bacterial resistance. Furthermore, it can be difficult to identify bacterial contamination early on before symptoms appear. The development of novel nanoparticles or the encapsulation of nano-vehicles with antibacterial components will be a good alternative for treating infected wounds, especially for clinical infection control, in order to overcome these drawbacks. [110]. The creation of multipurpose hydrogel-based wound dressings will be useful for diagnosing infections and dispensing medications. Controlling acute and chronic injuries brought on by trauma, surgery, or diabetes is made possible in large part by this system [111]. To ascertain physicochemical behavior and successfully carry out wound therapy, the researchers should create multifunctional systems.

Additionally, various substances like pH, toxins, enzymes, humidity, and inflammatory substances can identify early wound conditions. The most important and significant factor is the temperature. At the earliest stages of injury, there is an abnormal change in wound temperature, which is linked to infection and inflammation in the affected area. A wound dressing with flexible, multipurpose electronics is created for the early detection of pathological conditions by detecting temperature variations [112]. Given the variety and complexity of the wound healing process, it is necessary to concentrate future research efforts more on pathological systems and to create cutting-edge methods for treating wounds that combine pharmaceutics and sensing elements. Additionally, it is painful and impractical to change the wound dressings every day. As a result, there is a need to create novel wound dressings that are financially sound, capable of tracking the condition of the wound, and capable of applying the proper treatment when necessary. These dressings should be used in conjunction with the drug's sustained release in order to transfer the antibacterial agents precisely to the area of the wound [110]. It is challenging for a single dressing to treat all types of injuries because the wound and exudates are noticeably different. The best option is to create composite dressings, which combine the various properties of modern developments for faster, more effective wound healing [113]. Researchers should therefore concentrate on developing dressings with a variety of beneficial properties that promote wound healing and shield against microbial infection.

11.7 Conclusion

The complex process of wound healing involves a series of physiological actions that are dependent on the orderly action of cytokines. Most of the time, wounds heal on their own without any difficulty and in a timely manner. But occasionally, because some body parts are irreplaceable, defects brought on by injuries lower patients' quality of life. Any contamination brought on by microbes, such as bacteria, fungi, or viruses, delays the healing of the wound and delays the time at which it will heal. Although there are currently more than 3000 wound dressings on the market, certain restrictions like diabetes and venous leg ulcers make the process difficult. The development of biomaterials for the antibacterial activity in wound dressings is the subject of ongoing research on a global scale. Despite drug resistance and the absence of a set standard for assessing the controlled release of drugs to ensure the therapy's optimum effectiveness, antibiotics are still primarily used in clinical settings. Innovative antibacterial wound dressings that are somewhat toxic and difficult to remove are the result of ongoing research into more effective bioactive molecules and modifications of nanoparticle structures with those compounds. The natural sources used to make the wound dressing have excellent antibacterial qualities without producing any biological waste or poisonous effects. It is more biocompatible and degradable. But the limitation is the complexity of the process involved in its extraction and purification.

Moreover, the main research hotspots that can be further exploited for their antibacterial activity are the emerging nanomaterials, the hybrid of organic and inorganic materials, and the combination of natural and synthetic materials. Globally, the use of dressings containing antimicrobial agents for direct application to infected wounds is gaining popularity. They work well, are safe, economical, and biocompatible. However, because dressings cannot treat all types of wounds due to their unique structures, it is challenging to compare the dressings based on their morphological characteristics for their effectiveness in wound healing. Therefore, instead of focusing on one wound model for all dressings, researchers need to develop more trustworthy models. Additionally, the goal should be to produce inexpensive dressings that require little synthesis and generate little toxic waste. It is clear from this that more clinical trials based on empirical evidence are needed in order to choose the proper dressings for patients with wound infections. Additionally, interdisciplinary research is required to create antimicrobial dressings that are efficient at reducing infection and accelerating wound healing.

References

[1] P.H. Wang, B.S. Huang, H.C. Horng, C.C. Yeh, Y.J. Chen, Wound healing, J. Chin. Med. Assoc. 81 (2018) 94–101.

[2] C.K. Sen, Human wounds and its burden: an updated compendium of estimates, Adv. Wound Care 8 (2019) 39.

[3] A.C.D.O. Gonzalez, Z.D.A. Andrade, T.F. Costa, A.R.A.P. Medrado, Wound healing—a literature review, An. Bras. Dermatol. 91 (2016) 614.

[4] H.A. Wallace, B.M. Basehore, P.M. Zito, Wound healing phases, StatPearls (2021).

[5] V. Coger, N. Million, C. Rehbock, B. Sures, M. Nachev, S. Barcikowski, et al., Tissue concentrations of zinc, iron, copper, and magnesium during the phases of full thickness wound healing in a rodent model, Biol. Trace Elem. Res. 191 (2019) 167–176.

[6] T. Shaw, P. Martin, Wound repair at a glance, J. Cell. Sci. 122 (2009) 3209–3213.

[7] A. RAP Medrado, L.S. Pugliese, S.A. Regina Reis, Z.A. Andrade, Influence of low level laser therapy on wound healing and its biological action upon myofibroblasts, Lasers Surg. Med. 32 (2003) 239–244.

[8] J. Li, J. Chen, R. Kirsner, Pathophysiology of acute wound healing, Dermatol. Cutan. Surg. 25 (2007) 9–18.

[9] R. José De Mendonça, J. Coutinho-Netto, Cellular aspects of healing, An. Bras. Dermatol. 84 (2009) 257–262.

[10] M. Calin, T. Coman, M. Calin, The effect of low level laser therapy on surgical wound healing. 62 (2010), 617–627.

[11] A. Medrado, T. Costa, T. Prado, S. Reis, Z. Andrade, Phenotype characterization of pericytes during tissue repair following low-level laser therapy, Photodermatol. Photoimmunol. Photomed. 26 (2010) 192–197.

[12] L.G. Bowden, H.M. Byrne, P.K. Maini, D.E. Moulton, A morphoelastic model for dermal wound closure, Biomech. Model. Mechanobiol. 15 (2016) 663–681.

[13] R. Cotran, V. Kumar, R. Stanley, Robbins Pathologic Basis Of Disease, 2004.

[14] P. Martin, Wound healing—aiming for perfect skin regeneration, Science 276 (1997) 75–81.

[15] M. Xue, C. Jackson, Extracellular matrix reorganisation during wound healing nad its impact on abnormal scarring, Adv. Wound Care. 4 (3) (2015) 119–136.

[16] N.B. Menke, K.R. Ward, T.M. Witten, D.G. Bonchev, R.F. Diegelmann, Impaired wound healing, Clin. Dermatol. 25 (2007) 19–25.

[17] R. Edwards, K.G. Harding, Bacteria and wound healing, Curr. Opin. Infect. Dis. 17 (2) (2004) 91–96.

[18] P. Kaur, D. Choudhury, Insulin promotes wound healing by inactivating NFkβP50/P65 and activating protein and lipid biosynthesis and alternating pro/anti-inflammatory cytokines dynamics, Biomol. Concepts 10 (2019) 11–24.

[19] N. Woodford, D.M. Livermore, Infections caused by gram-positive bacteria: a review of the global challenge, J. Infect. 59 (Suppl 1) (2009).

[20] K. Sandy-Hodgetts, K. Carville, G. Leslie, Determining risk factors for surgical wound dehiscence: a literature review, Int. Wound J. 12 (2013) 265–275.

[21] K. Findley, E.A. Grice, The skin microbiome: a focus on pathogens and their association with skin disease, PLoS Pathog. (2014) 10.

[22] G. Suleyman, G. Alangaden, A.C. Bardossy, The role of environmental contamination in the transmission of nosocomial pathogens and healthcare-associated infections, Curr. Infect. Dis. Rep. (2018) 20.

[23] W. Norbury, D.N. Herndon, J. Tanksley, M.G. Jeschke, C.C. Finnerty, Infection in burns, Surg. Infect. (Larchmt). 17 (2016) 250–255.

[24] G. Dow, A. Browne, R. Sibbald, Infection in chronic wounds: controversies in diagnosis and treatment, Ostomy Wound Manag. 45 (8) (1999) 29–40.

[25] R.A. Cooper, T. Bjarnsholt, M. Alhede, Biofilms in wounds: a review of present knowledge, J. Wound Care 23 (2014) 570–582.

[26] O. Sarheed, A. Ahmed, D. Shouqair, J. Boateng, Antimicrobial Dressings for Improving Wound Healing. Wound Heal. - New insights into Anc. Challenges, 2016.

[27] K.G. Barki, A. Das, S. Dixith, P. Ghatak, Das, S. Mathew-Steiner, et al., Electric field based dressing disrupts mixed-species bacterial biofilm infection and restores functional wound healing, Ann. Surg. 269 (2019) 756–766.

[28] K. Ganesh, M. Sinha, S.S. Mathew-Steiner, A. Das, S. Roy, C.K. Sen, Chronic wound biofilm model, Adv. Wound Care 4 (2015) 382–388.

[29] A.V. Kostarnoy, P.G. Gancheva, D.Y. Logunov, L.V. Verkhovskaya, M.A. Bobrov, D.V. Scheblyakov, et al., Topical bacterial lipopolysaccharide application affects inflammatory response and promotes wound healing, J. Interferon Cytokine Res. 33 (2013) 514–522.

[30] S. Guo, L.A. DiPietro, Factors affecting wound healing, J. Dent. Res. 89 (2010) 219–229.

[31] A.A. Tandara, T.A. Mustoe, Oxygen in wound healing—more than a nutrient, World J. Surg. 28 (2004) 294—300.

[32] D.G. Greenhalgh, Wound healing and diabetes mellitus, Clin. Plast. Surg. 30 (2003) 37—45.

[33] P.L. Faries, V.J. Teodorescu, N.J. Morrissey, L.H. Hollier, M.L. Marin, The role of surgical revascularization in the management of diabetic foot wounds, Am. J. Surg. 187 (2004) S34—S37.

[34] J.Y. Park, K.S. Seo, Staphylococcus aureus, Food Microbiol. Fundam. Front. (2022) 555—584.

[35] A.N. Athanasopoulos, M. Economopoulou, V.V. Orlova, A. Sobke, D. Schneider, H. Weber, et al., The extracellular adherence protein (Eap) of Staphylococcus aureus inhibits wound healing by interfering with host defense and repair mechanisms, Blood 107 (2006) 2720—2727.

[36] R.M.N. Fard, M.D. Barton, M.W. Heuzenroeder, Novel bacteriophages in enterococcus spp, Curr. Microbiol. 60 (2010) 400—406.

[37] J.H. Ch'ng, K.K.L. Chong, L.N. Lam, J.J. Wong, K.A. Kline, Biofilm-associated infection by enterococci, Nat. Rev. Microbiol. 17 (2019) 82—94.

[38] R. Serra, R. Grande, L. Butrico, A. Rossi, U.F. Settimio, B. Caroleo, et al., Chronic wound infections: the role of Pseudomonas aeruginosa and Staphylococcus aureus, Expert. Rev. Anti. Infect. Ther. 13 (2015) 605—613.

[39] S.O. Oyibo, E.B. Jude, I. Tarawneh, H.C. Nguyen, D.G. Armstrong, L.B. Harkless, et al., The effects of ulcer size and site, patient's age, sex and type and duration of diabetes on the outcome of diabetic foot ulcers, Diabet. Med. 18 (2001) 133—138.

[40] H.J. Doughari, P.A. Ndakidemi, I.S. Human, S. Benade, The ecology, biology and pathogenesis of acinetobacter spp.: an overview, Microbes Env. 26 (2011) 101—112.

[41] 6th edition IIDF Diabetes Atlas [WWW Document]. https://diabetesatlas.org/atlas/sixth-edition/?msclkid = 9a1ecf3dc00711ecb21d2c6c7a44e5a5, n.d (accessed 4.19.22).

[42] M.P. Khanolkar, S.C. Bain, J.W. Stephens, The diabetic foot, QJM 101 (2008) 685—695.

[43] P. Van Battum, N. Schaper, L. Prompers, J. Apelqvist, E. Jude, A. Piaggesi, et al., Differences in minor amputation rate in diabetic foot disease throughout Europe are in part explained by differences in disease severity at presentation, Diabet. Med. 28 (2011) 199—205.

[44] G.W. Gibbons, The diabetic foot: amputations and drainage of infection, J. Vasc. Surg. 5 (1987) 791—793.

[45] S. Noor, M. Zubair, J. Ahmad, Diabetic foot ulcer—a review on pathophysiology, classification and microbial etiology, Diabetes Metab. Syndr. Clin. Res. Rev. 9 (2015) 192—199.

[46] C.N. Dang, Y.D.M. Prasad, A.J.M. Boulton, E.B. Jude, Methicillin-resistant Staphylococcus aureus in the diabetic foot clinic: a worsening problem, Diabet. Med. 20 (2003) 159—161.

[47] I.B. Wall, C.E. Davies, K.E. Hill, M.J. Wilson, P. Stephens, K.G. Harding, et al., Potential role of anaerobic cocci in impaired human wound healing, Wound Repair. Regen. 10 (2002) 346—353.

[48] D. Church, S. Elsayed, O. Reid, B. Winston, R. Lindsay, Burn wound infections, Clin. Microbiol. Rev. 19 (2006) 403—434.

[49] N.S. Gibran, D.M. Heimbach, Current status of burn wound pathophysiology, Clin. Plast. Surg. 27 (2000) 11—21.

[50] J.P. Barret, D.N. Herndon, Effects of burn wound excision on bacterial colonization and invasion, Plast. Reconstr. Surg. 111 (2003) 744—750.

[51] S. Erol, U. Altoparlak, M.N. Akcay, F. Celebi, M. Parlak, Changes of microbial flora and wound colonization in burned patients, Burns 30 (2004) 357—361.

[52] R.L. Bang, R.K. Gang, S.C. Sanyal, E.M. Mokaddas, A.R.A. Lari, Beta-haemolytic Streptococcus infection in burns, Burns 25 (1999) 242—246.

[53] U. Altoparlak, S. Erol, M.N. Akcay, F. Celebi, A. Kadanali, The time-related changes of antimicrobial resistance patterns and predominant bacterial profiles of burn wounds and body flora of burned patients, Burns 30 (2004) 660—664.

[54] C. Harrison-Balestra, A.L. Cazzaniga, S.C. Davis, P.M. Mertz, A wound-isolated Pseudomonas aeruginosa grows a biofilm in vitro within 10 hours and is visualized by light microscopy, Dermatol. Surg. 29 (2003) 631−635.

[55] M. Zubair, A. Malik, J. Ahmad, Clinico-microbiological study and antimicrobial drug resistance profile of diabetic foot infections in North India, Foot 21 (1) (2011) 6−14.

[56] B. Aderibigbe, B. Buyana, Alginate in wound dressings, Pharmaceutics 10 (2018) 42.

[57] A. Paduraru, C. Ghitulica, R. Trusca, V.A. Surdu, I.A. Neacsu, A.M. Holban, et al., Antimicrobial wound dressings as potential materials for skin tissue regeneration, Mater 12 (2019) 1859.

[58] D. Simões, S.P. Miguel, M.P. Ribeiro, P. Coutinho, A.G. Mendonça, I.J. Correia, Recent advances on antimicrobial wound dressing: a review, Eur. J. Pharm. Biopharm. 127 (2018) 130−141.

[59] J. Kurczewska, P. Pecyna, M. Ratajczak, M. Gajecka, C. Schroeder, Halloysite nanotubes as carriers of vancomycin in alginate-based wound dressing, Saudi Pharm. J. 25 (6) (2017) 911−920.

[60] H. Pawar, J. Tetteh, J.S. Boateng, Preparation, optimisation and characterisation of novel wound healing film dressings loaded with streptomycin and diclofenac, Colloids Surf. B Biointerfaces 102 (2013) 102−110.

[61] M. Sabitha, S. Rajiv, Preparation and characterization of ampicillin-incorporated electrospun polyurethane scaffolds for wound healing and infection control, Polym. Eng. Sci. 55 (2014) 541−548.

[62] N. Adhirajan, N. Shanmugasundaram, S. Shanmuganathan, M. Babu, Collagen-based wound dressing for doxycycline delivery: in-vivo evaluation in an infected excisional wound model in rats, J. Pharm. Pharmacol. 61 (2010) 1617−1623.

[63] M. Mohseni, A. Shamloo, Z. Aghababaei, M. Vossoughi, H. Moravvej, Antimicrobial wound dressing containing silver sulfadiazine with high biocompatibility: in vitro study, Artif. Organs 40 (2016) 765−773.

[64] N. Pásztor, E. Rédai, Z. Szabo, E. Sipos, Preparation and characterization of levofloxacin-loaded nanofibers as potential wound dressings, Acta Med. Marisiensis 63 (2) (2017) 66−69.

[65] M. Kohanski, D. Dwyer, J. Collins, How antibiotics kill bacteria: from targets to networks, Nat. Rev. Microbiol. 8 (2010) 423−435.

[66] E. Etebu, I. Arikekpar, Antibiotics: classification and mechanisms of action with emphasis on molecular perspectives, Int. J. Appl. Microbiol. Biotechnol. Res. 4 (2016) 90−101.

[67] W. Hong, J. Zeng, J. Xie, Antibiotic drugs targeting bacterial RNAs, Acta Pharm. Sin. 4 (4) (2014) 258−265.

[68] A. Bermingham, J. Derrick, The folic acid biosynthesis pathway in bacteria: evaluation of potential for antibacterial drug discovery, BioEssays 24 (7) (2002) 637−648.

[69] G. Patrick, An Introduction to Medicinal Chemistry, 2013.

[70] R. Singh, M. Smitha, S. Singh, The role of nanotechnology in combating multi-drug resistant bacteria, J. Nanosci. Nanotechnol. 14 (2014) 4745−4756.

[71] M. Berthet, Y. Gauthier, C. Lacroix, B. Verrier, C. Monge, Nanoparticle-based dressing: the future of wound treatment? Trends Biotechnol. 35 (8) (2017) 770−784.

[72] Y. Yang, Z. Qin, W. Zeng, T. Yang, Y. Cao, C. Mei, et al., Toxicity assessment of nanoparticles in various systems and organs, Nanotechnol. Rev. 6 (2017) 279−289.

[73] V. Kandi, S. Kandi, Antimicrobial properties of nanomolecules: potential candidates as antibiotics in the era of multi-drug resistance, Epidemiol. Health (2015) 37.

[74] N. Sanvicens, M. Marco, Multifunctional nanoparticles—properties and prospects for their use in human medicine, Trends Biotechnol. 26 (8) (2008) 425−433.

[75] L. Wang, C. Hu, L. Shao, The antimicrobial activity of nanoparticles: present situation and prospects for the future, Int. J. Nanomed. 14 (12) (2017) 1227−1249.

[76] H. Bahadar, F. Maqbool, K. Niaz, M. Abdollahi, Toxicity of nanoparticles and an overview of current experimental models, Iran. Biomed. J. 20 (1) (2016) 1−11.

[77] C.S. Yah, G.S. Simate, Nanoparticles as potential new generation broad spectrum antimicrobial agents, DARU J. Pharm. Sci. (2015) 23.

[78] R. Gharibi, S. Kazemi, H. Yeganeh, V. Tafakori, Utilizing dextran to improve hemocompatibility of antimicrobial wound dressings with embedded quaternary ammonium salts, Int. J. Biol. Macromol. 131 (2019) 1044−1056.

[79] A. Arora, A. Mishra, Antibacterial polymers−a mini review, Mater. Today: Proc. 5 (9) (2018) 17156−17161.

[80] I. Bano, M. Arshad, T. Yasin, M.A. Ghauri, M. Younus, Chitosan: a potential biopolymer for wound management, Int. J. Biol. Macromol. 102 (2017) 380−383.

[81] A. Idrees, P. Varela, F. Ruini, J.M. Vasquez, J. Salber, U. Greiser, et al., Drug-free antibacterial polymers for biomedical applications, Biomed. Sci. Eng. 2 (1) (2018).

[82] K. Kalantari, A. Afifi, H. Jahangirian, T. Webster, Biomedical applications of chitosan electrospun nanofibers as a green polymer−review, Carbohydr. Polym. 207 (2019) 588−601.

[83] R. Goy, D. Britto, O. Assis, A review of the antimicrobial activity of chitosan, Polimeros 19 (3) (2009).

[84] M. Arkoun, F. Daigle, M. Heuzey, A. Ajji, Mechanism of action of electrospun chitosan-based nanofibers against meat spoilage and pathogenic bacteria, Molecules 22 (4) (2017).

[85] M. Essa, A. Manickavasagan, E. Sukumar, Natural products and their active compounds on disease prevention, 2012.

[86] M. Saleem, M. Nazir, M. Ali, H. Hussain, Y. Lee, N. Riaz, et al., Antimicrobial natural products: an update on future antibiotic drug candidates, Nat. Prod. Rep. 27 (2010) 238−254.

[87] A. Scagnelli, Therapeutic review: Manuka honey, J. Exotic Pet. Med. (2016) 25.

[88] J. Packer, J. Irish, B. Herbert, C. Hill, M. Padula, S. Blair, et al., Specific non-peroxide antibacterial effect of manuka honey on the Staphylococcus aureus proteome, Int. J. Antimicrob. 40 (2012) 43−50.

[89] P.H.S. Kwakman, A.A. Te Velde, L. De Boer, D. Speijer, C.M.J.E. Vandenbroucke-Grauls, S.A.J. Zaat, How honey kills bacteria, FASEB J. 24 (2010) 2576−2582.

[90] K. Cutting, Honey and contemporary wound care: an overview, Ostomy Wound Manag. 53 (11) (2007) 49−54.

[91] P.C. Molan, The evidence supporting the use of honey as a wound dressing, Int. J. Low. Extrem. Wounds 5 (2006) 40−54.

[92] A. Simon, K. Traynor, K. Santos, G. Blaser, U. Bode, P. Molan, Medical honey for wound care—still the "latest resort"? Evid. Based Complement. Altern. Med. 6 (2008) 165−173.

[93] Y. Xin Seow, C. Rou Yeo, H. Ling Chung, Yuk Hyun-Gyun, Plant essential oils as active antimicrobial agents, Crit. Rev. Food Sci. Nutr. 54 (2014) 625−644.

[94] S. Bulman, G. Tronci, P. Goswami, C. Carr, J. Russell, Antibacterial properties of nonwoven wound dressings coated with manuka honey or methylglyoxal, Materials 10 (8) (2017).

[95] O. Akhavan, E. Ghaderi, Toxicity of graphene and graphene oxide nanowalls against bacteria, ACS Nano 4 (2010) 5731−5736.

[96] S. Liu, T.H. Zeng, M. Hofmann, E. Burcombe, J. Wei, R. Jiang, et al., Antibacterial activity of graphite, graphite oxide, graphene oxide, and reduced graphene oxide: membrane and oxidative stress, ACS Nano 5 (2011) 6971−6980.

[97] P. Wick, A.E. Louw-Gaume, M. Kucki, H.F. Krug, K. Kostarelos, B. Fadeel, et al., Classification framework for graphene-based materials, Angew. Chem. Int. Ed. 53 (2014) 7714−7718.

[98] H.E. Karahan, C. Wiraja, C. Xu, J. Wei, Y. Wang, L. Wang, et al., Graphene Materials in antimicrobial nanomedicine: current status and future perspectives, Adv. Healthc. Mater. 7 (2018) 1701406.

[99] R. Li, N.D. Mansukhani, L.M. Guiney, Z. Ji, Y. Zhao, C.H. Chang, et al., Identification and optimization of carbon radicals on hydrated graphene oxide for ubiquitous antibacterial coatings, ACS Nano 10 (2016) 10966−10980.

[100] G. Daeschlein, K. Mumcuoglu, O. Assadian, B. Hoffmeister, A. Kramer, In vitro antibacterial activity of Lucilia sericata maggot secretions, Skin. Pharmacol. Physiol. 20 (2) (2007) 112−115.

[101] G. Daeschlein, S. Scholz, A. Arnold, S. Von Podewils, H. Haase, S. Emmert, et al., In vitro susceptibility of important skin and wound pathogens against low temperature atmospheric pressure plasma jet (APPJ) and dielectric barrier discharge plasma, Plasma Process. Polym. 9 (2012) 380−389.

[102] G. Daeschlein, Antimicrobial and antiseptic strategies in wound management, Int. Wound J. 10 (2013) 9−14.

[103] M. Yao, M. Fabbi, H. Hayashi, N. Park, K. Attala, G. Gu, et al., A retrospective cohort study evaluating efficacy in high-risk patients with chronic lower extremity ulcers treated with negative pressure wound therapy, Int. Wound J. 11 (2012) 483−488.

[104] C.M. Mouës, M.C. Vos, G.J.C.M. Van Den Bemd, T. Stijnen, S.E.R. Hovius, Bacterial load in relation to vacuum-assisted closure wound therapy: a prospective randomized trial, Wound Repair. Regen. 12 (2004) 11−17.

[105] B. Lehner, W. Fleischmann, R. Becker, G.N. Jukema, First experiences with negative pressure wound therapy and instillation in the treatment of infected orthopaedic implants: a clinical observational study, Int. Orthop. 35 (2011) 1415−1420.

[106] Yuqing Liang, Yongping Liang, H. Zhang, B. Guo, Antibacterial biomaterials for skin wound dressing, Asian J. Pharm. Sci. (2022).

[107] N. Tang, Y. Zheng, H. Haick, X. Jiang, C. Zhou, H. Jin, et al., Wearable sensors and systems for wound healing-related pH and temperature detection, Micromachines (2021) 12.

[108] F. Kajzar, E. Pearce, N. Turovskij, O. Mukbaniani, Key Engineering Materials: Interdisciplinary Concepts and Research, 2014.

[109] P. Yandell, H. Korzendorfer, H. Hettrick, M. Vaughn, C. Gokoo, The use of collagen dressings in long-term care: a retrospective case series, Wounds 23 (8) (2011) 243−251.

[110] A. Bal-Öztürk, B. Özkahraman, Z. Özbaş, G. Yaşayan, E. Tamahkar, E. Alarçin, Advancements and future directions in the antibacterial wound dressings—a review, J. Biomed. Mater. Res. Part. B Appl. Biomater. 109 (2021) 703−716.

[111] B. Mirani, E. Pagan, B. Currie, M.A. Siddiqui, R. Hosseinzadeh, P. Mostafalu, et al., An advanced multifunctional hydrogel-based dressing for wound monitoring and drug delivery, Adv. Healthc. (2017) 6.

[112] Q. Pang, D. Lou, S. Li, G. Wang, B. Qiao, S. Dong, et al., Smart flexible electronics-integrated wound dressing for real-time monitoring and on-demand treatment of infected wounds, Adv. Sci. 7 (6) (2020).

[113] E. Rezvani Ghomi, S. Khalili, S. Nouri Khorasani, R. Esmaeely Neisiany, S. Ramakrishna, Wound dressings: current advances and future directions, J. Appl. Polym. Sci. 136 (2019) 47738.

Index

Note: Page numbers followed by "*f*" and "*t*" refer to figures and tables, respectively.

CPI Antony Rowe
Eastbourne, UK
March 21, 2023